Y0-EDG-365

Current Developments in Mathematics 2000

Harvard University

B. Mazur
W. Schmid
S. T. Yau

Massachusetts Institute of Technology

J. de Jong
D. Jerison
G. Lustig

International Press
PO Box 43502
Somerville, MA 02143

Current Developments in Mathematics, 2000, final edition,
ISBN 1-57146-079-9

Contents

Current Developments in Mathematics, 2000
Pages 1–107

Reflection Groups, Braid Groups, Hecke Algebras, Finite Reductive Groups

Michel Broué

November 2000

Abstract. Finite subgroups of $\mathrm{GL}_n(\mathbb{Q})$ generated by reflections, known as Weyl groups, classify simple complex Lie Groups as well as simple algebraic groups. They are also building stones for many other significant mathematical objects like braid groups and Hecke algebras.

Through recent work on representations of reductive groups over finite fields based upon George Lusztig's fundamental work, and motivated by conjectures about modular representations of general finite groups, it has become clearer and clearer that finite subgroups of $\mathrm{GL}_n(\mathbb{C})$ generated by pseudo–reflections ("complex reflection groups") behave very much like Weyl groups, and might even be as important.

We present here a concatenation of some recent work (mainly by D. Bessis, G. Malle, J. Michel, R. Rouquier and the author) on complex reflection groups, their braid groups and Hecke algebras, emphasizing the general properties which generalize basic properties of Weyl groups.

By many aspects, the family of finite groups $\mathbf{G}(q)$ over finite fields with q elements behave as if they were the specialisations at $x = q$ of an object depending on an indeterminate x. Convincing indices tend to show that, although complex reflection groups which are not Weyl groups do not define finite groups over finite fields , they might be associated to similar mysterious objects. We present here some aspects of the machinery allowing to emphasize this point of view. We use this machinery to state the general conjectures about representations of finite reductive groups over ℓ–adic rings which, ten years ago, originated this work .

1991 *Mathematics Subject Classification.* Primary 16G99, 20C20, 20C30, 20G05 ; Secondary 17B67, 05E10.

I thank Gunter Malle, Jean Michel and Raphaël Rouquier for years of permanent cooperation, and for having allowed me to use freely our common works [**BrMa1**], [**BrMa2**], [**BMM2**] and [**BMR**]. I thank David Bessis, Meinolf Geck and George Lusztig for fruitful conversations.

Contents

INTRODUCTION

Weyl groups.

Weyl groups are finite groups acting as a reflection group on a rational vector space. These rational reflection groups appear as the "skeleton" of many important mathematical objects, like algebraic groups, Hecke algebras, braid groups, etc.

A particularly fascinating area is the study of algebraic reductive groups over finite fields ("finite reductive groups"). In the study of subgroup structure, ordinary and modular representation theory of finite reductive groups, it turns out that many properties behave in a generic manner, *i.e.,* the results can be phrased in terms independent of the order q of the field of definition. Following in particular [**BrMa1**], [**BMM1**], [**BMM2**], we shall present here (chap. 4) a framework in which this generic behaviour can be conveniently formulated (and partly also proved).

More precisely, assume we are given a family of groups of Lie type, like the groups $\mathrm{GL}_n(q)$ for fixed n but varying prime power q. Then a crucial role in the description of this family of groups is played by the Weyl group, which is the same for all members in the family, together with the action induced by the Frobenius morphism (the twisting). The Weyl group occurs in a natural way as a reflection group on the vector space generated by the coroots, and the Frobenius morphism induces an automorphism of finite order. This leads to the concept of reflection datum $\mathbb{G}$. The orders of the groups attached to a reflection datum $\mathbb{G}$ can be obtained as values of one single polynomial, the order polynomial $|\mathbb{G}|$ of $\mathbb{G}$. In analogy to the concept of ℓ-subgroup of a finite group, for any cyclotomic polynomial Φ_d dividing the order polynomial $|\mathbb{G}|$ one can develop a theory of "Φ_d–subdata" of $\mathbb{G}$. These satisfy a complete analogue of the Sylow theorems for finite groups [**BrMa1**].

Complex reflection groups.

By extension of base field, Weyl groups may be viewed as particular finite *complex reflection groups.* Such groups have been characterized by Shephard–Todd and Chevalley as those linear finite groups whose ring of invariants in the corresponding symmetric algebra is still a regular ring. The irreducible finite complex reflection groups have been classified by Shephard–Todd.

It has been recently discovered that complex reflection groups (and not only Weyl groups) play a key role in the structure as well as in representation theory of finite reductive groups.

In the representation theory of finite reductive groups, the right generic objects are the unipotent characters. It follows from results of Lusztig that they can be parametrized in terms only depending on the reflection datum $\mathbb{G}$. Moreover, the functor of Deligne–Lusztig induction can be shown to be generic. It turns out that for any d such that Φ_d divides $|\mathbb{G}|$ there holds a d–Harish–Chandra theory for unipotent characters. This gives rise to a whole family of generalized Harish–Chandra theories which contains the usual Harish–Chandra theory for complex characters as the case $d = 1$. The resulting generalized Harish–Chandra series allow a description of the subdivision into ℓ-blocks of the unipotent characters of a finite reductive group, at least for large primes ℓ, thus giving an application of the generic formalism to the ℓ-modular representation theory of groups of Lie type. The generalized Harish–Chandra series can be indexed by irreducible characters of suitable relative Weyl groups such that the decomposition of the functor of Deligne–Lusztig induction can be compared to ordinary induction in these relative Weyl groups [**BMM1**]. The most surprising fact is that these relative Weyl groups turn out to be (complex) reflection groups.

These discoveries led naturally to more work on complex reflection groups, which in turn unvailed the fact that complex reflection groups (as well as the associated braid groups, Hecke algebras, etc.) have many unexpected properties in common with Weyl groups.

In the course of his classification of unipotent characters of finite reductive groups (which are ultimately determined by Weyl groups), Lusztig [**Lu2**] observed that similar sets can also be attached to those finite real reflection groups which are not Weyl groups (namely, the finite Coxeter groups), just as if there existed a "fake algebraic group" whose Weyl group is non–crystallographic.

It was at the conference held on the little Greek island named *Spetses*, during the summer of 1993, that Gunter Malle, Jean Michel and the author realized that these constructions might also exist for *non real* reflection groups. Since then, this was shown to be true ([**Ma1**], [**BMM3**]) for certain types of non–real irreducible reflection groups (called "spetsial") by constructing sets of "unipotent characters", Fourier transform matrices, and "eigenvalues of Frobenius", which satisfy suitable generalizations of combinatorial properties of actual finite reductive groups. These were announced in the report [**Ma3**] and will be developed in [**BMM3**]. The mysterious objects which ought to be there, behind the scene, have be named "'spetses" (singular : "spets").

ℓ–blocks of finite reductive groups.

With what precedes in mind, we present the form taken by the *abelian defect group conjecture* (a conjecture about ℓ–adic representations of abstract finite groups, see [**Bro1**]) for the particular case of finite reductive groups. The whole machinery of complex reflection groups, their braid groups and Hecke algebras, is relevant here. Moreover, here again, the phenomenon which we predict seems to be "generic", and it will probably generalized, some day, to the mysterious spetses.

Notation

The complex conjugate of a complex number λ is denoted by λ^*.

If $P(t_1, t_2, \ldots, t_m)$ is a Laurent polynomial in the indeterminates $t_1, t_2, \ldots, t_m$ with complex coefficients, $P(t_1, t_2, \ldots, t_m)^*$ denotes the Laurent polynomial whose coefficients are the complex conjugate of the coefficients of $P(t_1, t_2, \ldots, t_m)$.

We denote by $\boldsymbol{\mu}_\infty$ the group of all roots of unity in $\mathbb{C}$. For d an integer ($d \geq 1$), we denote by $\boldsymbol{\mu}_d$ the subgroup of $\boldsymbol{\mu}_\infty$ consisting of all d-th roots of unity, and we set $\zeta_d := \exp(2\pi i/d)$.

The letter K will denote a subfield of the field of complex numbers $\mathbb{C}$. We denote by $\boldsymbol{\mu}(K)$ the group of roots of unity in K.

For V a K–vector space, we denote by $V^\vee$ the dual space $V^\vee = \operatorname{Hom}(V, K)$.

CHAPTER I

COMPLEX REFLECTION GROUPS

For most of the results quoted here, we refer the reader to the classical literature on complex reflection groups, such as [**Bou1**], [**Ch**], [**OrSo**j] ($j = 1, 2, 3$), [**OrTe**], [**ShTo**], [**Sp**].

We have made constant use of the papers [**BMM2**] and [**BMR**].

Let V be a K–vector space of dimension r.

Let W be a finite subgroup of $\mathrm{GL}(V)$ We denote by S the symmetric algebra of V, by $R = S^W$ the algebra of invariants of W, by R_+ the ideal of R consisting of elements of positive degree. We set $S_W := S/(R_+.S)$, and call it the *coinvariant algebra* of (V, W).

A *pseudo–reflection* (or simply *reflection*) of $\mathrm{GL}(V)$ is a non trivial element s of $\mathrm{GL}(V)$ which acts trivially on a hyperplane, called the *reflecting hyperplane* of s.

The pair (V, W) is called a K*–reflection group* (or simply a reflection group if we omit to mention the ground field) of rank r whenever W is a finite subgroup of $\mathrm{GL}(V)$ generated by pseudo–reflections.

Characterization

After Shephard and Todd discovered the following theorem as a consequence of their classification theorem (see 1.5 below), Chevalley [**Ch**] gave an *a priori* (*i.e.*, classification free) proof of it.

1.1. THEOREM. *Let V be an r–dimensional complex vector space and let W be a finite subgroup of* $\mathrm{GL}(V)$. *The following assertions are equivalent :*

(i) *(V, W) is a (complex) reflection group.*

(ii) *The (graded) algebra $R = S^W$ is regular (i.e., it is a polynomial algebra over r algebraically independant homogeneous elements).*

(iii) *S is free as an R–module.*

Suppose that (V, W) is a complex reflection group. Let $f_1, f_2, \ldots, f_r$ be a family of algebraically independant homogeneous elements of S, with degrees respectively $(d_1, d_2, \ldots, d_r)$, such that $R = \mathbb{C}[f_1, f_2, \ldots, f_r]$. The family $(d_1, d_2, \ldots, d_r)$ depends only on (V, W). It is called the family of *degrees* of (V, W).

The Poincaré series $\mathrm{grdim}R(x)$ of R (an element of $\mathbb{Z}[[x]]$, also called the "graded dimension of R") satifies the following identity :

$$\mathrm{grdim}R(x) = \frac{1}{|W|} \sum_{w \in W} \frac{1}{\det(1 - xw)} = \frac{1}{\prod_{j=1}^{j=r}(1 - x^{d_j})} .$$

A consequence of the preceding identity is the following formula :

$$|W| = d_1 d_2 \cdots d_r .$$

Reflection groups inside reflection groups.

The next two theorems, which provide a way of constructing subgroups of a complex reflection group which are again complex reflection groups, both rely on the above characterization 1.1.

Parabolic subgroups and Steinberg theorem.

For X a subset of V, let us denote by W_X the pointwise stabilizer of X in W.

DEFINITION. The parabolic subgroups of W are the subgroups W_X where X runs over the set of subsets of V.

The next result is due to Steinberg ([**St1**], Thm. 1.5) — cf. also exercises 7 and 8 in [**Bou**], Ch. v, §6.

1.2. THEOREM. *Let X be a subset of V. Then the parabolic subgroup W_X, consisting of all elements of W which fix X pointwise, is generated by the pseudo-reflections in W whose reflecting hyperplane contains X. In particular (V, W_X) is a reflection group.*

Let $\mathcal{A}$ be the set of all reflecting hyperplanes of reflections in W. Let us denote by $I(\mathcal{A})$ the set of all intersections of elements of $\mathcal{A}$. The following result is a consequence of 1.2.

1.3. COROLLARY. *The map $X \mapsto W_X$ is a (non-increasing) bijection from the set $I(\mathcal{A})$ onto the set of all parabolic subgroups of W.*

Regular elements and Springer theorem.

An element of V is said to be *regular* if it does not belong to any reflection hyperplane of W. We introduce the *regular variety*

$$\mathcal{M} := V - \bigcup_{H \in \mathcal{A}} H .$$

Thus $\mathcal{M}$ is the set of regular elements.

DEFINITION. Let d a be positive integer and let $\zeta \in \boldsymbol{\mu}_d$.

• An element w of W is said to be ζ–regular if it has a regular ζ–eigenvector, *i.e.*, if $\ker(w - \zeta \mathrm{Id}) \cap \mathcal{M} \neq \emptyset$. If $\zeta = \zeta_d$, we also say that w is d–regular.

• If there exists a ζ–regular element in W, we say that d is a regular number for W.

Notice (see [**Bes2**]) that d is a regular number for W if and only if the set $(\mathcal{M}/W)^{\boldsymbol{\mu}_d}$ (set of fixed points of $\mathcal{M}/W$ under multiplications by d–the roots of unity) is not empty.

The next result is essentially due to Springer [**Sp**] (see [**Le**], 5.8, or [**DeLo**] for the third assertion). It has been generalized in [**LeSp2**] and in [**BrMa2**] (see below 5.7).

1.4. THEOREM. *Let d be a positive integer and let $\zeta \in \boldsymbol{\mu}_d$. Let w be a ζ-regular element of W. Let $W(w)$ denote the centralizer of w in W, and let $V(w) :=$* $\ker(w - \zeta \mathrm{Id})$.

(1) *The pair $(V(w), W(w))$ is a complex reflection group.*
(2) *Its family of degrees consists of the degrees of W which are divisible by d.*

(3) *Let $\mathcal{A}(w)$ be the set of reflecting hyperplanes of $(V(w), W(w))$ and let $\mathcal{M}(w)$ be the corresponding regular variety. Then the elements of $\mathcal{A}(w)$ are the intersections with $V(w)$ of the elements of $\mathcal{A}$ and we have*

$$\mathcal{M}(w) = \mathcal{M} \cap V(w).$$

Classification

For more details about the classification and the exceptional groups, one may refer for example to [**ShTo**] and to [**OrTe**].

The general infinite family $G(de, e, r)$.

Let d, e and r be three positive integers.

- Let $D_r(de)$ be the set of diagonal complex matrices with diagonal entries in the group $\boldsymbol{\mu}_{de}$ of all de–th roots of unity.
- The d–th power of the determinant defines a surjective morphism

$$\det{}^d \colon D_r(de) \twoheadrightarrow \boldsymbol{\mu}_e .$$

  Let $A(de, e, r)$ be the kernel of the above morphism. In particular we have $|A(de, e, r)| = (de)^r/e$.
- Identifying the symmetric group $\mathfrak{S}_r$ with the usual $r \times r$ permutation matrices, we define

$$G(de, e, r) := A(de, e, r) \rtimes \mathfrak{S}_r .$$

We have $|G(de, e, r)| = (de)^r r!/e$, and $G(de, e, r)$ is the group of all monomial $r \times r$ matrices, with entries in $\boldsymbol{\mu}_{de}$, and product of all non-zero entries in $\boldsymbol{\mu}_d$.

Let $(x_1, x_2, \ldots, x_r)$ be a basis of V. Let us denote by $(\Sigma_j(x_1, x_2, \ldots, x_r))_{1 \le j \le r}$ the family of fundamental symmetric polynomials. Let us set

$$\begin{cases} f_j := \Sigma_j(x_1^{de}, x_2^{de}, \ldots, x_r^{de}) & \text{for } 1 \le j \le r-1, \\ f_r := (x_1 x_2 \cdots x_r)^d . \end{cases}$$

Then we have

$$\mathbb{C}[x_1, x_2, \ldots, x_r]^{G(de,e,r)} = \mathbb{C}[f_1, f_2, \ldots, f_r] .$$

EXAMPLES.

- $G(e, e, 2)$ is the dihedral group of order $2e$.
- $G(d, 1, r)$ is isomorphic to the wreath product $\boldsymbol{\mu}_d \wr \mathfrak{S}_r$. For $d = 2$, it is isomorphic to the Weyl group of type B_r (or C_r).
- $G(2, 2, r)$ is isomorphic to the Weyl group of type D_r.

About the exceptional groups.

There are 34 exceptional irreducible complex reflection groups, of ranks from 2 to 8, denoted $G_4, G_5, \ldots, G_{37}$.

The rank 2 groups are connected with the finite subgroups of $\mathrm{SL}_2(\mathbb{C})$ (the *binary polyhedral groups*).

1.5. Theorem. (Shephard–Todd) *Let (V, W) be an irreducible complex reflection group. Then one of the following assertions is true :*

- *There exist integers d, e, r, with $de \geq 2$, $r \geq 1$ such that $(V, W) \simeq G(de, e, r)$.*
- *There exists an integer $r \geq 1$ such that $(V, W) \simeq (\mathbb{C}^{r-1}, \mathfrak{S}_r)$.*
- *(V, W) is isomorphic to one of the 34 exceptional groups G_n $(n = 4, \ldots, 37)$.*

Field of definition

The following theorem has been proved (using a case by case analysis) by Bessis [**Bes1**] (see also [**Ben**]), and generalizes a well known result on Weyl groups.

1.6. Theorem–Definition.

Let (V, W) be a reflection group. Let K be the field generated by the traces on V of all elements of W. Then all irreducible KW-representations are absolutely irreducible.

The field K is called the field of definition of the reflection group W.

- If $K \subseteq \mathbb{R}$, the group W is a (finite) *Coxeter group*.
- If $K = \mathbb{Q}$, the group W is a Weyl group.

Differential forms and derivations

Here we follow closely Orlik and Solomon [**OrSo2**].

Generalities.

Let us denote by Δ_1 the S-module of derivations of the K-algebra S, and by Ω^1 the S-dual of Δ_1 ("module of 1-forms"). We have

$$\Delta_1 = S \otimes V^\vee \quad \text{and} \quad \Omega^1 = S \otimes V\,,$$

and there is an obvious duality

$$\langle\ ,\ \rangle \colon \Omega^1 \times \Delta_1 \longrightarrow S\,.$$

Let $(x_1, x_2, \ldots, x_r)$ be a basis of V. Note that the family $(\frac{\partial}{\partial x_1}, \frac{\partial}{\partial x_2}, \ldots, \frac{\partial}{\partial x_r})$ of elements of Δ_1 is the dual basis. Denote by $d \colon \Omega^1 \longrightarrow \Omega^1$ the derivation of the S-module $\Omega^1 = S \otimes V$ defined by $d(x \otimes 1) := 1 \otimes x$ for all $x \in V$. Then

$$\begin{cases} (\frac{\partial}{\partial x_1}, \frac{\partial}{\partial x_2}, \ldots, \frac{\partial}{\partial x_r}) \text{ is a basis of the } S\text{-module } \Delta_1\,. \\ (dx_1\,, dx_2\,, \ldots\,, dx_r) \text{ is a basis of the } S\text{-module } \Omega^1\,. \end{cases}$$

Let us endow Δ_1 and Ω^1 respectively with the graduations defined by

$$\begin{cases} \Delta_1 = \bigoplus_{n=-1}^{\infty} \Delta_1^{(n)} \text{ where } \Delta_1^{(n)} := S^{n+1} \otimes V^\vee \\ \Omega^1 = \bigoplus_{n=1}^{\infty} \Omega^{1,(n)} \text{ where } \Omega^{1,(n)} := S^{n-1} \otimes V \end{cases}$$

In other words, we have

$$\begin{cases} (\delta \in \Delta_1^{(n)}) \Longleftrightarrow (\delta(S^m) \subseteq S^{n+m})\,, \\ (\omega \in \Omega^{1,(n)}) \Longleftrightarrow (\langle \omega, \Delta_1^{(m)} \rangle \subseteq S^{n+m})\,, \end{cases}$$

and so (with previous notation)

$$\begin{cases} \text{the elements } (\frac{\partial}{\partial x_1}, \frac{\partial}{\partial x_2}, \dots, \frac{\partial}{\partial x_r}) \text{ have degree } -1, \\ \text{while the elements } (dx_1\,,\, dx_2\,,\, \dots\,,\, dx_r) \text{ have degree } 1. \end{cases}$$

We denote by Δ and Ω the exterior algebras of respectively the S–modules Δ_1 and Ω^1. We have

$$\Delta = S \otimes \Lambda(V^\vee) \quad \text{and} \quad \Omega = S \otimes \Lambda(V)\,.$$

We endow Δ and Ω respectively with the bi–graduations extending the graduations of Δ_1 and Ω^1 :

$$\begin{cases} \Delta^{(m,n)} := S^m \otimes \Lambda^{-n}(V^\vee) \\ \Omega^{(m,n)} := S^m \otimes \Lambda^{n}(V) \end{cases}$$

The Poincaré series of the preceding bigraded modules are defined as follows :

$$\begin{cases} \operatorname{grdim} \Delta(t,u) := \displaystyle\sum_{m=0}^{\infty} \sum_{n=0}^{n=r} \dim \Delta^{(m,-n)} t^m (-u)^{-n} \\ \operatorname{grdim} \Omega(t,u) := \displaystyle\sum_{m=0}^{\infty} \sum_{n=0}^{n=r} \dim \Omega^{(m,-n)} t^m (-u)^{n}\,. \end{cases}$$

Fixed points under W.

Since S is a free graded R–module, the fixed points modules $(\Omega^1)^W$ and $(\Delta_1)^W$ are free graded R–modules.

- *Degrees again* : If $f_1, \dots, f_r$ is a family of algebraically independant homogeneous elements of S such that $R = \mathbb{C}[f_1, f_2, \dots, f_r]$, then $df_1, \dots, df_r$ is a basis of $(\Omega^1)^W$ over R (thus consisting in a family of homogeneous elements with degrees respectively $(d_1, d_2, \dots, d_r)$).
- *Codegrees* : If $\delta_1, \delta_2, \dots, \delta_r$ is a basis of $(\Delta_1)^W$ over R consisting of homogeneous elements of degrees $(d_1^\vee, d_2^\vee, \dots, d_r^\vee)$, the family $(d_1^\vee, d_2^\vee, \dots, d_r^\vee)$ is called the family of codegrees of (V, W).

The Poincaré series of $(\Omega^1)^W$ and $(\Delta_1)^W$ are respectively

$$\begin{cases} \operatorname{grdim}\,(\Omega^1)^W(q) := \dfrac{q^{-1}\sum_{j=1}^{j=r} q^{d_j}}{\prod_{j=1}^{j=r}(1-q^{d_j})} \\ \operatorname{grdim}\,(\Delta_1)^W(q) := \dfrac{q\sum_{j=1}^{j=r} q^{d_j^\vee}}{\prod_{j=1}^{j=r}(1-q^{d_j})}\,. \end{cases}$$

Let us denote by $\Delta^W := (S \otimes \Lambda(V^\vee))^W$ and $\Omega^W := (S \otimes \Lambda(V))^W$ respectively the subspaces of fixed points under the action of W.

1.7. Theorem. (Orlik–Solomon [**OrSo2**]) *The R–modules Δ^W and Ω^W are the exterior algebras of respectively the R–modules $(\Delta_1)^W$ and $(\Omega^1)^W$:*

$$\begin{cases} \Delta^W = \Lambda_R((\Delta_1)^W) \\ \Omega^W = \Lambda_R((\Omega^1)^W) \end{cases}$$

Let us denote by $\operatorname{grdim}\Delta(t,u)$ and $\operatorname{grdim}\Omega(t,u)$ respectively the generalized Poincaré series of these modules defined by

$$\begin{cases} \operatorname{grdim}\Delta^W(t,u) := \sum_{m=0}^{\infty}\sum_{n=0}^{n=r} \dim\,(\Delta^{(m,-n)})^W t^m(-u)^{-n} \\ \operatorname{grdim}\Omega^W(t,u) := \sum_{m=0}^{\infty}\sum_{n=0}^{n=r} \dim\,(\Omega^{(m,-n)})^W t^m(-u)^{n}\,. \end{cases}$$

1.8. Corollary. *We have*

$$\begin{cases} \operatorname{grdim}\Delta^W(t,u) = \dfrac{1}{|W|}\sum_{w\in W}\dfrac{\det(1-w^{-1}u^{-1})}{\det(1-wt)} = \prod_{j=1}^{j=r}\dfrac{1-u^{-1}t^{d_j^\vee+1}}{1-t^{d_j}} \\ \operatorname{grdim}\Omega^W(t,u) = \dfrac{1}{|W|}\sum_{w\in W}\dfrac{\det(1-wu)}{\det(1-wt)} = \prod_{j=1}^{j=r}\dfrac{1-ut^{d_j-1}}{1-t^{d_j}} \end{cases}$$

Complement. The previous corollary implies a multiplicative identity for the Poincaré series which has been conjectured by Jean Michel from computations.

1.9. Lemma. *We have*

$$\left(\prod_{w\in W}\det_V(1-tw)\right)^{1/|W|} = \prod_{j=1}^{j=r}(1-t^{d_j})^{1/d_j}\,.$$

Proof of 1.9. By 1.8 we have

$$\frac{1}{|W|}\sum_{w\in W}\frac{\det(1-wu)}{\det(1-wt)} = \prod_{j=1}^{j=r}\frac{1-ut^{d_j-1}}{1-t^{d_j}}\,.$$

Let us differentiate with respect to u both sides of the preceding equality. We get

$$\frac{1}{|W|}\sum_{w\in W}\frac{\frac{d}{du}\det_V(1-wu)}{\det_V(1-tw)} = \sum_{j=1}^{j=r} -t^{d_j-1}\left(\frac{\prod_{k\neq j}(1-ut^{d_k-1})}{\prod_{k=1}^{k=r}(1-t^{d_k})}\right)\,.$$

Now specialize the preceding equality at $u=t$. We get

$$\frac{d}{dt}\mathrm{Log}\left(\prod_{w\in W}\det_V(1-tw)\right)^{1/|W|} = \frac{d}{dt}\mathrm{Log}\left(\prod_{j=1}^{j=r}(1-t^{d_j})^{1/d_j}\right)\,,$$

thus proving the identity announced in 1.9. □

We recall that $\mathcal{A}$ denotes the set of reflecting hyperplanes of (V,W), and we set $N := |\mathcal{A}|$. We denote by $N^\vee$ the number of pseudo-reflections in W (note that for real reflection groups we have $N = N^\vee$).

For $H\in\mathcal{A}$, we denote by $e_H := |W_H|$ the order of the pointwise stabilizer of H. The group W_H is a minimal non trivial parabolic subgroup of W. All its non trivial elements are pseudo-reflections. The group W_H is cyclic : if s_H denotes the element of W_H with determinant ζ_{e_H}, we have $W_H = \langle s_H\rangle$, the group generated by s_H. Such a pseudo–reflection as s_H is called a *distinguished* pseudo–reflection in W.

The centralizer $C_W(W_H)$ of W_H in W is also its normalizer, as well as the normalizer (set-wise stabilizer) of H.

For $\mathcal{C} \in \mathcal{A}/W$ an orbit of hyperplanes, we denote by $N_{\mathcal{C}}$ its cardinality. We have $N_{\mathcal{C}} = |W : C_W(W_H)|$ for $H \in \mathcal{C}$. We also set $e_{\mathcal{C}} := e_H$ for $H \in \mathcal{C}$.

The following identities are all consequences of 1.8.

- We have

$$(x + d_1 - 1)(x + d_2 - 1) \cdots (x + d_r - 1) = \sum_{w \in W} x^{\dim V^{\langle w \rangle}}$$

(where $V^{\langle w \rangle}$ denotes the space of fixed points of w). It follows that

$$\sum_{j=1}^{j=r} (d_j - 1) = \sum_{H \in \mathcal{A}} (e_H - 1) = \sum_{\mathcal{C} \in \mathcal{A}/W} N_{\mathcal{C}}(e_{\mathcal{C}} - 1) = N^\vee . \tag{1.10}$$

- We have

$$(x - d_1^\vee - 1)(x - d_2^\vee - 1) \cdots (x - d_r^\vee - 1) = \sum_{w \in W} \det_V(w) x^{\dim V^{\langle w \rangle}} .$$

It follows that

$$\sum_{j=1}^{j=r} (d_j^\vee + 1) = \sum_{H \in \mathcal{A}} 1 = \sum_{\mathcal{C} \in \mathcal{A}/W} N_{\mathcal{C}} = N , \tag{1.11}$$

and so

$$N + N^\vee = \sum_{j=1}^{j=r} (d_j + d_j^\vee) = \sum_{\mathcal{C} \in \mathcal{A}/W} N_{\mathcal{C}} e_{\mathcal{C}} . \tag{1.12}$$

Regular numbers, degrees and codegrees.

Lehrer and Springer ([**LeSp2**], 5.1) have noticed the following characterization of regular numbers.

1.13. Proposition. *An integer d is regular for W if and only if it divides as many degrees as codegrees.*

It should be noticed that the "if" implication is only known at the moment by case–by–case inspection.

Cohomology of the hyperplanes complement

For $H \in \mathcal{A}$, let us denote by α_H a linear form on V with kernel H, and let us define the holomorphic differential form ω_H on $\mathcal{M}$ by the formula

$$\omega_H := \frac{1}{2i\pi} \frac{d\alpha_H}{\alpha_H} ,$$

which we also write $\omega_H = \dfrac{1}{2i\pi} d\mathrm{Log}(\alpha_H)$. We denote by $[\omega_H]$ the corresponding de Rham cohomology class.

Brieskorn (cf. [**Br2**], Lemma 5) has proved the following result.

1.14. *Let $\mathbb{C}[(\omega_H)_{H\in\mathcal{A}}]$ (resp. $\mathbb{Z}[(\omega_H)_{H\in\mathcal{A}}]$) be the $\mathbb{C}$-subalgebra (resp. the $\mathbb{Z}$-subalgebra) of the $\mathbb{C}$-algebra of holomorphic differential forms on $\mathcal{M}$ which is generated by $\{\omega_H\}_{H\in\mathcal{A}}$. Then the map $\omega_H \mapsto [\omega_H]$ induces an isomorphism between $\mathbb{C}[(\omega_H)_{H\in\mathcal{A}}]$ and the cohomology algebra $\mathrm{H}^*(\mathcal{M},\mathbb{C})$ (resp. an isomorphism between $\mathbb{Z}[(\omega_H)_{H\in\mathcal{A}}]$ and the singular cohomology algebra $\mathrm{H}^*(\mathcal{M},\mathbb{Z})$).*

From now on, we write ω_H instead of $[\omega_H]$.

Orlik and Solomon (cf. [**OrSo1**]) give a description of the algebra $\mathrm{H}^*(\mathcal{M},\mathbb{C})$. Before stating their result, we need to introduce more notation.

- Let $\mathbb{C}\mathcal{A} := \bigoplus_{H\in\mathcal{A}} \mathbf{e}_H$ be the vector space with basis indexed by $\mathcal{A}$, and let $\Lambda\mathcal{A}$ be its exterior algebra, endowed with the usual Koszul differential map $\delta\colon \Lambda\mathcal{A} \to \Lambda\mathcal{A}$ of degree -1.
- For $\mathcal{B} = \{H_1, H_2, \dots, H_k\} \subset \mathcal{A}$, we denote by $D_\mathcal{B}$ the line generated by $\mathbf{e}_{H_1} \wedge \mathbf{e}_{H_2} \wedge \cdots \wedge \mathbf{e}_{H_k}$.
- We say that $\mathcal{B}$ is *dependent* if $\operatorname{codim}(\bigcap_{H\in\mathcal{B}} H) < |\mathcal{B}|$.
- We denote by $\mathrm{I}\Lambda\mathcal{A}$ the (graded) ideal of $\Lambda\mathcal{A}$ generated by the $\delta(D_\mathcal{B})$ where $\mathcal{B}$ runs overs the set of all dependent subsets of $\mathcal{A}$.

1.15. Theorem. (Orlik and Solomon) *The map $\mathbf{e}_H \mapsto \omega_H$ induces an isomorphism of graded algebras between $\Lambda\mathcal{A}/\mathrm{I}\Lambda\mathcal{A}$ and $\mathrm{H}^*(\mathcal{M},\mathbb{C})$.*

We recall that $I(\mathcal{A})$ denotes the set of intersections of elements of $\mathcal{A}$. For $X \in I(\mathcal{A})$, we set $\mathrm{H}^{(X)}(\mathcal{M},\mathbb{C}) := \sum D_\mathcal{B}$ where the summation is taken over all $\mathcal{B} \subset \mathcal{A}$, $|\mathcal{B}| = \operatorname{codim}(X)$, $\bigcap_{H\in\mathcal{B}} H = X$, and where $D_\mathcal{B}$ is the complex line generated by $\omega_{H_1} \wedge \omega_{H_2} \wedge \cdots \wedge \omega_{H_k}$ if $\mathcal{B} = (H_1, H_2, \dots, H_k)$.

Then it follows from Theorem 1.15 that

1.16. Corollary. *for any integer n, we have*

$$\mathrm{H}^n(\mathcal{M},\mathbb{C}) = \bigoplus_{\substack{(X\in I(\mathcal{A}))\\ (\operatorname{codim}(X)=n)}} \mathrm{H}^{(X)}(\mathcal{M},\mathbb{C})\,.$$

Moreover, we see that

1.17. Corollary.

(1) *the family $(\omega_H)_{H\in\mathcal{A}}$ is a basis of $\mathrm{H}^1(\mathcal{M},\mathbb{C})$,*

(2) *for X an element of $I(\mathcal{A})$ with codimension 2, if H_X denotes a fixed element of $\mathcal{A}$ which contains X,*
   - *whenever H and H' are two elements of $\mathcal{A}$ which contain X, we have* $\omega_H \wedge \omega_{H'} = \omega_{H_X} \wedge \omega_{H'} - \omega_{H_X} \wedge \omega_H$,
   - *the family $(\omega_{H_X} \wedge \omega_H)_{(H\supset X)(H\neq H_X)}$ is a basis of $\mathrm{H}^{(X)}(\mathcal{M},\mathbb{C})$.*

The codegrees are determined by the arrangement $\mathcal{A}$, by the following consequence of Theorem 1.15.

1.18. Corollary. *The Poincaré polynomial*

$$P_\mathcal{M}(q) := \sum_n q^n \dim(\mathrm{H}^n(\mathcal{M},\mathbb{C}))$$

of the cohomology algebra $\mathrm{H}^*(\mathcal{M},\mathbb{C})$ is given by the following formulae :

$$P_\mathcal{M}(q) = (1 + (1 + d_1^\vee)q)\cdots(1 + (1 + d_r^\vee)q) = \sum_{w\in W} \det_V(w)(-q)^{\operatorname{codim}(V^{\langle w\rangle})}\,.$$

Coinvariant algebra and fake degrees

The coinvariant algebra $S_W = S/(S.(S_+^W)$, viewed as a KW–module, is isomorphic to the regular representation of W (see [**Bou**], chap V, 5.2, th. 2, (ii), or [**Ch**], p.779). Moreover we have an isomorphism of graded KW–modules (cf. for example [**Bou**], chap. V, §5, th. 2) :

$$S \simeq S_W \otimes_K R .$$

Thus the graded dimension $\sum_n \dim S_W^n x^n$ of S_W is

$$\mathrm{grdim} S_W(x) = \prod_{j=1}^{j=r} (1 + x + \cdots + x^{d_j - 1}) .$$

Since $\sum_{j=1}^{j=r}(d_j - 1) = N^\vee$, it follows that

$$S_W = \bigoplus_{n=0}^{n=N^\vee} S_W^n \quad \text{with} \quad \dim S_W^0 = \dim S_W^{N^\vee} = 1 . \tag{1.19}$$

Let χ be a character of W. *The fake degree of* χ is the element of $\mathbb{Z}[x]$ defined as follows:

$$\mathrm{Feg}_\chi(x) := \sum_n \langle \mathrm{tr}(\cdot, S_W^n), \chi \rangle \, x^n . \tag{1.20}$$

In particular, if χ is (absolutely) irreducible, its fake degree is the "graded" multiplicity of an irreducible representation with character χ in the graded module S_W.

1.21. Example. Assume that W is a cyclic group of order e. The set of irreducible characters consists of the characters $\det_V^j$ for $0 \leq j < e$. Then the fake degree of $\det_V^j$ is x^j.

The following formulae are well known and they may be found, for example, in [**Sp**] (see also chap. 4 below).

(1) We have

$$\mathrm{Feg}_\chi(x) = \frac{1}{|W|} \sum_{w \in W} \frac{\chi(w)}{\det(1 - xw)^*} \left(\prod_{j=1}^{j=r} (1 - x^{d_j}) \right) .$$

(2) Since the coinvariant algebra S_W, viewed as a KW–module, is isomorphic to the regular representation of W, we have

$$\mathrm{Feg}_\chi(1) = \chi(1) . \tag{1.22}$$

The integer $N(\chi)$ is defined as

$$N(\chi) := \frac{d}{dx} \mathrm{Feg}_\chi(x)_{|x=1} .$$

One writes sometimes

$$\mathrm{Feg}_\chi(x) = x^{e_1(\chi)} + \cdots + x^{e_{\chi(1)}(\chi)} ,$$

where the integers $e_j(\chi)$ are called the *exponents of* χ. Then we have

$$N(\chi) = e_1(\chi) + \cdots + e_{\chi(1)}(\chi)\,.$$

The following result is a reformulation of a theorem of Gutkin (see [**Gu**]).

1.23. PROPOSITION. *We have*

$$N(\chi) = \sum_{H\in\mathcal{A}} N(\mathrm{Res}^W_{W_H}\chi)\,,$$

where in the right hand side N is interpreted with respect to the reflection group (V, W_H).

PROOF OF 1.23. Let us consider the element $S_\chi(x) \in K(x)$ defined by

$$S_\chi(x) := \frac{1}{|W|}\sum_{w\in W}\frac{\chi(w)}{\det(1-xw)^*} = \left(\prod_{j=1}^{j=r}(1-x^{d_j})\right)^{-1}\mathrm{Feg}_\chi(x)\,.$$

We define $\psi(\chi)$ by writing the first terms of the expansion of $S_\chi(x)$ as a Laurent series in $(1-x)$ as follows:

$$S_\chi(x) = \frac{\chi(1)}{|W|}\frac{1}{(1-x)^r} + \frac{\psi(\chi)}{|W|}\frac{1}{(1-x)^{r-1}} + \cdots.$$

Let us compute $\psi(\chi)$ in two different ways.

- We have $\psi(\chi) = |W|\dfrac{d}{dx}((1-x)^r S_\chi(x))_{|x=1}$. Since

$$\prod_{j=1}^{j=r} d_j = |W| \quad\text{and}\quad \sum_{j=1}^{j=r}(d_j-1) = N^\vee\,,$$

an easy computation gives

$$\psi(\chi) = \frac{N^\vee}{2} - N(\chi)\,.$$

- On the other hand, if $\mathrm{Ref}(W)$ denotes the set of all reflections in W, we have

$$\psi(\chi) = \sum_{\rho\in\mathrm{Ref}(W)}\frac{\chi(\rho)}{1-\det(\rho)^*}\,.$$

Thus we get

$$N(\chi) = \frac{N^\vee}{2} - \sum_{\rho\in\mathrm{Ref}(W)}\frac{\chi(\rho)}{1-\det(\rho)^*} = \sum_{\rho\in\mathrm{Ref}(W)}\frac{1-\det(\rho)^*+2\chi(\rho)}{2(1-\det(\rho)^*)}\,. \tag{1.24}$$

Since $\mathrm{Ref}(W)$ is the disjoint union of the sets $\mathrm{Ref}(W_H)$ for $H\in\mathcal{A}$, 1.23 results now from the preceding formula. □

Let us denote by $m^\chi_{\mathcal{C},j}$ the multiplicity of $\det^j_V$ as a constituent of $\mathrm{Res}^W_{W_H}\chi$ (for $H\in\mathcal{C}$). In other words, we have, for all integers k,

$$\chi(s^k_H) = \sum_{j=0}^{j=e_\mathcal{C}-1} m^\chi_{\mathcal{C},j}\det{}^j_V(s^k_H)\,.$$

We denote by χ^* the complex conjugate of the character χ, which is then the character of the contragredient representation of a representation defining χ.

The following properties may be found in [**BrMi2**], 4.1 and 4.2.

1.25. PROPOSITION. *Whenever* $\chi \in \mathrm{Irr}(W)$, *we have*

(1) $N(\chi) = \sum_{C\in\mathcal{A}/W} \sum_{j=0}^{j=e_C-1} j\, N_C\, m_{C,j}^{\chi}$,

(2) $N(\chi) + N(\chi^*) = \sum_{C\in\mathcal{A}/W} \sum_{j=1}^{j=e_C-1} N_C e_C m_{C,j}^{\chi}$,

(3) $\chi(1)(N + N^{\vee}) - (N(\chi) + N(\chi^*)) = \sum_{C\in\mathcal{A}/W} N_C e_C m_{C,0}^{\chi}$.

Coxeter–like presentations

Coxeter–like presentations and minimal number of "generating reflections".

One says that W has a Coxeter–like presentation (cf. [**Op2**], 5.2) if it has a presentation of the form

$$\langle s \in S \mid \{v_i = w_i\}_{i\in I}, \{s^{e_s} = 1\}_{s\in S}\rangle$$

where S is a finite set of distinguished reflections, and I is a finite set of relations which are multi–homogeneous, *i.e.*, such that (for each i) v_i and w_i are positive words with the same length in elements of S.

Bessis ([**Bes3**], Thm. 0.1) has recently proven *a priori* (without using the Shephard–Todd classification) a general result about presentations of braid groups (see below 2.27) which implies that any complex reflection group has a Coxeter–like presentation.

The following property follows partially from the main theorem of [**Bes3**], and partially from a case–by–case analysis (see *loc.cit.*, 4.2).

The first two assertions are by–products of analogous results about associated braid groups (see below 2.28). The third assertion is proved in [**BMR**] through a case–by–case analysis.

1.26. THEOREM. *Let* (V, W) *be a complex reflection group. We set* $r := \dim(V)$. *Let* $(d_1, d_2, \ldots, d_r)$ *be the family of its invariant degrees, ordered to that* $d_1 \le d_2 \le \cdots \le d_r$. *Let* $g(W)$ *denote the minimal number of reflections needed to generate* W.

(1) *We have* $g(W) = \lceil (N + N^{\vee})/d_r \rceil$.

(2) *We have either* $g_W = r$ *or* $g_W = r + 1$.

(3) *The group* W *has a Coxeter–like presentation by* g_W *reflections.*

The tables in Appendix 1 provide a complete list of the irreducible finite pseudo–reflection groups, as classified by Shephard and Todd, together with Coxeter–like presentations of these groups symbolized by diagrams "à la Coxeter", as well as some of the data attached to these groups.

Many of these presentations were previously known. This is the case of the rank r groups which are generated by r reflections, studied by Coxeter [**Cx**]. Some others (the ones corresponding to the infinite series) occurred in [**BrMa**] or were inspired by [**Ari**].

The reader may refer to Appendix 1 to understand what follows.

Isomorphisms between diagrams.

We may notice that the only isomorphisms between the diagrams of our tables are between the diagrams of $G(2,1,2)$ and $G(4,4,2)$, between the diagrams of $\mathfrak{S}_4$ and $G(2,2,3)$, between the diagrams of $\mathfrak{S}_3$ and $G(3,3,2)$, and between the diagrams of $\mathfrak{S}_2$ and $G(2,1,1)$.

Coxeter groups.

- $G(e,e,2)$ $(e \geq 3)$ is the dihedral group of order $2e$,
- G_{28} is the Coxeter group of type F_4,
- G_{35} is the Coxeter group of type E_6,
- G_{36} is the Coxeter group of type E_7,
- G_{37} is the Coxeter group of type E_8,
- G_{23} is the Coxeter group of type H_3,
- G_{30} is the Coxeter group of type H_4.

Admissible subdiagrams and parabolic subgroups.

Let $\mathcal{D}$ be one of the diagrams. Let us define an equivalence relation between nodes by $s \sim s$ and, for $s \neq t$,

$$s \sim t \iff s \text{ and } t \text{ are not in a homogeneous relation with support } \{s,t\}\,.$$

Then we see that the equivalence classes have 1 or 3 elements, and that there is at most one class with 3 elements.

If there is no class with 3 elements, the rank r of the group is the number of nodes of the diagram, while it is this number minus 1 in case there is a class with 3 elements. Thus s ② ② t 5 ③ u has rank 2, as well as ④ ③ s t .

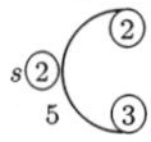

Remark. One must point out that, in the first of the preceding two diagrams, s, t and u must be considered as linked by a line (so t and u *do not* commute).

An **admissible subdiagram** is a full subdiagram of the same type, namely a diagram with 1 or 3 elements per class.

Thus, the diagram s ② ② t 5 ③ u has five admissible subdiagrams, namely the empty diagram, the three diagrams consisting of one node, and the whole diagram.

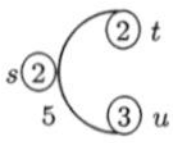

1.27. Fact. *Let $\mathcal{D}$ be the diagram of W as given in tables 1 to 4 in Appendix 2 below.*

(1) *If $\mathcal{D}'$ is an admissible subdiagram of $\mathcal{D}$, it gives a presentation of the corresponding subgroup $W(\mathcal{D}')$ of W. This subgroup is a parabolic subgroup.*
(2) *Assume W is neither G_{27}, G_{29}, G_{33} nor G_{34}. If $P_1 \subseteq P_2 \subseteq \cdots P_n$ is a chain of parabolic subgroups of W, there exist $g \in W$ and a chain $\mathcal{D}_1 \subseteq \mathcal{D}_2 \subseteq \cdots \mathcal{D}_n$ of admissible subdiagrams of $\mathcal{D}$ such that*

$$(P_1, P_2, \ldots, P_n) = {}^g(W(\mathcal{D}_1), W(\mathcal{D}_2), \ldots, W(\mathcal{D}_n))\,.$$

Remark.

For groups G_{27} and G_{29}, all isomorphism classes of parabolic subgroups are represented by admissible subdiagrams of our diagrams, but not all *conjugacy classes* of parabolics subgroups are represented by admissible subdiagrams, as noticed by Orlik.

For groups G_{33} and G_{34}, not all isomorphism classes of parabolic subgroups are represented by admissible subdiagrams of our diagrams.

CHAPTER II

ASSOCIATED BRAID GROUPS

Notation

For X a topological space, we denote by $\mathcal{P}(X)$ its fundamental groupoid, where the composition of (classes of) paths is defined so that, if γ_1 is a path going from x_0 to x_1 and γ_2 is a path going from x_1 to x_2, then the composite map going from x_0 to x_2 is denoted by $\gamma_2 \cdot \gamma_1$.

Given a point $x_0 \in X$, we denote by $\pi_1(X, x_0)$ (or $\pi_1(X)$ if the choice of x_0 is clear) the fundamental group with base point x_0. So we have $\pi_1(X, x_0) = \mathrm{End}_{\mathcal{P}(X)}(x_0)$. If $f\colon X \to Y$ is a continuous map, we denote by $\mathcal{P}(f)$ the corresponding functor from $\mathcal{P}(X)$ to $\mathcal{P}(Y)$. We also denote by $\pi_1(f, x_0)$ (or $\pi_1(f)$) the group homomorphism from $\pi_1(X, x_0)$ to $\pi_1(Y, f(x_0))$ induced by $\mathcal{P}(f)$.

We choose, once for all, a square root of (-1) in $\mathbb{C}$, which is denoted by i. Moreover, for every $z \in \mathbb{C}^\times$, we identify $\pi_1(\mathbb{C}^\times, z)$ with $\mathbb{Z}$ by sending onto 1 the loop $\lambda_z\colon [0,1] \to \mathbb{C}^\times$ defined by $\lambda_z(t) := z\exp(2i\pi t)$.

Generalities about hyperplane complements

What follows is probably well known to specialists of hyperplane complements and topologists. We include it for the convenience of the reader, and because of the lack of convenient references.

Distinguished braid reflections around an irreducible divisor.

We define here what we mean by a "*distinguished braid reflection around an irreducible divisor*", usually called "generator of the monodromy" (around this divisor), and we recall some well known properties.

Let Y be a smooth connected complex algebraic variety, I a finite family of irreducible codimension 1 closed subvarieties (irreducible divisors) and $Z := \cup_{D \in I} D$. Let $X := Y - Z$ and $x_0 \in X$.

For $D \in I$, let D_s be the smooth part of D and let us define $\tilde{D} := D_s - \left(D_s \cap \bigcup_{D' \in I, D' \neq D} D'\right)$.

A "*path from x_0 to D in X*" is by definition a path γ in Y such that $\gamma(0) = x_0$, $\gamma(1) \in \tilde{D}$ and $\gamma(t) \in X$ for $t \neq 1$.

Let γ' be another path from x_0 to D in X. We say that γ and γ' are *D-homotopic* if there is a continuous map $T : [0,1] \times [0,1] \to Y$ such that $T(t,0) = \gamma(t)$ and $T(t,1) = \gamma'(t)$ for $t \in [0,1]$, $T(0,u) = x_0$ and $T(1,u) \in \tilde{D}$ for all $u \in [0,1]$ and $T(t,u) \in X$ for $t \in [0,1[$ and $u \in [0,1]$. We denote by $[\gamma]$ the D-homotopy class of γ.

Given a path γ from x_0 to D in X, let B be a connected open neighbourhood of $\gamma(1)$ in $X \cup \tilde{D}$ such that $B \cap X$ has a fundamental group free abelian of rank 1. Let $u \in [0,1[$ such that $\gamma(t) \in B$ for $t \geq u$. Put $x_1 := \gamma(u)$. The orientation of $B \cap X$ coming from the orientation of X gives an isomorphism $f\colon \pi_1(B \cap X, x_1) \xrightarrow{\sim} \mathbb{Z}$. Let λ be a loop in $B \cap X$ from x_1 such that $f([\lambda]) = 1$.

Let γ_u be the "restriction" of γ to $[0,u]$, defined by $\gamma_u(t) := \gamma(ut)$ for all $t \in [0,1]$. Define $\rho_{\gamma,\lambda} := \gamma_u{}^{-1} \cdot \lambda \cdot \gamma_u$. Then, the homotopy class of $\rho_{\gamma,\lambda}$ in $\pi_1(\mathcal{M}, x_0)$ depends only on the D-homotopy class of γ and is denoted by $\rho_{[\gamma]}$. We call it *the distinguished braid reflection around D associated to* $[\gamma]$.

Given two paths γ and γ' from x_0 to D, the distinguished braid reflections $\rho_{[\gamma]}$ and $\rho_{[\gamma']}$ are conjugate.

2.1. PROPOSITION. *Let i be the injection of an irreducible divisor D in a smooth connected complex variety Y and $x_0 \in Y - D$. Then, the kernel of the morphism $\pi_1(i)\colon \pi_1(Y - D, x_0) \to \pi_1(Y, x_0)$ is generated by all the distinguished braid reflections around D.*

Indeed, note that the singular points of D form a closed subvariety D_{sing} of D, distinct from D, hence of (complex) codimension at least 2 in Y. Therefore the natural morphism $\pi_1(Y - D - D_{\text{sing}}, x_0) \to \pi_1(Y - D, x_0)$ is an isomorphism, and in order to prove 2.1 we may assume D is smooth, which we do now.

The lemma then follows from the fact that given a locally constant sheaf $\mathcal{F}$ over $Y - D$, its extension $i_*\mathcal{F}$ to Y is locally constant if and only if every distinguished braid reflection around D acts trivially on $\mathcal{F}$.

2.2. PROPOSITION. *Suppose that Y is simply connected. The fundamental group $\pi_1(X, x_0)$ is generated by all the distinguished braid reflections around the divisors $D \in I$.*

Indeed, this follows immediately from Proposition 2.1 by induction on $|I|$.

Lifting distinguished braid reflections.

Let $p\colon Y \to \overline{Y}$ be a finite covering between two smooth connected complex varieties. Let D be the branch locus of p and $\overline{D} = p(D)$. We assume $\overline{D}$ is an irreducible divisor. We set $\overline{X} := \overline{Y} - \overline{D}$ and $X := Y - D$.

We shall see that a distinguished braid reflection around $\overline{D}$ (associated to a path $\overline{\gamma}$ from $\overline{x}_0$ to $\overline{D}$ in $\overline{Y}$) may be naturally lifted to an element of $\mathcal{P}(X)$ (which depends only on the $\overline{D}$-homotopy class of $\overline{\gamma}$).

Indeed, let γ be the path from x_0 to an irreducible component, say D_γ, of D, which lifts $\overline{\gamma}$. Let $\overline{B}$ be an open neighbourhood of $\overline{x}_0$ in $\overline{Y}$ such that the fundamental group of $\overline{B} \cap \overline{X}$ is free abelian of rank 1 and $B \cap (X \cup \tilde{D}_\gamma) \to \overline{B}$ is unramified outside D_γ. Let $u \in [0,1[$ such that $\overline{\gamma}(t) \in \overline{B}$ for $t \geq u$. Let $\overline{\lambda}$ be a loop in $\overline{B} \cap \overline{X}$ with origin $\overline{\gamma}(u)$ which is a positive generator of $\pi_1(\overline{B} \cap \overline{X}, \overline{\gamma}(u))$.

Let λ be the path from $\gamma(u)$ which lifts $\overline{\lambda}$. Let γ_u be the restriction of γ to $[0,u]$. Let $\gamma_u^\vee$ be the path from $\lambda(1)$ which lifts $(\overline{\gamma}_u)^{-1}$, where $\overline{\gamma}_u$ is the "restriction" of $\overline{\gamma}$ to $[0,u]$.

The proof of the following proposition is left to the reader.

2.3. PROPOSITION. *We define $\rho_\gamma := \gamma_u^\vee \cdot \lambda \cdot \gamma_u$.*

(1) *The homotopy class of ρ_γ in $\mathcal{P}(X)$ depends only on the $\overline{D}$-homotopy class of $\overline{\gamma}$.*
(2) *Let e_D denote the ramification index of p on $\tilde{D}$. Then $\rho_\gamma^{e_D}$ is the distinguished braid reflection around D_γ associated to γ.*

Hyperplane complements.

Let $\mathcal{A}$ be a finite set of affine hyperplanes (*i.e.*, affine subspaces of codimension one) in a finite dimensional complex vector space V. We set $\mathcal{M} := V - \bigcup_{H \in \mathcal{A}} H$.

Let $x_0 \in \mathcal{M}$. We shall give now some properties of the fundamental group $\pi_1(\mathcal{M}, x_0)$.

Distinguished braid reflections around the hyperplanes.

For $H \in \mathcal{A}$, let α_H be an affine map $V \to \mathbb{C}$ such that $H = \{x \in V \mid \alpha_H(x) = 0\}$. Its restriction to $\mathcal{M} \to \mathbb{C}^\times$ induces a functor $\mathcal{P}(\alpha_H) \colon \mathcal{P}(\mathcal{M}) \to \mathcal{P}(\mathbb{C}^\times)$, and in particular a group homomorphism $\pi_1(\alpha_H, x_0) \colon \pi_1(\mathcal{M}, x_0) \to \mathbb{Z}$

2.4. LEMMA. *For $H, H' \in \mathcal{A}$ and γ a path from x_0 to H, we have*

$$\pi_1(\alpha_{H'})(\rho_{[\gamma]}) = \delta_{H,H'} \,.$$

Indeed, let us set $\mathcal{M}_H := H - \bigcup_{\substack{H' \in \mathcal{A} \\ H' \neq H}} H'$. Let $x_\gamma := \gamma(1)$ and let B be an open ball with center x_γ contained in $\mathcal{M} \cup \mathcal{M}_H$. Let $u \in [0,1[$ such that $\gamma(t) \in B$ for $t \geq u$. We set $x_1 := \gamma(u)$. Then, the restriction of α_H to $B \cap \mathcal{M}$ induces an isomorphism $\pi_1(\alpha_H) \colon \pi_1(B \cap \mathcal{M}, x_1) \to \mathbb{Z}$. Let λ be a loop in $B \cap \mathcal{M}$, with origin x_1, whose image under $\pi_1(\alpha_H)$ is 1. Let γ_u be the "restriction" of γ to $[0,u]$, defined by $\gamma_u(t) := \gamma(ut)$ for all $t \in [0,1]$. Define $\rho_{\gamma,\lambda} := {\gamma_u}^{-1} \cdot \lambda \cdot \gamma_u$. Then the loop $\rho_{\gamma,\lambda}$ induces the distinguished braid reflection $\rho_{[\gamma]}$, and

$$\pi_1(\alpha_{H'})(\rho_{\gamma,\lambda}) = \pi_1(\alpha_{H'})(\lambda) = \delta_{H,H'} \,.$$

The following proposition is immediate.

2.5. PROPOSITION.

(1) *The fundamental group $\pi_1(\mathcal{M}, x_0)$ is generated by all the distinguished braid reflections around the affine hyperplanes $H \in \mathcal{A}$.*

(2) *Let $\pi_1(\mathcal{M}, x_0)^{\mathrm{ab}}$ denote the largest abelian quotient of $\pi_1(\mathcal{M}, x_0)$. For $H \in \mathcal{A}$, we denote by ρ_H^{ab} the image of $\rho_{H,\gamma}$ in $\pi_1(\mathcal{M}, x_0)^{\mathrm{ab}}$. Then*

$$\pi_1(\mathcal{M}, x_0)^{\mathrm{ab}} = \prod_{H \in \mathcal{A}} \langle \rho_H^{\mathrm{ab}} \rangle \,,$$

where each $\langle \rho_H^{\mathrm{ab}} \rangle$ is infinite cyclic. Dually, we have

$$\mathrm{Hom}(\pi_1(\mathcal{M}, x_0), \mathbb{Z}) = \prod_{H \in \mathcal{A}} \langle \pi_1(\alpha_H) \rangle \,.$$

REMARK. Let us recall that we have natural isomorphisms

$$\pi_1(\mathcal{M}, x_0)^{\mathrm{ab}} \xrightarrow{\sim} \mathrm{H}_1(\mathcal{M}, \mathbb{Z}) \quad \text{and} \quad \mathrm{Hom}(\pi_1(\mathcal{M}, x_0), \mathbb{Z}) \xrightarrow{\sim} \mathrm{H}^1(\mathcal{M}, \mathbb{Z}) \,.$$

Moreover, the duality between $\pi_1(\mathcal{M}, x_0)^{\mathrm{ab}}$ and $\mathrm{H}^1(\mathcal{M}, \mathbb{Z})$ may be seen as follows. For γ a loop in $\mathcal{M}$ with origin x_0 and for ω a holomorphic differential 1-form on $\mathcal{M}$, we set $\langle \gamma, \omega \rangle := \int_\gamma \omega$. It is then clear that, under the isomorphism

$$\mathrm{Hom}(\pi_1(\mathcal{M}, x_0), \mathbb{Z}) \xrightarrow{\sim} \mathrm{H}^1(\mathcal{M}, \mathbb{Z}) \,,$$

the element $\pi_1(\alpha_H)$ is sent onto the class of the 1-form $\omega_H = \dfrac{1}{2i\pi} \dfrac{d\alpha_H}{\alpha_H}$ (see 1.14 for more details).

About the center of the fundamental group.

In this part, we assume the hyperplanes in $\mathcal{A}$ to be *linear*.

2.6. NOTATION. *We denote by* $\boldsymbol{\pi}$ *the loop* $[0,1] \to \mathcal{M}$ *defined by*

$$\boldsymbol{\pi} : t \mapsto x_0 \exp(2i\pi t) .$$

2.7. LEMMA.

(1) $\boldsymbol{\pi}$ *belongs to the center* $Z(\pi_1(\mathcal{M}, x_0))$ *of the fundamental group* $\pi_1(\mathcal{M}, x_0)$.

(2) *For all* $H \in \mathcal{A}$, *we have* $\pi_1(\alpha_H)(\boldsymbol{\pi}) = 1$.

Generalities about the braid groups

More notation.

We go back to notation introduced in chap. 1. In particular, $\mathcal{A}$ is now the set of reflecting hyperplanes of a finite subgroup W of $\mathrm{GL}(V)$ generated by pseudo-reflections. We denote by $p\colon \mathcal{M} \to \mathcal{M}/W$ the canonical surjection.

Let $x_0 \in \mathcal{M}$. We introduce the following notation for the fundamental groups:

$$P := \pi_1(\mathcal{M}, x_0) \quad \text{and} \quad B := \pi_1(\mathcal{M}/W, p(x_0)) ,$$

and we call B and P respectively the *braid group* (at x_0) and the *pure braid group* (at x_0) associated to W. We shall often write $\pi_1(\mathcal{M}/W, x_0)$ for $\pi_1(\mathcal{M}/W, p(x_0))$.

The covering $\mathcal{M} \to \mathcal{M}/W$ is Galois by Steinberg's theorem (see Theorem 1.2 above), hence the projection p induces a surjective map $B \twoheadrightarrow W$, $\sigma \mapsto \overline{\sigma}$, as follows :

Let $\tilde{\sigma}\colon [0,1] \to \mathcal{M}$ be a path in $\mathcal{M}$, such that $\tilde{\sigma}(0) = x_0$, which lifts σ. Then $\overline{\sigma}$ is defined by the equality $\overline{\sigma}(x_0) = \tilde{\sigma}(1)$.

Note that the map $\sigma \mapsto \overline{\sigma}$ is an *anti-morphism*.

Denoting by W^{op} the group opposite to W, we have the following short exact sequence :

$$1 \to P \to B \to W^{\mathrm{op}} \to 1 , \tag{2.8}$$

where the map $B \to W^{\mathrm{op}}$ is defined by $\sigma \mapsto \overline{\sigma}$.

The spaces $\mathcal{M}$ and $\mathcal{M}/W$ are conjectured to be $K(\pi,1)$-spaces.

The following result is due to Fox and Neuwirth [**FoNe**] for the type A_n, to Brieskorn [**Br2**] for Coxeter groups of type different from H_3, H_4, E_6, E_7, E_8, to Deligne [**De2**] for general Coxeter groups. The case of the infinite series of complex reflection groups $G(de, e, r)$ has been solved by Nakamura [**Na**]. For the non-real Shephard groups (non-real groups with Coxeter braid diagrams), this has been proven by Orlik and Solomon [**OrSo3**]. Note that the rank 2 case is trivial.

2.9. THEOREM. *Assume* W *has no irreducible component of type* G_{24}, G_{27}, G_{29}, G_{31}, G_{33} *or* G_{34}. *Then,* $\mathcal{M}$ *and* $\mathcal{M}/W$ *are* $K(\pi,1)$*-spaces.*

Distinguished braid reflections around the hyperplanes.

For $H \in \mathcal{A}$, we set $\zeta_H := \zeta_{e_H}$, We recall that we denote by s_H and call *distinguished reflection* the pseudo-reflection in W with reflecting hyperplane H and determinant ζ_H. We set

$$L_H := \mathrm{im}(s_H - \mathrm{Id}_V) .$$

For $x \in V$, we set $x = \mathrm{pr}_H(x) + \mathrm{pr}_H^\perp(x)$ with $\mathrm{pr}_H(x) \in H$ and $\mathrm{pr}_H^\perp(x) \in L_H$.

Thus, we have $s_H(x) = \zeta_H \mathrm{pr}_H^\perp(x) + \mathrm{pr}_H(x)$.

If $t \in \mathbb{R}$, we set $\zeta_H^t := \exp(2i\pi t/e_H)$, and we denote by s_H^t the element of $\mathrm{GL}(V)$ (a pseudo–reflection if $t \neq 0$) defined by :

(2.10) $$s_H^t(x) = \zeta_H^t \mathrm{pr}_H^\perp(x) + \mathrm{pr}_H(x)\,.$$

For $x \in V$, we denote by $\sigma_{H,x}$ the path in V from x to $s_H(x)$, defined by :

$$\sigma_{H,x}\colon [0,1] \to V\ ,\ t \mapsto s_H^t(x)\,.$$

For any path γ in $\mathcal{M}$, with initial point x_0 and terminal point x_H, the path defined by $s_H(\gamma^{-1})\colon t \mapsto s_H(\gamma^{-1}(t))$ is a path in $\mathcal{M}$ going from $s_H(x_H)$ to $s_H(x_0)$.

Whenever γ is a path in $\mathcal{M}$, with initial point x_0 and terminal point x_H, we define the path $\sigma_{H,\gamma}$ from x_0 to $s_H(x_0)$ as follows :

(2.11) $$\sigma_{H,\gamma} := s_H(\gamma^{-1}) \cdot \sigma_{H,x_H} \cdot \gamma\,.$$

It is not difficult to see that, provided x_H is chosen "close to H and far from the other reflecting hyperplanes", the path $\sigma_{H,\gamma}$ is in $\mathcal{M}$, and its homotopy class does not depend on the choice of x_H, and the element it induces in the braid group B is actually a distinguished braid reflection around the image of H in $\mathcal{M}/W$.

The following properties are immediate.

2.12. Lemma.

(1) *The image of* $\mathbf{s}_{H,\gamma}$ *in* W *is* s_H.

(2) *Whenever* γ' *is a path in* $\mathcal{M}$*, with initial point* x_0 *and terminal point* x_H*, if* τ *denotes the loop in* $\mathcal{M}$ *defined by* $\tau := {\gamma'}^{-1}\gamma$*, one has*

$$\sigma_{H,\gamma'} = \tau \cdot \sigma_{H,\gamma} \cdot \tau^{-1}$$

and in particular $\mathbf{s}_{H,\gamma}$ *and* $\mathbf{s}_{H,\gamma'}$ *are conjugate in* P.

(3) *The path* $\prod_{j=e_H-1}^{j=0} \sigma_{H,s_H^j(\gamma)}$*, a loop in* $\mathcal{M}$*, induces the element* $\mathbf{s}_{H,\gamma}^{e_H}$ *in the braid group* B*, and belongs to the pure braid group* P*. It is homotopy equivalent, as a loop in* $\mathcal{M}$*, to the distinguished braid reflection* $\rho_{[\gamma]}$ *around* H *in* P.

2.13. Definition. *Let* s *be a distinguished pseudo–reflection in* W*, with reflecting hyperplane* H*. An* s*–distinguished braid reflection is a distinguished braid reflection* $\mathbf{s}$ *around the image of* H *in* $\mathcal{M}/W$ *such that* $\overline{\mathbf{s}} = s$.

Discriminants and length

Let $\mathcal{C}$ be an orbit of W on $\mathcal{A}$. Recall that we denote by $e_\mathcal{C}$ the (common) order of the pointwise stabilizer W_H for $H \in \mathcal{C}$. We call *discriminant at* $\mathcal{C}$ and we denote by $\delta_\mathcal{C}$ the element of the symmetric algebra of $V^\vee$ defined (up to a non zero scalar multiplication) by

$$\delta_\mathcal{C} := (\prod_{H\in\mathcal{C}} \alpha_H)^{e_\mathcal{C}}\,.$$

Since (see for example [**Co**], 1.8) $\delta_\mathcal{C}$ is W-invariant, it induces a continuous function $\delta_\mathcal{C}\colon \mathcal{M}/W \to \mathbb{C}^\times$, hence induces a functor $\mathcal{P}(\delta_\mathcal{C})\colon \mathcal{P}(\mathcal{M}/W) \to \mathcal{P}(\mathbb{C}^\times)$, and in particular it induces a group homomorphism $\pi_1(\delta_\mathcal{C})\colon B \to \mathbb{Z}$.

2.14. PROPOSITION. *For any $H \in \mathcal{A}$, we have*

$$\pi_1(\delta_{\mathcal{C}})(\mathbf{s}_{H,\gamma}) = \begin{cases} 1 & \text{if } H \in \mathcal{C}\,, \\ 0 & \text{if } H \notin \mathcal{C}\,. \end{cases}$$

What precedes allows us to define length functions on B.

• There is a unique length function $\ell\colon B \to \mathbb{Z}$ defined as follows (see [**BMR**], Prop. 2.19): if $b = \mathbf{s}_1^{n_1} \cdot \mathbf{s}_2^{n_2} \cdots \mathbf{s}_m^{n_m}$ where (for all j) $n_j \in \mathbb{Z}$ and $\mathbf{s}_j$ is a distinguished braid reflection around an element of $\mathcal{A}$ in B, then

$$\ell(b) = n_1 + n_2 + \cdots + n_m\,.$$

Indeed, we set $\ell := \pi_1(\delta)$. Let $b \in B$. By Theorem 2.15 above, there exists an integer k and for $1 \leq j \leq k$, $H_j \in \mathcal{A}$, a path γ_j from x_0 to H_j and an integer n_j such that

$$b = \mathbf{s}_{H_1,\gamma_1}^{n_1} \mathbf{s}_{H_2,\gamma_2}^{n_2} \cdots \mathbf{s}_{H_k,\gamma_k}^{n_k}\,.$$

From Proposition 2.14 above, it then results that we have $\ell(b) = \sum_{j=1}^{j=k} n_j$.

If $\{\mathbf{s}\}$ is a set of distinguished braid reflections around hyperplanes which generates B, let us denote by B^+ the sub–monoid of B generated by $\{\mathbf{s}\}$. Then for $b \in B^+$, its length $\ell(b)$ coincide with its length on the distinguished set of generators $\{\mathbf{s}\}$ of the monoid B^+.

• More generally, given $\mathcal{C} \in \mathcal{A}/W$, there is a unique length function $\ell_{\mathcal{C}}\colon B \to \mathbb{Z}$ (this is the function denoted by $\pi_1(\delta_{\mathcal{C}})$ in [**BMR**], see Prop. 2.16 in *loc.cit.*) defined as follows: if $b = \mathbf{s}_1^{n_1} \cdot \mathbf{s}_2^{n_2} \cdots \mathbf{s}_m^{n_m}$ where (for all j) $n_j \in \mathbb{Z}$ and $\mathbf{s}_j$ is a distinguished braid reflection around an element of $\mathcal{C}_j$, then

$$\ell_{\mathcal{C}}(b) = \sum_{\{j \mid (\mathcal{C}_j = \mathcal{C})\}} n_j\,.$$

Thus we have, for all $b \in B$,

$$\ell(b) = \sum_{\mathcal{C} \in \mathcal{A}/W} \ell_{\mathcal{C}}(b)\,.$$

Generators and abelianization of B

2.15. THEOREM.

(1) *The group B is generated by the generators $\{\mathbf{s}_{H,\gamma}\}$ (for all hyperplanes $H \in \mathcal{A}$ and all paths γ from x_0 to H in $\mathcal{M}$) of the monodromy (in B) around the elements of $\mathcal{A}$.*

(2) *We denote by B^{ab} the largest abelian quotient of B. For $\mathcal{C} \in \mathcal{A}/W$, we denote by $\mathbf{s}_{\mathcal{C}}^{\mathrm{ab}}$ the image of $\mathbf{s}_{H,\gamma}$ in B^{ab} for $H \in \mathcal{C}$. Then*

$$B^{\mathrm{ab}} = \prod_{\mathcal{C} \in \mathcal{A}/W} \langle \mathbf{s}_{\mathcal{C}}^{\mathrm{ab}} \rangle\,,$$

where each $\langle \mathbf{s}_{\mathcal{C}}^{\mathrm{ab}} \rangle$ is infinite cyclic. Dually, we have

$$\mathrm{Hom}(B, \mathbb{Z}) = \prod_{\mathcal{C} \in \mathcal{A}/W} \langle \pi_1(\delta_{\mathcal{C}}) \rangle\,.$$

Indeed, the second assertion is immediate by the first one and by Proposition 2.14. Let us sketch a proof of (1).

Since W is generated by the set $\{s_H\}_{(H\in\mathcal{A})}$ and since we have the exact sequence (2.8), it is enough to prove that the pure braid group P is generated by all the conjugates in P of the elements $\mathbf{s}_{H,\gamma}^{e_H}$. This is a consequence of Proposition 2.5, (1).

REMARK. We have natural isomorphisms

$$B^{\mathrm{ab}} \xrightarrow{\sim} \mathrm{H}_1(\mathcal{M}/W,\mathbb{Z}) \quad \text{and} \quad \mathrm{Hom}(B,\mathbb{Z}) \xrightarrow{\sim} \mathrm{H}^1(\mathcal{M}/W,\mathbb{Z}),$$

and, under the second isomorphism, we have

$$\pi_1(\delta_C) \mapsto e_C \sum_{H\in C} \frac{1}{2i\pi}\frac{d\alpha_H}{\alpha_H} = \frac{1}{2i\pi}\mathrm{dLog}(\delta_C)\,.$$

Let us denote by $\mathrm{Gen}(B)$ the set of all distinguished braid reflections in B (see Definition 2.13 above). For $\mathbf{s} \in \mathrm{Gen}(B)$, we denote by $e_{\mathbf{s}}$ the order of $\overline{\mathbf{s}}$.

In other words, if s is a distinguished braid reflection around the reflecting hyperplane $H \in \mathcal{A}$, we set now (using the notation of Definition 2.13) : $e_{\mathbf{s}} := e_H$.

The following property is a consequence of general results recalled at the beginning of this section.

2.16. PROPOSITION.

(1) *The pure braid group P is generated by $\{\mathbf{s}^{e_{\mathbf{s}}}\}_{\mathbf{s}\in\mathrm{Gen}(B)}$.*

(2) *We have*

$$W \simeq B/\langle \mathbf{s}^{e_{\mathbf{s}}}\rangle_{\mathbf{s}\in\mathrm{Gen}(B)}\,.$$

About the center of B

2.17. NOTATION. *We denote by $\boldsymbol{\beta}$ the path $[0,1] \to \mathcal{M}$ defined by*

$$\boldsymbol{\beta}\colon t \mapsto x_0 \exp(2i\pi t/|Z(W)|)\,.$$

From now on, we assume that W *acts irreducibly on* V. Note that, since W is irreducible on V, it results from Schur's lemma that

$$Z(W) = \{\zeta_{|Z(W)|}^k \mid (k \in \mathbb{Z})\}\,,$$

and so in particular $\boldsymbol{\beta}$ defines an element of B, which we will still denote by $\boldsymbol{\beta}$.

2.18. LEMMA.

(1) *The image $\overline{\boldsymbol{\beta}}$ of $\boldsymbol{\beta}$ in W is the scalar multiplication by $\zeta_{|Z(W)|}$. It is a generator of the center $Z(W)$ of W.*

(2) *We have $\boldsymbol{\beta} \in Z(B)$, $\boldsymbol{\pi} \in Z(P)$, and $\boldsymbol{\pi} = \boldsymbol{\beta}^{|Z(W)|}$.*

The following proposition (see [**BMR**] 2.23) is now easy.

2.19. PROPOSITION. *Let $\overline{\mathcal{M}}$ be the image of $\mathcal{M}$ in $(V-\{0\})/\mathbb{C}^\times$. Then, we have a commutative diagram, where all short sequences are exact :*

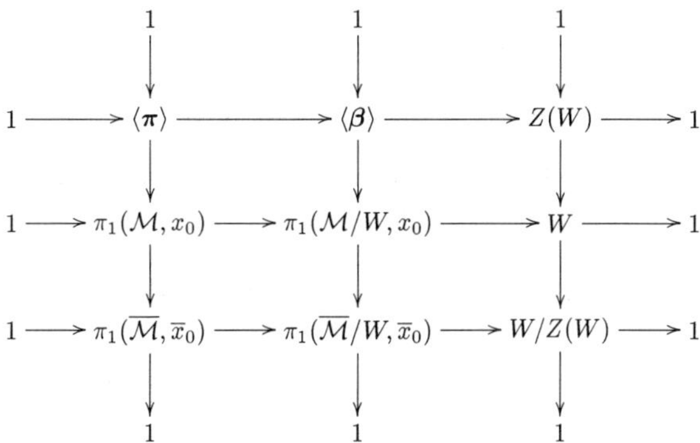

The following statement is known for Coxeter groups (see [**Del**] or [**BrSa**]). The result holds as well for G_{25}, G_{26}, G_{32}, since the corresponding braid groups are the same as braid groups of Coxeter groups. The proof for the particular case of groups in dimension 2 is easy (see [**BMR**]. A proof for all the infinite series is given in [**BMR**], §3. We conjecture it is still true in the case of G_{31}, as well as for G_{24}, G_{27}, G_{29}, G_{33}, G_{34}.

As for all results proved for all but a finite number of irreducible complex reflection groups and conjectured to be true in general, we call it a "*Theorem–Assumption*".

2.20. THEOREM–ASSUMPTION. *The center $Z(B)$ of B is infinite cyclic and generated by $\boldsymbol{\beta}$, the center $Z(P)$ of P is infinite cyclic and generated by $\boldsymbol{\pi}$, and the short exact sequence (2.8) induces a short exact sequence*

$$1 \to Z(P) \to Z(B) \to Z(W) \to 1\,.$$

2.21. COROLLARY. *Let $\boldsymbol{\beta}^{\mathrm{ab}}$ be the image in B^{ab} of the central element $\boldsymbol{\beta}$ of B. Then we have*

$$\boldsymbol{\beta}^{\mathrm{ab}} = \prod_{\mathcal{C}\in\mathcal{A}/W} (\mathbf{s}_{\mathcal{C}}^{\mathrm{ab}})^{e_{\mathcal{C}} N_{\mathcal{C}}/|Z(W)|}\,.$$

Indeed, it suffices to prove that, if $\mathcal{C}\in\mathcal{A}/W$, then $\pi_1(\delta_{\mathcal{C}})(\boldsymbol{\beta}^{\mathrm{ab}}) = e_{\mathcal{C}}N_{\mathcal{C}}/|Z(W)|$. This is immediate :

$$\begin{aligned}\pi_1(\delta_{\mathcal{C}})(\boldsymbol{\beta}^{\mathrm{ab}}) &= \sum_{H\in\mathcal{C}} \frac{e_{\mathcal{C}}}{2i\pi}\int_0^1 \frac{d\alpha_H(x_0\exp(2i\pi t/|Z(W)|))}{\alpha_H(x_0\exp(2i\pi t/|Z(W)|))} \\ &= \frac{e_{\mathcal{C}}}{2i\pi}\sum_{H\in\mathcal{C}} \frac{2i\pi}{|Z(W)|}\int_0^1 dt = e_{\mathcal{C}}N_{\mathcal{C}}/|Z(W)|\,.\end{aligned}$$

□

The following result is a consequence of Corollary 2.21. Notice that it generalizes a result of Deligne [**Del**], (4.21) (see also [**BrSa**]), from which it follows that if W is a Coxeter group, then $\ell(\boldsymbol{\pi}) = 2N$. It was noticed "experimentally" in [**BrMi2**], (4.8).

2.22. Corollary. *We have $\ell_{\mathcal{C}}(\boldsymbol{\beta}) = N_{\mathcal{C}}e_{\mathcal{C}}/|Z(W)|$ and $\ell_{\mathcal{C}}(\boldsymbol{\pi}) = N_{\mathcal{C}}e_{\mathcal{C}}$. In particular, we have $\ell(\boldsymbol{\beta}) = (N + N^\vee)/|Z(W)|$ and $\ell(\boldsymbol{\pi}) = N + N^\vee$.*

Roots of $\boldsymbol{\pi}$ and Bessis theorem

Here we follow [**BrMi**] and [**Bes2**]. The results of this paragraph should be viewed as the "lifting" up to braid groups level of Springer theory of regular elements.

Let d be a positive integer, let $\zeta \in \boldsymbol{\mu}_d$ and let w a ζ–regular element of W. Recall (see chap. 1 above) that we denote by $W(w)$ the centralizer of w in W, we set $V(w) := \ker(w - \zeta\mathrm{Id})$, $\mathcal{M}(w) := \mathcal{P} \cap V(w)$. By 1.4 we know that $(V(w), W(w))$ is a complex reflection group with regular variety $\mathcal{M}(w)$.

Let us choose a base point x_0 in $\mathcal{M}(w)$, and let us denote by $P(w)$ and $B(w)$ respectively the corresponding pure braid group and braid group.

Let us denote by $\mathbf{w}$ the element of B defined by the path

$$\mathbf{w}(t) \colon t \mapsto x_0 \exp(2\pi i t/d)\,.$$

The following proposition tells that a regular element can be lifted up to a root of $\boldsymbol{\pi}$ in B.

2.23. Proposition.
(1) *The image of $\mathbf{w}$ through the natural morphism $B \twoheadrightarrow W$ is w.*
(2) *We have $\mathbf{w}^d = \boldsymbol{\pi}$.*

Roots of $\boldsymbol{\pi}$ (at least, for the case where W is a Weyl group) seem to play a key role in representation theory of the corresponding finite reductive groups (see below chap. VI).

Let us notice the following consequence of 2.23 and 2.22.

2.24. Corollary. *Let d be a regular number for W. Whenever $\mathcal{C} \in \mathcal{A}/W$, then d divides $N_{\mathcal{C}}e_{\mathcal{C}}$. In particular, d divides $N + N^\vee$.*

[Indeed, we have $\ell_{\mathcal{C}}(\mathbf{w}) = N_{\mathcal{C}}e_{\mathcal{C}}/d$ and $\ell(\mathbf{w}) = (N + N^\vee)/d$.]

The natural maps

$$\mathcal{M}(w) \hookrightarrow \mathcal{M} \quad \text{and} \quad \mathcal{M}(w)/W(w) \hookrightarrow \mathcal{M}/W(w) \twoheadrightarrow \mathcal{M}/W$$

induce the following commutative diagram, where the lines are exact :

$$\begin{array}{ccccccccc}
1 & \longrightarrow & P(w) & \longrightarrow & B(w) & \longrightarrow & W(w)^{\mathrm{op}} & \longrightarrow & 1 \\
 & & \downarrow & & \downarrow & & \downarrow & & \\
1 & \longrightarrow & P & \longrightarrow & B & \longrightarrow & W^{\mathrm{op}} & \longrightarrow & 1
\end{array}$$

Answering positively (for almost all cases of complex reflection groups) a question raised in [**BrMi2**] (3.5), Bessis [**Bes2**] has proven the following theorem for the case where W is any irreducible complex reflection group different from the exceptional groups G_n for $n \in \{28, 30, 31, 33, 35, 36, 37\}$. Since we conjecture that the results holds in all cases, it is again a "theorem–assumption".

2.25. Theorem–Assumption. *The vertical arrows in the preceding diagram are all injective. In particular, the map $B(w) \longrightarrow B$ identifies $B(w)$ with a subgroup of the centralizer $C_B(\mathbf{w})$ of $\mathbf{w}$ in B.*

Remark. In the course of his proof, Bessis notices that the image of $\mathcal{M}(w)$ in $\mathcal{M}/W$ through any path of the following commutative diagram

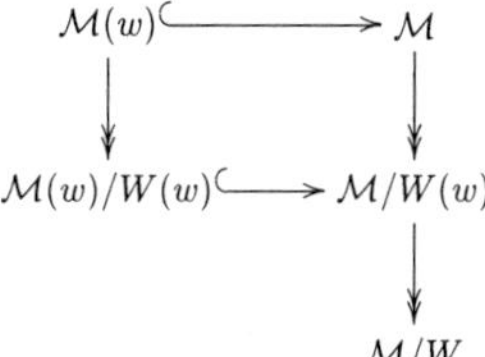

$$\begin{array}{ccc} \mathcal{M}(w) & \hookrightarrow & \mathcal{M} \\ \downarrow & & \downarrow \\ \mathcal{M}(w)/W(w) & \hookrightarrow & \mathcal{M}/W(w) \\ & & \downarrow \\ & & \mathcal{M}/W \end{array}$$

is actually contained in the variety $(\mathcal{M}/W)^{\boldsymbol{\mu}_d}$ of fixed points of the group $\boldsymbol{\mu}_d$ acting on $\mathcal{M}/W$, and that the natural map $\mathcal{M}(w)/W(w) \longrightarrow (\mathcal{M}/W)^{\boldsymbol{\mu}_d}$ is an homeomorphism. Then his result may be reformulated as follows : the image under π_1 of the natural injection $(\mathcal{M}/W)^{\boldsymbol{\mu}_d} \hookrightarrow \mathcal{M}/W$ is injective

A natural question (see [**BDM**], 0.1) consists then in asking whether $B(w)$ is isomorphic to $C_B(\mathbf{w})$. A partial positive answer to this question has been given in [**BDM**], 0.2.

2.26. Theorem. *Let (V, W) be an irreducible complex reflection group of the following types :*

$$\mathfrak{S}_r\,,\ G(d,1,r)\ (d>1)\,,\ G_n \text{ for } n \in \{4,5,8,10,16,18,25,26,32\}\,.$$

Then the natural injection $B(w) \hookrightarrow C_B(\mathbf{w})$ is an isomorphism

Notice that the braid diagrams (see below) of the groups listed in theorem 2.26 above are all braid diagrams of Coxeter groups.

Parabolic braid subgroups

Let X be an intersection of reflecting hyperplanes: $X = \bigcap_{\substack{H\in\mathcal{A}\\ X\subseteq H}} H$. We set

$$\mathcal{A}_X := \{H \in \mathcal{A} \mid (X \subseteq H)\} \quad \text{and} \quad \mathcal{M}_X := V - \bigcup_{H\in\mathcal{A}_X} H\,.$$

We recall (see 1.2 above) that the parabolic subgroup W_X, the pointwise stabilizer of X, is generated by all the cyclic reflection groups W_H for $H \in \mathcal{A}_X$.

For $x_0 \in \mathcal{M}$, we set

$$P_X := \pi_1(\mathcal{M}_X, x_0) \quad \text{and} \quad B_X := \pi_1(\mathcal{M}_X/W_X, p(x_0))\,.$$

We recall (see [**BMR**], Proposition 2.29) that there is a morphism of short exact sequences (see 2.8 above)

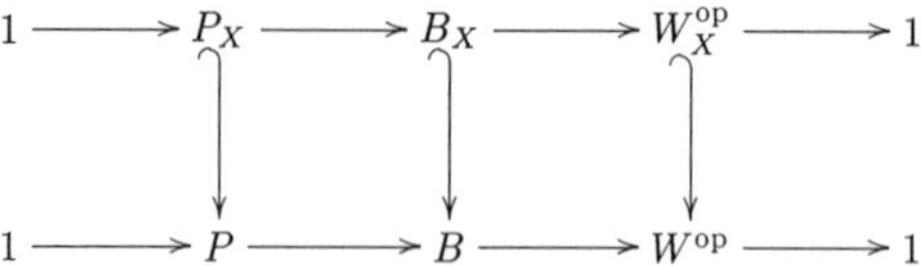

where all vertical arrows are injective, and where the corresponding injection $B_X \hookrightarrow B$ is well–defined up to P–conjugation.

Any subgroup of B which is the image of B_X by one of the preceding injections is called a *parabolic braid subgroup of B associated with X*. For a given X, such parabolic braid subgroups constitute a conjugacy class of subgroups under P.

Artin–like presentations and the braid diagrams

General results.

We say that B has an *Artin–like* presentation (cf. [**Op2**], 5.2) if it has a presentation of the form

$$\langle \mathbf{s} \in \mathbf{S} \mid \{\mathbf{v}_i = \mathbf{w}_i\}_{i \in I} \rangle$$

where $\mathbf{S}$ is a finite set of distinguished braid reflections, and I is a finite set of relations which are multi–homogeneous, *i.e.*, such that (for each i) $\mathbf{v}_i$ and $\mathbf{w}_i$ are positive words in elements of $\mathbf{S}$ (and hence, for each $\mathcal{C} \in \mathcal{A}/W$, we have $\ell_{\mathcal{C}}(\mathbf{v}_i) = \ell_{\mathcal{C}}(\mathbf{w}_i)$).

The following result, which has recently been proved by Bessis ([**Bes3**], Thm. 0.1), shows in particular that any braid group has an Artin–like presentation.

2.27. THEOREM. *Let W be a complex reflection group of rank r, with associated braid group B. Let d be one of the degrees of W. Assume that d is a regular number for W. Let $n := (N + N^\vee)/d$. Then there exists a subset $\mathbf{S} = \{\mathbf{s}_1, \dots, \mathbf{s}_n\}$ of B such that*

(1) *The elements $\mathbf{s}_1, \dots, \mathbf{s}_n$ are distinguished braid reflections, and therefore their images $s_1, \dots, s_n$ in W are distinguished reflections.*
(2) *The set $\mathbf{S}$ generates B, and therefore $S := \{s_1, \dots, s_n\}$ generates W.*
(3) *The product $(\mathbf{s}_1 \cdots \mathbf{s}_n)^d$ is central in B and belongs to the pure braid group P.*
(4) *The product $c := s_1, \dots, s_n$ is a ζ_d–regular element in W.*
(5) *There exists a set $\mathcal{R}$ of relations of the form $\mathbf{w}_1 = \mathbf{w}_2$, where $\mathbf{w}_1$ and $\mathbf{w}_2$ are positive words of equal length in the elements of $\mathbf{S}$, such that $\langle S \mid \mathcal{R} \rangle$ is a presentation of B.*
(6) *Viewing now $\mathcal{R}$ as a set of relations in S, the group W is presented by*

$$\langle S \mid \mathcal{R}; (\forall s \in S)(s^{e_s} = 1) \rangle$$

where e_s denotes the order of s in W.

The following result is mainly due to Bessis, who has shown that theorem 1.26 has a generalization in terms of braid groups (cf. [**Bes3**], 4.2).

The first assertion is due to Bessis. The second assertion is partially a consequence of *loc.cit.* and are partially proved in [**BMR**] through a case–by–case analysis.

2.28. THEOREM. *Let (V, W) be a complex reflection group. We set $r := \dim(V)$. Let $(d_1, d_2, \ldots, d_r)$ be the family of its invariant degrees, ordered to that $d_1 \leq d_2 \leq \cdots \leq d_r$.*

(1) *The following integers are equal.*
  (a) *The minimal number of reflections needed to generate W.*
  (b) *The minimal number of distinguished braid reflections around an hyperplane needed to generate B.*
  (c) $\lceil (N + N^\vee)/d_r \rceil$.

(2) *If $g(W)$ denotes the integer defined by properties (a) to (c) above, we have either $g_W = r$ or $g_W = r+1$, and the group B has an Artin-like presentation by g_W reflections.*

The braid diagrams.

Let us first introduce some more notation.

Let (V, W) be a finite irreducible complex reflection group. As previously, we set $\mathcal{M} := V - \bigcup_{H \in \mathcal{A}} H$, $B := \pi_1(\mathcal{M}/W, x_0)$, and we denote by $\sigma \mapsto \overline{\sigma}$ the antimorphism $B \twoheadrightarrow W$ defined by the Galois covering $\mathcal{M} \twoheadrightarrow \mathcal{M}/W$.

Let $\mathcal{D}$ be one of the diagrams given in tables 1, 2, 3 (Appendix 1 below) symbolizing a set of relations as described in Appendix 1.

• We denote by $\mathcal{D}_{\text{br}}$ and we call *braid diagram associated to* $\mathcal{D}$ the set of nodes of $\mathcal{D}$ subject to all relations of $\mathcal{D}$ but the orders of the nodes, and we represent the braid diagram $\mathcal{D}_{\text{br}}$ by the same picture as $\mathcal{D}$ where numbers insides the nodes are omitted. Thus, if $\mathcal{D}$ is the diagram s(a) e (b)t (c)u, then $\mathcal{D}_{\text{br}}$ is the diagram

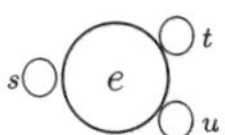

and represents the relations

$$\underbrace{stustu\cdots}_{e \text{ factors}} = \underbrace{tustus\cdots}_{e \text{ factors}} = \underbrace{ustust\cdots}_{e \text{ factors}}.$$

Note that this braid diagram for $e = 3$ is the braid diagram associated to $G(2d, 2, 2)$ $(d \geq 2)$, as well as G_7, G_{11}, G_{19}. Also, for $e = 4$, this is the braid diagram associated to G_{12} and for $e = 5$, the braid diagram associated to G_{22}. Similarly, the braid diagram s t 5 u is associated to the diagrams of both G_{15} and $G(4d, 4, 2)$.

• We denote by $\mathcal{D}^{\text{op}}$ and we call *opposite diagram associated to* $\mathcal{D}$ the set of nodes of $\mathcal{D}$ subject to all opposite relations (words in reverse order) of $\mathcal{D}$. Thus, if $\mathcal{D}$ is the diagram s(a) e (b)t (c)u, then $\mathcal{D}^{\text{op}}$ represents the relations

$$s^a = t^b = u^c = 1 \text{ and } \underbrace{utsuts\cdots}_{e \text{ factors}} = \underbrace{sutsut\cdots}_{e \text{ factors}} = \underbrace{tsutsu\cdots}_{e \text{ factors}}.$$

Note that $\mathcal{D}^{\mathrm{op}}$ is the diagram 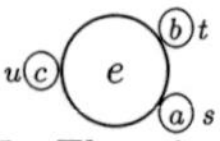 . Finally, we denote by $\mathcal{D}^{\mathrm{op}}_{\mathrm{br}}$ the braid diagram associated with $\mathcal{D}^{\mathrm{op}}$. Thus, in the above case, $\mathcal{D}^{\mathrm{op}}_{\mathrm{br}}$ is the diagram 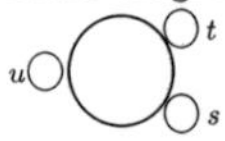 . Note that if $\mathcal{D}_{\mathrm{br}}$ is a Coxeter type diagram, then it is equal to $\mathcal{D}^{\mathrm{op}}_{\mathrm{br}}$.

The following statement is well known for Coxeter groups (see for example [**Br1**] or [**De2**]). It has been noticed by Orlik and Solomon (see [**OrSo3**], 3.7) for the case of non real Shephard groups (*i.e.*, non real complex reflection groups whose braid diagram – see above – is a Coxeter diagram). It has been proved for all the infinite series, as well as checked case by case for all the exceptional groups but G_{24}, G_{27}, G_{29}, G_{31}, G_{33}, G_{34} in [**BMM2**]. It is conjectured that it still holds for G_{31}.

2.29. THEOREM. *Let W be a finite irreducible complex reflection group, different from G_{24}, G_{27}, G_{29}, G_{33}, G_{34} – and also different from G_{31} for which the following assertions are still conjectural.*

Let $\mathcal{N}(\mathcal{D})$ be the set of nodes of the diagram $\mathcal{D}$ for W given in tables 1–3 below, identified with a set of pseudo-reflections in W. For each $s \in \mathcal{N}(\mathcal{D})$, there exists an s-distinguished braid reflection $\mathbf{s}$ in B such that the set $\{\mathbf{s}\}_{s\in\mathcal{N}(\mathcal{D})}$, together with the braid relations of $\mathcal{D}^{\mathrm{op}}_{\mathrm{br}}$, is a presentation of B.

Monoids.

Given a presentation of B as defined by a braid diagram $\mathcal{D}$, and since the relations symbolized by $\mathcal{D}$ only involve positive powers of the generators, one may consider the *monoid* $B^+_{\mathcal{D}}$ defined by "the same" presentation.

It is known (cf. for example [**De2**] or [**BrSa**]) that whenever B is the usual braid diagram associated with a Coxeter group, the follwoing two properties are satisfied.

1. The natural monoid morphism $B^+_{\mathcal{D}} \longrightarrow B$ is injective.
2. Identifying $B^+_{\mathcal{D}}$ with its image in B, we have $B = \{\boldsymbol{\pi}^n b \mid (n \in \mathbb{Z})(b \in B^+_{\mathcal{D}})\}$.

Ruth Corran ([**Cor**], chap. 8) has recently checked that

• the same properties hold for all the braid diagrams described in appendix 1 and associated with exceptional complex reflection groups but G_{24}, G_{27}, G_{29}, G_{31}, G_{33}, G_{34},

• but the injectivity of the morphism $B^+_{\mathcal{D}} \longrightarrow B$ fails for all braid diagrams of the infinite series which are not Coxeter diagrams.

CHAPTER III

GENERIC HECKE ALGEBRAS

In this chapter we follow essentially [**BMR**] and [**BMM2**].

A monodromy representation of the braid group

We extend to the case of complex reflection groups the construction of generalized Knizhnik–Zamolodchikov connections for Weyl groups due to Cherednik ([**Ch1**], [**Ch2**], [**Ch3**]; see also the constructions of Dunkl [**Du**], Opdam [**Op**] and Kohno [**Ko1**]).[1] This allows us to construct explicit isomorphisms between the group algebra of a complex reflection group and its Hecke algebra.

Background from differential equations and monodromy.

What follows is well known, and is introduced here at an elementary level for the convenience of the unexperienced reader, since we only need this elementary approach. For a more general approach, see for example [**De1**].

We go back to the setting of chap. 1. Let A be a finite dimensional complex vector space. We denote by 1 a chosen non zero point of A — in the applications, A we will be an algebra. We set $E := \mathrm{End}(A)$. Let ω be a holomorphic differential form on $\mathcal{M}$ with values in E, *i.e.*, a holomorphic map $\mathcal{M} \to \mathrm{Hom}(V, E)$, where $\mathrm{Hom}(V, E)$ denotes the space of linear maps from V into E, such that (see 1.14 and 1.17, (1)) we have

$$\omega = \sum_{H \in \mathcal{A}} f_H \omega_H \,,$$

with $\omega_H = \dfrac{1}{2i\pi} \dfrac{d\alpha_H}{\alpha_H}$, and $f_H \in E$. For $x \in \mathcal{M}$ and $v \in V$, we have $\omega(x)(v) = \dfrac{1}{2i\pi} \sum_{H \in \mathcal{A}} \dfrac{\alpha_H(v)}{\alpha_H(x)} f_H$.

We consider the following linear differential equation

$$(\mathrm{Eq}(\omega)) \qquad dF = \omega(F) \,,$$

where F is a holomorphic function defined on an open subset of $\mathcal{M}$ with values in A. In other words, for x in this open subset, we have $dF(x) \in \mathrm{Hom}(V, A)$, and we want F to satisfy, for all $v \in V$, $dF(x)(v) = \omega(x)(v)(F(x))$, or in other words $dF(x)(v) = \dfrac{1}{2i\pi} \sum_{H \in \mathcal{A}} \dfrac{\alpha_H(v)}{\alpha_H(x)} f_H(F(x))$.

For $y \in \mathcal{M}$, let us denote by $\mathcal{V}(y)$ the largest open ball with center y contained in $\mathcal{M}$. The existence and unicity theorem for linear differential equations shows that for each $y \in \mathcal{M}$, there exists a unique function

$$F_y \colon \mathcal{V}(y) \to A \,, \; x \mapsto F_y(x) \,,$$

---

[1] This construction has also been noticed independently by Opdam, who is able to deduce from it some important consequences concerning the "generalized fake degrees" of a complex reflection group.

solution of $\mathrm{Eq}(\omega)$ and such that $F_y(y) = 1$. From now on, we set

$$F(x,y) := F_y(x)\,.$$

Assume now that the finite group W acts linearly on A through a morphism $\varphi\colon W \to \mathrm{GL}(A)$. Then it induces an action of W on the space of differential forms on $\mathcal{M}$ with values in E, and an easy computation shows that ω is W-stable if and only if, for all $w \in W$,

(3.1) $$\omega(w(x)) = \varphi(w)(\omega(x)\cdot w^{-1})\varphi(w^{-1})\,,$$

which can also be written, for all $x \in \mathcal{M}$ and $v \in V$:

$$\sum_{H\in\mathcal{A}} \omega_H(wx)(v) f_H = \sum_{H\in\mathcal{A}} \omega_H(x)(w^{-1}(v))\varphi(w) f_H \varphi(w^{-1})\,.$$

An easy computation shows that this is equivalent to

(3.2) $$\sum_{H\in\mathcal{A}} f_{w(H)} \frac{d\alpha_{w(H)}}{\alpha_H} = \sum_{H\in\mathcal{A}} \varphi(w) f_H \varphi(w^{-1}) \frac{d\alpha_{w(H)}}{\alpha_H}\,.$$

In particular we see that

3.3. *If $f_{w(H)} = \varphi(w) f_H \varphi(w^{-1})$ for all $H \in \mathcal{A}$ and $w \in W$, then the form ω is W-stable.*

From (3.1) (and from the existence and unicity theorem), it follows that

3.4. *If ω is W-stable, then for all $y \in \mathcal{M}$, $x \in \mathcal{V}(y)$ and $w \in W$, the solution $x \mapsto F(x,y)$ satisfies*

$$\varphi(w)(F(x,y)) = F(w(x), w(y))\,.$$

The case of an interior W-algebra.

The following hypothesis and notation will be in force for the rest of this paragraph.

From now on, we assume that A is endowed with a structure of $\mathbb{C}$-algebra with unity, and that ω takes its values in the subalgebra of E consisting of the multiplications by the elements of A – which, by abuse of notation, we still denote by A. With this abuse of notation, we may assume that

$$\omega = \sum_{H\in\mathcal{A}} a_H \omega_H\,,$$

where $a_H \in A$, and the equation $\mathrm{Eq}(\omega)$ is written

$$dF = \omega F \text{ or } dF(x)(v) = \frac{1}{2i\pi} \sum_{H\in\mathcal{A}} \frac{\alpha_H(v)}{\alpha_H(x)} a_H F(x)\,.$$

Let γ be a path in $\mathcal{M}$. From the existence and unicity of local solutions of $\mathrm{Eq}(\omega)$, it results that the solution $x \mapsto F(x, \gamma(0))$ has an analytic continuation $t \mapsto (\gamma^* F)(t, \gamma(0))$ along γ, which satisfies the following properties.

Let us say that a sequence of real numbers $t_0 = 0 < t_1 < \ldots < t_{n-1} < t_n = 1$ is adapted to $(\gamma, \mathrm{Eq}(\omega))$ if for all $1 \le j \le n$, we have $\gamma([t_{j-1}, t_j]) \subset \mathcal{V}(\gamma(t_j))$.

Then :

(1) there exists $\varepsilon > 0$ such that $(\gamma^* F)(t, \gamma(0)) = F(\gamma(t), \gamma(0))$ for $0 \le t \le \varepsilon$,

(2) whenever $t_0 = 0 < t_1 < \ldots < t_{n-1} < t_n = 1$ is adapted to $(\gamma, \mathrm{Eq}(\omega))$, we have $(\gamma^* F)(t_j, \gamma(0)) = F(\gamma(t_j), \gamma(t_{j-1}))(\gamma^* F)(t_{j-1}, \gamma(0))$ for all $j > 0$.

We see that

$$(3.5) \qquad (\gamma^* F)(1,\gamma(0)) = \prod_{j=n}^{j=1} F(\gamma(t_j),\gamma(t_{j-1}))\,.$$

The case of an integrable form.

We recall that the form ω is said to be integrable if $d\omega + \omega \wedge \omega = 0$.

The following fact was noticed, for example, by Kohno (see [**Ko2**], 1.2). This is an immediate consequence of 1.17, (2).

3.6. LEMMA. *The form $\omega = \sum_{H\in\mathcal{A}} a_H\omega_H$ is integrable if and only if, for all subspaces X of V with codimension 2, and for all $H \in \mathcal{A}$ such that $X \subset H$, a_H commutes with $\sum_{\substack{(H'\in\mathcal{A})\\(H'\supset X)}} a_{H'}$.*

If ω is integrable, the value $(\gamma^* F)(1,\gamma(0))$ depends only on the homotopy class of γ. By (3.5), we see that we get a *covariant functor*

$$S\colon \begin{cases} \mathcal{P}(\mathcal{M}) \to A^\times \\ \gamma \mapsto (\gamma^* F)(1,\gamma(0)) \end{cases}$$

Action of W.

Assume now that A is an *interior* W-algebra, *i.e.*, that there is a group morphism $W \to A^\times$ (through which the image of $w \in W$ is still denoted by w), which defines a linear operation φ of W on A by composition with the injection $A^\times \hookrightarrow \mathrm{GL}(A)$. So, with our convention, for $w \in W$ and $a \in A$ we have $\varphi(w)(a) = wa$.

The form ω is then W-stable if and only if, for all $w \in W$ and $x \in \mathcal{M}$,

$$\omega(w(x)) = w(\omega(x)\cdot w^{-1})w^{-1}\,,$$

which can also be written, for all $x \in \mathcal{M}$ and $v \in V$:

$$\sum_{H\in\mathcal{A}} \omega_H(wx)(v)a_H = \sum_{H\in\mathcal{A}} \omega_H(x)(w^{-1}(v))wa_H w^{-1}\,.$$

By 3.3, we have the following criterion.

3.7. *If, for all $H \in \mathcal{A}$ and $w \in W$, we have $a_{w(H)} = wa_H w^{-1}$, then the form ω is W-stable.*

By 3.4, the solution F of $\mathrm{Eq}(\omega)$ then satisfies

$$wF(x,y)w^{-1} = F(w(x),w(y))\,.$$

3.8. DEFINITION–PROPOSITION. *Assuming that ω is W-stable, we define a group morphism*

$$T\colon \pi_1(\mathcal{M}/W,x_0) \to (A^\times)^{\mathrm{op}}$$

(or, in other words, a group anti-morphism $T\colon \pi_1(\mathcal{M}/W,x_0) \to A^\times$), called the monodromy morphism associated with ω, as follows.

For $\sigma \in B$, with image $\overline{\sigma}$ in W through the natural anti-morphism $B \to W$, we denote by $\tilde{\sigma}$ a path in $\mathcal{M}$ from x_0 to $\overline{\sigma}(x_0)$ which lifts σ. Then we set

$$T(\sigma) := S(\tilde{\sigma}^{-1})\overline{\sigma}\,.$$

Dependence of parameters.

Suppose the form ω depends holomorphically on m parameters $t_1, \dots, t_m$. Denoting by $\mathcal{O}$ the ring of holomorphic functions of the variables $t_1, \dots, t_m$, we have $\omega = \sum_{H\in\mathcal{A}} f_H \omega_H$ where $f_H \in \mathcal{O} \otimes_{\mathbb{C}} E$. Then, for $y \in \mathcal{M}$, the function F_y is a holomorphic function of $t_1, \dots, t_m$, *i.e.*, F_y has values in $\mathcal{O} \otimes_{\mathbb{C}} A$.

Then, given a path γ in $\mathcal{M}$, the analytic continuation $t \mapsto (\gamma^* F)(t, \gamma(0))$ depends holomorphically of $t_1, \dots, t_m$.

If ω is integrable and W-stable, then the monodromy morphism depends holomorphically on the parameters $t_1, \dots, t_m$. It follows that we have a monodromy morphism

$$T : \pi_1(\mathcal{M}/W, x_0)^{\mathrm{op}} \to (\mathcal{O} \otimes_{\mathbb{C}} A)^\times .$$

The main theorem.

From now on, we assume that $A = \mathbb{C}W$.

We denote by $\mathcal{O}$ the ring of holomorphic functions of a set of $\sum_{\mathcal{C}\in\mathcal{A}/W} e_{\mathcal{C}}$ variables

$$\mathbf{z} := (z_{\mathcal{C},j})_{(\mathcal{C}\in\mathcal{A}/W)(0\le j\le e_{\mathcal{C}}-1)} .$$

For $H \in \mathcal{C}$, we set $z_{H,j} := z_{\mathcal{C},j}$.

We put

$$q_{\mathcal{C},j} = \exp(-2\pi i z_{\mathcal{C},j}/e_{\mathcal{C}}) \quad \text{for } \mathcal{C} \in \mathcal{A}/W \, , \, 0 \le j \le e_{\mathcal{C}} - 1 \, .$$

For $H \in \mathcal{C}$, we set $q_{H,j} := q_{\mathcal{C},j}$.

Let $\mathcal{C} \in \mathcal{A}/W$ and let $H \in \mathcal{C}$. For $0 \le j \le e_{\mathcal{C}} - 1$, we denote by $\varepsilon_j(H)$ the primitive idempotent of the group algebra $\mathbb{C}W_H$ associated with the character $\det_V^j$ of the group W_H. Thus we have

$$\varepsilon_j(H) = \frac{1}{e_{\mathcal{C}}} \sum_{k=0}^{k=e_{\mathcal{C}}-1} \zeta_{e_{\mathcal{C}}}^{jk} \, s_H^k \, .$$

We set

$$a_H := \sum_{j=0}^{j=e_H-1} z_{H,j}\varepsilon_j(H) \quad \text{and} \quad \omega := \sum_{H\in\mathcal{A}} a_H \frac{d\alpha_H}{\alpha_H} \, .$$

In other words, we have

$$\omega = \sum_{\mathcal{C}\in\mathcal{A}/W} \sum_{j=0}^{j=e_{\mathcal{C}}-1} \sum_{H\in\mathcal{C}} z_{\mathcal{C},j}\varepsilon_j(H) 2\pi i \omega_H \, .$$

The following lemma is clear.

3.9. Lemma. *The map $\mathcal{A} \to A$, $H \mapsto a_H$ has the following properties :*
(1) *it is W-stable, i.e., for all $w \in W$ and $H \in \mathcal{A}$, we have $a_{w(H)} = w a_H w^{-1}$,*
(2) *for all $H \in \mathcal{A}$, a_H belongs to the image of $\mathbb{C}W_H$ in A.*

The following property follows from 3.9.

3.10. Lemma. *The form ω is W-stable and integrable.*

The proof of the following result may be found in [**BMR**], §4.

3.11. THEOREM. *We denote by* $T\colon B^{\mathrm{op}} \to (\mathbb{C}W)^\times$ *the monodromy morphism associated with the differential form* ω *on* $\mathcal{M}$. *For all* $H \in \mathcal{C}$, *we have*

$$\prod_{j=0}^{j=e_{\mathcal{C}}-1} (T(\mathbf{s}_{H,\gamma}) - q_{H,j}\mathrm{det}_V(s_H)^j) = 0\,.$$

Furthermore, T *depends holomorphically on the parameters* $z_{\mathcal{C},j}$, *i.e., arises by specialization from a morphism* $\mathbf{T}\colon B^{\mathrm{op}} \to (\mathcal{O}W)^\times$.

Hecke algebras : first properties

We define a set

$$\mathbf{u} = (u_{\mathcal{C},j})_{(\mathcal{C}\in\mathcal{A}/W)(0\le j\le e_{\mathcal{C}}-1)}$$

of $\sum_{\mathcal{C}\in\mathcal{A}/W}(e_{\mathcal{C}})$ indeterminates. We denote by $\mathbb{Z}[\mathbf{u},\mathbf{u}^{-1}]$ the ring of Laurent polynomials in the indeterminates $\mathbf{u}$.

Let $\mathfrak{J}$ be the ideal of the group algebra $\mathbb{Z}[\mathbf{u},\mathbf{u}^{-1}]B$ generated by the elements

$$(\mathbf{s}_{H,\gamma} - u_{\mathcal{C},0})(\mathbf{s}_{H,\gamma} - u_{\mathcal{C},1})\cdots(\mathbf{s}_{H,\gamma} - u_{\mathcal{C},e_{\mathcal{C}}-1})$$

where $\mathcal{C} \in \mathcal{A}/W$, $H \in \mathcal{C}$, $\mathbf{s}_{H,\gamma}$ is a distinguished braid reflection around H in B and s is the image of $\mathbf{s}_{H,\gamma}$ in W.

3.12. DEFINITION. *The Hecke algebra, denoted by* $\mathcal{H}_{\mathbf{u}}(W)$, *is the* $\mathbb{Z}[\mathbf{u},\mathbf{u}^{-1}]$-*algebra* $\mathbb{Z}[\mathbf{u},\mathbf{u}^{-1}]B/\mathfrak{J}$.

Notice that, since all the distinguished braid reflections around any hyperplane $H \in \mathcal{C}$ are conjugate under B, it suffices to take one of the $\mathbf{s}_H$ in the above relations to generate the ideal $\mathfrak{J}$.

From now on, whenever R is a ring endowed with a ring morphism $\mathbb{Z}[\mathbf{u},\mathbf{u}^{-1}] \longrightarrow R$, we will write $R\mathcal{H}(W,\mathbf{u})$ for the implied tensor product $R\otimes_{\mathbb{Z}[\mathbf{u},\mathbf{u}^{-1}]}\mathcal{H}(W,\mathbf{u})$.

Now assume that W is a finite irreducible complex reflection group. Let $\mathcal{D}$ be the diagram of W, and let $\mathbf{s} \in \mathcal{N}(\mathcal{D})$ be a node of $\mathcal{D}$. We set $u_{\mathbf{s},j} := u_{\mathcal{C},j}$ for $j = 0, 1, \ldots, e_{\mathcal{C}} - 1$, where $\mathcal{C}$ denotes the orbit under W of the reflecting hyperplane of s.

The following proposition is an immediate consequence of Theorem 2.29 :

3.13. PROPOSITION. *Assume* W *is different from* G_{24}, G_{27}, G_{29}, G_{33}, G_{34} – *and also different from* G_{31} *for which the following assertion is still conjectural. The Hecke algebra* $\mathcal{H}_{\mathbf{u}}(W)$ *is isomorphic to the* $\mathbb{Z}[\mathbf{u},\mathbf{u}^{-1}]$-*algebra generated by elements* $(\mathbf{s})_{\mathbf{s}\in\mathcal{N}(\mathcal{D})}$ *such that*

- *the elements* $\mathbf{s}$ *satisfy the braid relations defined by* $\mathcal{D}_{\mathrm{br}}^{\mathrm{op}}$,
- *we have* $(\mathbf{s} - u_{\mathbf{s},0})(\mathbf{s} - u_{\mathbf{s},1})\cdots(\mathbf{s} - u_{\mathbf{s},e_{\mathbf{s}}-1}) = 0\,.$

Note that the specialization $u_{\mathcal{C},j} \mapsto \zeta_{e_{\mathcal{C}}}^j$ takes the generic Hecke algebra $\mathcal{H}(W,\mathbf{u})$ to the group algebra of W^{op} over a cyclotomic extension of $\mathbb{Z}$. Any specialization of the generic Hecke algebra $\mathcal{H}(W,\mathbf{u})$ through which the previous one factorizes is called *admissible*.

We shall denote by $\mathbf{h}\colon \mathbb{Z}[\mathbf{u},\mathbf{u}^{-1}]B \to \mathcal{H}(W,\mathbf{u})$ the natural epimorphism. For a $\mathbb{Z}[\mathbf{u},\mathbf{u}^{-1}]$-linear map t defined on $\mathcal{H}(W,\mathbf{u})$, we shall often omit the letter $\mathbf{h}$ by writing $t(b)$ instead of $t(\mathbf{h}(b))$.

Hecke algebras and monodromy representations.

By Theorem 3.11, we see that the monodromy representation $\mathbf{T}$ factors through $\mathcal{H}_{\mathbf{u}}(W)$. Indeed, let us set

$$\begin{cases} u_{\mathcal{C},j} := \exp\left(\dfrac{2\pi i(j - z_{\mathcal{C},j}}{e_{\mathcal{C}}}\right) & \text{for all } (\mathcal{C}, j)\,, \\ \mathbf{z} := (z_{\mathcal{C},j})_{(\mathcal{C}\in\mathcal{A}/W)(0\leq j\leq e_{\mathcal{C}}-1)}\,, \end{cases}$$

and let us denote by $\mathcal{O}$ the ring of holomorphic functions of the set of variables $\mathbf{z}$. Then we have the following commutative diagram :

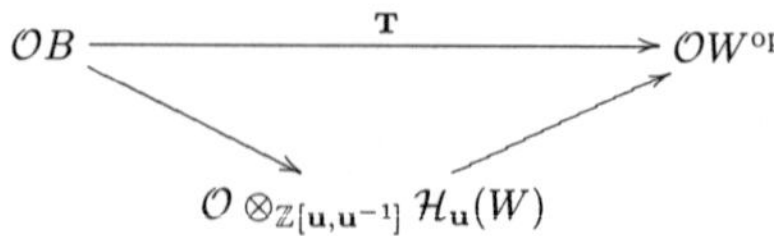

Let $\mathcal{K}$ be the field of fractions of $\mathcal{O}$.

The following lemma is a key point to understand the structure of $\mathcal{H}_{\mathbf{u}}(W)$. It is well-known to hold for Coxeter groups. For the infinite series of complex reflection groups, see [**ArKo**] for $G(d,1,r)$, [**BrMa**], (4.12) for $G(2d,2,r)$ and [**Ari**], Proposition 1.4 for the general case (it has been also checked for many of the remaining groups of small rank — see for example [**BrMa**], Satz 4.7). We conjecture it is true for all complex reflection groups.

3.14. Lemma–Assumption. *The* $\mathbb{Z}[\mathbf{u},\mathbf{u}^{-1}]$*-module* $\mathcal{H}_{\mathbf{u}}(W)$ *can be generated by* $|W|$ *elements.*

From this lemma, we can now deduce the following

3.15. Theorem–Assumption. *The monodromy representation* $\mathbf{T}$ *induces an isomorphism of* $\mathcal{K}$*-algebras*

$$\mathcal{K}\otimes_{\mathbb{Z}[\mathbf{u},\mathbf{u}^{-1}]}\mathcal{H}_{\mathbf{u}}(W)\xrightarrow{\sim}\mathcal{K}W^{\mathrm{op}}\,.$$

Furthermore, $\mathcal{H}_{\mathbf{u}}(W)$ *is a free* $\mathbb{Z}[\mathbf{u},\mathbf{u}^{-1}]$*-module of rank* $|W|$*.*

Indeed, by Lemma 3.14, there is a surjective morphism of $\mathbb{Z}[\mathbf{u},\mathbf{u}^{-1}]$-modules

$$\phi\colon \mathbb{Z}[\mathbf{u},\mathbf{u}^{-1}]^{|W|}\to\mathcal{H}_{\mathbf{u}}(W)\,.$$

Let $\mathfrak{m}$ be the ideal of $\mathcal{O}$ of the functions vanishing at the point $(t_{\mathcal{C},j}=1)$. The morphism $\mathcal{O}_{\mathfrak{m}}\otimes_{\mathbb{Z}[\mathbf{u},\mathbf{u}^{-1}]}\mathcal{H}_{\mathbf{u}}\to\mathcal{O}_{\mathfrak{m}}W$ induced by the monodromy is surjective by Nakayama's lemma, since it becomes an isomorphism after tensoring by $(\mathcal{O}_{\mathfrak{m}})/\mathfrak{m}$. Composing with $1_{\mathcal{O}_{\mathfrak{m}}}\otimes\phi$, we obtain an epimorphism $\mathcal{O}_{\mathfrak{m}}^{|W|}\to\mathcal{O}_{\mathfrak{m}}W$: this must be an isomorphism. Hence, $\ker\phi=0$, *i.e.*, ϕ is an isomorphism and $\mathcal{H}_{\mathbf{u}}$ is free of rank $|W|$ over $\mathbb{Z}[\mathbf{u},\mathbf{u}^{-1}]$.

Since the morphism $K\otimes_{\mathbb{Z}[\mathbf{u},\mathbf{u}^{-1}]}\mathcal{H}_{\mathbf{u}}\to KW$ is a surjective morphism between two K-modules with same dimensions, it is an isomorphism and Theorem 3.15 follows. □

Grading the Hecke algebra.

The algebra $\mathbb{Z}[\mathbf{u},\mathbf{u}^{-1}]$ is multi–graded over $\mathbb{Z}$: for each $\mathcal{C}\in\mathcal{A}/W$, we define the grading

$$\deg_{\mathcal{C}}\mathbb{Z}[\mathbf{u},\mathbf{u}^{-1}]\to\mathbb{Z}$$

by setting

$$\deg_{\mathcal{C}}(u_{\mathcal{C}',j}) := \begin{cases} 1 & \text{if } \mathcal{C}' = \mathcal{C}\,, \\ 0 & \text{if } \mathcal{C}' \neq \mathcal{C}\,. \end{cases}$$

If we set $\deg_{\mathcal{C}}(b) := \ell_{\mathcal{C}}(b)$ for $b \in B$ and $\mathcal{C} \in \mathcal{A}/W$, we see then that the group algebra $\mathbb{Z}[\mathbf{u}, \mathbf{u}^{-1}]B$ is endowed with a grading on $\mathbb{Z}^{\mathcal{A}/W}$. Since the ideal $\mathfrak{J}$ (see above 3.12) is homogeneous for this grading, it follows that the maps $\deg_{\mathcal{C}}$ endow the generic Hecke algebra $\mathcal{H}(W, \mathbf{u})$ with a structure of $\mathbb{Z}^{\mathcal{A}/W}$-graded algebra.

Parabolic Hecke sub–algebras.

Let X be an intersection of reflecting hyperplanes, and let W_X be the corresponding parabolic subgroup of W (see chap. 1 above). We recall that we denote by $\mathcal{A}_X$ the set of reflecting hyperplanes of W_X. Let $a : \mathcal{A}_X/W_X \to \mathcal{A}/W$ be the map sending a W_X-orbit of hyperplanes to the corresponding W-orbit.

Let $\mathbf{u}' = (u'_{\mathcal{C},j})_{(\mathcal{C} \in \mathcal{A}_X/W_X)(0 \leq j \leq e_{\mathcal{C}}-1)}$ be a set of $\sum_{\mathcal{C} \in \mathcal{A}_X/W_X} e_{\mathcal{C}}$ indeterminates. We have a morphism

$$\mathbb{Z}[\mathbf{u}', \mathbf{u}'^{-1}] \to \mathbb{Z}[\mathbf{u}, \mathbf{u}^{-1}]\,,\; u'_{\mathcal{C},j} \mapsto u_{a(\mathcal{C}),j}\,.$$

The injection $B_{W_X} \hookrightarrow B_W$, defined up to P–conjugation (see above), induces an inclusion, defined up to P–conjugation (see [**BMR**], §4)

$$\mathcal{H}(W_X, \mathbf{u}') \otimes_{\mathbb{Z}[\mathbf{u}', \mathbf{u}'^{-1}]} \mathbb{Z}[\mathbf{u}, \mathbf{u}^{-1}] \to \mathcal{H}(W, \mathbf{u})\,,$$

whose image is called the generic *parabolic Hecke sub–algebra of* $\mathcal{H}(W, \mathbf{u})$ *associated with* X, and is denoted by $\mathcal{H}(W_X, W, \mathbf{u})$ (note that only the P–conjugacy class of this subalgebra is well defined).

Semi–simple Hecke algebras and absolutely irreducible characters.

The following assertion is known to hold

- for all infinite families of finite irreducible reflection groups by [**ArKo**], [**BrMa2**], [**Ar**],
- for all Coxeter groups (see for example [**Bou**], chap. IV, §2, ex. 23),
- for the groups G_4, G_5, G_8 and G_{25} which occur in [**BrMa2**].

We conjecture it to be true in all cases (see [**BMR**], §4).

It shows in particular that the algebra $\mathcal{H}(W, \mathbf{u})$ is a free graded $\mathbb{Z}[\mathbf{u}, \mathbf{u}^{-1}]$–module of rank $|W|$, since it has a basis over $\mathbb{Z}[\mathbf{u}, \mathbf{u}^{-1}]$ consisting of homogeneous elements.

3.16. Theorem–Assumption. *There exists a family $(b_w)_{w \in W}$ of $|W|$ elements of the braid group B with*

- *b_w maps to w through the natural morphism $B \twoheadrightarrow W^{\mathrm{op}}$,*
- *$b_1 = 1$,*

such that the family $(\mathbf{h}(b_w))_{w \in W}$ is a basis of $\mathcal{H}(W, \mathbf{u})$ over $\mathbb{Z}[\mathbf{u}, \mathbf{u}^{-1}]$.

By Appendix 2, 8.4, (1) below, it follows that

3.17. Proposition. *For W satisfying Theorem-Assumption above, the $\mathbb{Q}(\mathbf{u})$–algebra $\mathbb{Q}(\mathbf{u})\mathcal{H}(W, \mathbf{u})$ is trace symmetric, hence separable and in particular semi–simple.*

Let $\mathbb{Q}(\boldsymbol{\mu}_\infty)$ be the subfield of $\mathbb{C}$ generated by all roots of unity. Assume that (V, W) is an irreducible K–reflection group, where K is a subfield of $\mathbb{Q}(\boldsymbol{\mu}_\infty)$ of finite degree over $\mathbb{Q}$. We denote by $K(\mathbf{u})$ the field of fractions of the Laurent polynomial ring $K[\mathbf{u},\mathbf{u}^{-1}]$, where as above $\mathbf{u}=(u_{\mathcal{C},j})_{(\mathcal{C}\in\mathcal{A}/W)(0\leq j\leq e_{\mathcal{C}}-1)}$.

The following theorem is proved for *any* reflection group in [**Ma4**] (see theorem 5.2. in *loc.cit.*), provided that the corresponding Hecke algebra is defined by generators and relations symbolized by the diagrams of Appendix 1 below. Since this last fact is unknown for the six exceptional groups G_{24}, G_{27}, G_{29}, G_{31}, G_{33}, G_{34}, we must call it a "theorem–assumption": it is proved for all irreducible groups but the six previous ones.

3.18. THEOREM–ASSUMPTION. *Let W be a complex reflection group satisfying Theorem-Assumption 3.16. Let $\mathbf{t}=(t_{\mathcal{C},j})_{(\mathcal{C}\in\mathcal{A}/W)(0\leq j\leq e_{\mathcal{C}}-1)}$ be a set of $\sum_{\mathcal{C}\in\mathcal{A}/W}e_{\mathcal{C}}$ indeterminates such that, for all $\mathcal{C},j$, we have $t_{\mathcal{C},j}^{|\mu(K)|}=\zeta_{e_{\mathcal{C}}}^{-j}u_{\mathcal{C},j}$. The field $K(\mathbf{t})$ is a splitting field for the Hecke algebra $\mathbb{Q}(\mathbf{u})\mathcal{H}(W,\mathbf{u})$.*

In what follows the field $K(\mathbf{t})$ above is called a *good splitting field*.

Let us denote by $\mathbb{Z}_K$ the ring of integers of K. By Appendix 2, 8.2 below, the specialization

$$t_{\mathcal{C},j}\mapsto 1$$

induces a bijection $\chi\mapsto\chi_{\mathbf{t}}$ from the set $\mathrm{Irr}(W^{\mathrm{op}})$ of absolutely irreducible characters of W^{op} onto the set $\mathrm{Irr}(\mathcal{H}(W,\mathbf{u}))$ of absolutely irreducible characters of the Hecke algebra $\mathcal{H}(W,\mathbf{u})$, such that the following diagram is commutative

$$\begin{array}{ccc}\mathcal{H}(W,\mathbf{u}) & \xrightarrow{\chi_{\mathbf{t}}} & \mathbb{Z}_K[\mathbf{t},\mathbf{t}^{-1}]\\ \downarrow & & \downarrow\\ \mathbb{Z}_KW^{\mathrm{op}} & \xrightarrow{\chi} & \mathbb{Z}_K\,.\end{array}$$

3.19. CONVENTION. As functions on the underlying set of W, the characters of W and of W^{op} coincide. We identify $\mathrm{Irr}(W)$ and $\mathrm{Irr}(W^{\mathrm{op}})$. Thus we get a bijection

$$\mathrm{Irr}(W)\xrightarrow{\sim}\mathrm{Irr}(\mathcal{H}(W,\mathbf{u}))\ ,\ \chi\mapsto\chi_{\mathbf{t}}$$

such that, for $b\in B$ with image $\bar{b}\in W$, we have

$$(\chi_{\mathbf{t}}(b))_{|\mathbf{t}=1}=\chi(\bar{b})\,.$$

Linear characters.

Let $\mathrm{Hom}(B,\mathbb{Q}(\mathbf{u})^\times)$ denote the group of linear characters of B with values in $\mathbb{Q}(\mathbf{u})^\times$. We have an isomorphism (see for example [**BMR**], prop. 2.16)

$$\mathrm{Hom}(B,\mathbb{Q}(\mathbf{u})^\times)\simeq\underbrace{\mathbb{Q}(\mathbf{u})^\times\times\mathbb{Q}(\mathbf{u})^\times\times\cdots\times\mathbb{Q}(\mathbf{u})^\times}_{|\mathcal{A}/W|\text{ times}}\,,$$

given by the map

$$\theta\mapsto(\theta(\mathbf{s}_{\mathcal{C}}))_{\mathcal{C}\in\mathcal{A}/W}\,,$$

where $\mathbf{s}_{\mathcal{C}}$ denotes a distinguished braid reflection around a hyperplane belonging to $\mathcal{C}$.

It follows from what precedes that the linear characters of $\mathcal{H}(W,\mathbf{u})$ are induced by those linear characters of B described as follows: there exists a family of integers $\mathbf{j}:=(j_{\mathcal{C}})_{\mathcal{C}\in\mathcal{A}/W}$ where $j_{\mathcal{C}}\in\{0,1,\ldots,e_{\mathcal{C}}-1\}$, such that $\mathbf{s}\mapsto u_{\mathcal{C},j_{\mathcal{C}}}$ if $\mathbf{s}$ is a distinguished braid reflection around an element of $\mathcal{C}$. We denote by $\theta_{\mathbf{j}}$ this linear character of $\mathcal{H}(W, \mathbf{u})$.

We denote by $\det_{\mathcal{C}}$ the linear character of W such that

$$\det_{\mathcal{C}}(s_H) = \begin{cases} \det_V(s_H) & \text{if } H \in \mathcal{C}\,, \\ 1 \quad \text{if } H \notin \mathcal{C}\,, \end{cases}$$

where s_H is any reflection with reflecting hyperplane H.

Then the character $(\det_{\mathcal{C},j})_{\mathbf{t}}$ of the Hecke algebra corresponding to $\det_{\mathcal{C}}^j$ is defined by

$$(\det_{\mathcal{C},j})_{\mathbf{t}}(\mathbf{s}_H) = \begin{cases} u_{\mathcal{C},j} & \text{if } H \in \mathcal{C}\,, \\ u_{\mathcal{C}',0} & \text{if } H \in \mathcal{C}'\,,\ \mathcal{C}' \neq \mathcal{C}\,, \end{cases}$$

where $\mathbf{s}_H$ is a distinguished braid reflection around H.

We see in particular that $(\det_{\mathcal{C},j})_{\mathbf{t}}$ is rational over $\mathbb{Q}(\mathbf{u})$: we have

$$(\det_{\mathcal{C},j})_{\mathbf{t}} \colon \mathcal{H}(W, \mathbf{u})^{\times} \to \mathbb{Z}[\mathbf{u}, \mathbf{u}^{-1}]^{\times}\,. \tag{3.20}$$

Notice that (with notation introduced above), for $\mathbf{j} = (j_{\mathcal{C}})_{\mathcal{C} \in \mathcal{A}/W}$ we have

$$\theta_{\mathbf{j}} = \prod_{\mathcal{C} \in \mathcal{A}/W} (\det_{\mathcal{C},j_{\mathcal{C}}})_{\mathbf{t}}\,.$$

Central morphisms associated with irreducible characters.

More generally, let $\chi \in \operatorname{Irr}(W)$. Recall that, for $\mathcal{C} \in \mathcal{A}/W$, we denote by $m_{\mathcal{C},j}^{\chi}$ the multiplicity of $\det_V^j$ in the restriction of χ to the cyclic group W_H. It results from [**BrMi2**], 4.17, that

$$\frac{m_{\mathcal{C},j}^{\chi} N_{\mathcal{C}} e_{\mathcal{C}}}{\chi(1)} \in \mathbb{N}\,.$$

Since $K(\mathbf{t})$ is a splitting field for the algebra $K(\mathbf{t})\mathcal{H}(W, \mathbf{u})$, the irreducible character $\chi_{\mathbf{t}}$ defines an algebra morphism from the center of $K(\mathbf{t})\mathcal{H}(W, \mathbf{u})$ onto $K(\mathbf{t})$ which we denote by

$$\omega_{\chi_{\mathbf{t}}} \colon Z(K(\mathbf{t})\mathcal{H}(W, \mathbf{u})) \to K(\mathbf{t})\,.$$

By [**BrMi2**], prop. 4.16, its value on the image of the central element $\boldsymbol{\pi}$ equals

$$\omega_{\chi_{\mathbf{t}}}(\boldsymbol{\pi}) = \zeta_{\chi(1)}^{N(\chi)} \prod_{\mathcal{A}/W} \prod_{j=0}^{j=e_{\mathcal{C}}-1} u_{\mathcal{C},j}^{\frac{m_{\mathcal{C},j}^{\chi} N_{\mathcal{C}} e_{\mathcal{C}}}{\chi(1)}}\,.$$

Now using the equalities $t_{\mathcal{C},j}^{|\mu(K)|} = \zeta_{e_{\mathcal{C}}}^{j} u_{\mathcal{C},j}$ and 1.25 above, we get the equivalent formula

$$\omega_{\chi_{\mathbf{t}}}(\boldsymbol{\pi}) = \prod_{\mathcal{C} \in \mathcal{A}/W} \prod_{j=0}^{j=e_{\mathcal{C}}-1} t_{\mathcal{C},j}^{|\mu(K)| \frac{m_{\mathcal{C},j}^{\chi} N_{\mathcal{C}} e_{\mathcal{C}}}{\chi(1)}}\,. \tag{3.21}$$

A proof like in [**BrMi2**], 4.16, provides the following generalization of (3.21).

3.22. PROPOSITION. *Assume that* $\mathbf{w} \in B$ *is such that* $\mathbf{w}^d = \boldsymbol{\pi}$ *(for some positive integer d). Let w denote its image in W. Then we have*

$$\chi_{\mathbf{t}}(\mathbf{w}) = \chi(w) \prod_{\mathcal{C} \in \mathcal{A}/W} \prod_{j=0}^{j=e_{\mathcal{C}}-1} t_{\mathcal{C},j}^{\frac{|\boldsymbol{\mu}(K)|}{d} \frac{m_{\mathcal{C},j}^{\chi} N_{\mathcal{C}} e_{\mathcal{C}}}{\chi(1)}} .$$

Characters as homogeneous functions.

The following proposition results from Appendix 2 below, 8.1, (2).

3.23. PROPOSITION. *Assume W satisfies Theorem-Assumption 3.16. Then for all* $\chi \in \operatorname{Irr}(W)$ *and* $b \in B$, *the value* $\chi_{\mathbf{t}}(b)$ *(as an element of* $\mathbb{Z}_K[\mathbf{t}, \mathbf{t}^{-1}]$*) is multi-homogeneous of degree* $|\boldsymbol{\mu}(K)|\ell_{\mathcal{C}}(b)$ *in the indeterminates* $(t_{\mathcal{C},j})_{0 \leq j < e_{\mathcal{C}}}$ *(for all* $\mathcal{C} \in \mathcal{A}/W$*).*

Galois operations.

Both extensions $K(\mathbf{t})/K(\mathbf{u})$ and $K(\mathbf{t})/\mathbb{Q}(\mathbf{t})$ are Galois (although in general the extension $K(\mathbf{t})/\mathbb{Q}(\mathbf{u})$ is not Galois). Since $\mathcal{H}(W, \mathbf{u})$ is defined over $\mathbb{Q}(\mathbf{u})$, both groups $\operatorname{Gal}(K(\mathbf{t})/K(\mathbf{u}))$ and $\operatorname{Gal}(K(\mathbf{t})/\mathbb{Q}(\mathbf{t}))$ act on the set of irreducible characters of the algebra $K(\mathbf{t})\mathcal{H}(W, \mathbf{u})$, hence, through the preceding bijection, both groups act on $\operatorname{Irr}(W)$.

- For $g \in \operatorname{Gal}(K(\mathbf{t})/K(\mathbf{u}))$ and $\chi \in \operatorname{Irr}(W)$, we denote by $g(\chi)$ the irreducible character of W defined by the condition

$$g(\chi_{\mathbf{t}}) = (g(\chi))_{\mathbf{t}} .$$

- The group $\operatorname{Gal}(K(\mathbf{t})/\mathbb{Q}(\mathbf{t}))$ is isomorphic (through the restriction map from $K(\mathbf{t})$ to K) to the Galois group $\operatorname{Gal}(K/\mathbb{Q})$. Through this isomorphism we get an action of $\operatorname{Gal}(K/\mathbb{Q})$ on $\operatorname{Irr}(W)$. It is easy to see that this action coincides with the usual action of $\operatorname{Gal}(K/\mathbb{Q})$ on $\operatorname{Irr}(W)$. For $g \in \operatorname{Gal}(K/\mathbb{Q})$ and $\chi \in \operatorname{Irr}(W)$, we still denote by $g(\chi)$ the image of χ under g.

The following proposition is 4.1 and 4.2 in [**BrMi2**].

3.24. PROPOSITION.

(1) *The linear characters of W are fixed under* $\operatorname{Gal}(K(\mathbf{t})/K(\mathbf{u}))$.

(2) *Whenever* $\chi \in \operatorname{Irr}(W)$ *and* $g \in \operatorname{Gal}(K(\mathbf{t})/K(\mathbf{u}))$, *we have*
   - (a) $\chi(1) = g(\chi)(1)$,
   - (b) $\chi(s) = g(\chi)(s)$ *whenever s is a reflection in W,*
   - (c) $N(\chi) = N(g(\chi))$.

Some anti-automorphisms of the generic Hecke algebra.

We denote by $\alpha \mapsto \alpha^{\vee}$ the automorphism of $\mathbb{Z}[\mathbf{u}, \mathbf{u}^{-1}]$ consisting of the simultaneous inversion of the indeterminates.

The following proposition is an immediate consequence of the definition of the Hecke algebra.

3.25. PROPOSITION. *There is a unique anti-automorphism of* $\mathcal{H}(W, \mathbf{u})$, *denoted by* $\mathbf{a}_1$, *such that:*

(a1) *For* $b \in B$, *we have*

$$\mathbf{a}_1(\mathbf{h}(b)) = \mathbf{h}(b^{-1}) .$$

(a2) *For* $\alpha \in \mathbb{Z}[\mathbf{u}, \mathbf{u}^{-1}]$, *we have*

$$\mathbf{a}_1(\alpha) = \alpha^\vee .$$

The group

$$\mathfrak{S}_W := \prod_{\mathcal{C} \in \mathcal{A}/W} \mathfrak{S}_{e_\mathcal{C}}$$

permutes the set of indeterminates $\mathbf{u}$ in an obvious way, thus acts on the $\mathbb{Z}$–algebra $\mathbb{Z}[\mathbf{u}, \mathbf{u}^{-1}]$. For $\sigma \in \mathfrak{S}_W$, we still denote by σ the corresponding automorphism of $\mathbb{Z}[\mathbf{u}, \mathbf{u}^{-1}]$, as well as the $\mathbb{Z}$–algebra automorphism of $\mathcal{H}(W, \mathbf{u})$ which extends the automorphism σ of $\mathbb{Z}[\mathbf{u}, \mathbf{u}^{-1}]$ by the condition $\sigma(\mathbf{h}(b)) = \mathbf{h}(b)$ for all $b \in B$. We define

$$\mathbf{a}_\sigma := \sigma\, \mathbf{a}_1 ; \tag{3.26}$$

then $\mathfrak{S}_W$ acts regularly on the set of all the anti-automorphisms $\mathbf{a}_\sigma$.

Anti–automorphisms and characters of the generic Hecke algebra.

A linear character θ is transformed into another linear character by an anti-automorphism $\mathbf{a}$ through the formula

$$^{\mathbf{a}}\theta := \mathbf{a}\theta\mathbf{a}^{-1} .$$

It is immediate to check that, for $\sigma \in \mathfrak{S}_W$, we have

$$^{\mathbf{a}_\sigma}\theta_{\mathbf{j}} = \theta_{\sigma(\mathbf{j})} .$$

Notation.

- We extend the automorphism $\alpha \mapsto \alpha^\vee$ of the ring $\mathbb{Z}[\mathbf{u}, \mathbf{u}^{-1}]$ to the automorphism (still denoted by $\alpha \mapsto \alpha^\vee$) of the ring $\mathbb{Z}_K[\mathbf{t}, \mathbf{t}^{-1}]$ such that

$$t_{\mathcal{C},j}^\vee := t_{\mathcal{C},j}^{-1} \text{ (for all } \mathcal{C} \text{ and } j), \text{ and } \lambda^\vee := \lambda^* \text{ (for all } \lambda \in \mathbb{Z}_K) .$$

- We extend the anti–automorphism $\mathbf{a}_1$ to the algebra $\mathbb{Z}_K[\mathbf{t}, \mathbf{t}^{-1}]\mathcal{H}(W, \mathbf{u})$ and to the algebra $K(\mathbf{t})\mathcal{H}(W, \mathbf{u})$ by stipulating that $\mathbf{a}_1$ induces the automorphism $\alpha \mapsto \alpha^\vee$ on the ring $\mathbb{Z}_K[\mathbf{t}, \mathbf{t}^{-1}]$. Thus, for $\lambda \in K$, $b \in B$, $\mathcal{C} \in \mathcal{A}/W$ and $0 \leq j < e_\mathcal{C}$, we have

$$\mathbf{a}_1(\lambda t_{\mathcal{C},j}\mathbf{h}(b)) = \lambda^* t_{\mathcal{C},j}^{-1}\mathbf{h}(b^{-1}) .$$

- For any function $f \colon \mathcal{H}(W, \mathbf{u}) \to \mathbb{Z}_K[\mathbf{t}, \mathbf{t}^{-1}]$ we set (for $h \in \mathcal{H}$)

$$f^\vee(h) := f(\mathbf{a}_1(h))^\vee . \tag{3.27}$$

Thus, in particular, for $b \in B$ we have

$$f^\vee(\mathbf{h}(b)) := f(\mathbf{h}(b^{-1}))^\vee .$$

3.28. Lemma. *For all* $\chi \in \mathrm{Irr}(W)$, *we have* $\chi_{\mathbf{t}}^\vee = \chi_{\mathbf{t}}$.

[It suffices to check that $\chi_{\mathbf{t}}^\vee$ specializes to χ for the specialization $t_{\mathcal{C},j} \mapsto 1$, which is clear.]

The opposite algebra and the Lusztig involution.

The opposite algebra $\mathcal{H}(W,\mathbf{u})^{\mathrm{op}}$ is defined as usual:

- as a $\mathbb{Z}[\mathbf{u},\mathbf{u}^{-1}]$-module, we have $\mathcal{H}(W,\mathbf{u})^{\mathrm{op}} = \mathcal{H}(W,\mathbf{u})$;
- the multiplication in $\mathcal{H}(W,\mathbf{u})^{\mathrm{op}}$ is defined by $(h,h') \mapsto h'h$.

It is easy to check the following two properties.

3.29. *The opposite algebra $\mathcal{H}(W,\mathbf{u})^{\mathrm{op}}$ is*

(a) *isomorphic to $\mathbb{Z}[\mathbf{u},\mathbf{u}^{-1}]B^{\mathrm{op}}/\mathfrak{I}^{\mathrm{op}}$, where $\mathfrak{I}^{\mathrm{op}}$ denotes the ideal of the algebra $\mathbb{Z}[\mathbf{u},\mathbf{u}^{-1}]B^{\mathrm{op}}$ generated by all the elements*

$$(\mathbf{s}_H - u_{\mathcal{C},0})(\mathbf{s}_H - u_{\mathcal{C},1})\cdots(\mathbf{s}_H - u_{\mathcal{C},e_{\mathcal{C}}-1})$$

where $\mathcal{C} \in \mathcal{A}/W$, $H \in \mathcal{C}$, $\mathbf{s}_H$ is a distinguished braid reflection around H in B,

(b) *semi-linearly isomorphic to the algebra $\mathcal{H}(W,\mathbf{u})$ through the map $h \mapsto h^\vee$.*

We denote by

$$h \mapsto {}^r h\,,\ \mathcal{H}(W,\mathbf{u}) \longrightarrow \mathcal{H}(W,\mathbf{u})^{\mathrm{op}}$$

the (linear) isomorphism such that ${}^r\mathbf{s} = \mathbf{s}$ whenever $\mathbf{s}$ is the image in $\mathcal{H}(W,\mathbf{u})$ of a distinguished braid reflection around an hyperplane.

Definition. We call *Lusztig involution* of $\mathcal{H}(W,\mathbf{u})$ the involutive semi-linear automorphism ι of $\mathcal{H}(W,\mathbf{u})$ defined by $\iota(h) := {}^r h^\vee$.

Let $\mathbf{s}_1^{n_1}\mathbf{s}_2^{n_2}\cdots\mathbf{s}_l^{n_l} \in B$, where (for $1 \le j \le l$) $\mathbf{s}_j$ is a distinguished braid reflection and $n_j \in \mathbb{Z}$. Then for all $\lambda \in \mathbb{Z}[\mathbf{u},\mathbf{u}^{-1}]$, we have

$$\iota(\lambda \mathbf{s}_1^{n_1}\mathbf{s}_2^{n_2}\cdots\mathbf{s}_l^{n_l}) = \lambda^\vee \mathbf{s}_1^{-n_1}\mathbf{s}_2^{-n_2}\cdots\mathbf{s}_l^{-n_l}\,.$$

Generic Hecke algebras as symmetric algebras

3A. The canonical trace form.

An element $P(\mathbf{u}) \in \mathbb{Z}[\mathbf{u},\mathbf{u}^{-1}]$ is called "multi–homogeneous" if, for each $\mathcal{C} \in \mathcal{A}/W$, it is homogeneous as a Laurent polynomial in the indeterminates $u_{\mathcal{C},j}$ for $j = 0,\ldots,e_{\mathcal{C}}-1$.

The following assertion is conjectured to be true for all reflection groups.

- It is proved in general ([**BMM2**], §2A) that if there exists a form satisfying all conditions (1),(a), (b), (c) of 3.30, then it is unique.
- It is proved in general for W a Coxeter group (*loc.cit.*, §2A).
- It is proved for all infinite families of non Coxeter complex reflection groups

  [Indeed, it is proved in *loc.cit.*, §4, under Malle's conjecture [**Mal**] about the Schur elements, which is now proved in [**GIM**]].

- Finally, Theorem–Assumption 2 is only partly checked for the non Coxeter exceptional groups. A "good" candidate for $t_{\mathbf{u}}$ is known for all groups: it satisfies all properties below but property (1)(a). Property (1)(a) is probably easily checkable on computer for some of the small groups, but it is still open for large exceptional non Coxeter reflection groups.

3.30. THEOREM–ASSUMPTION.

(0) *W satisfies Theorem-Assumption 3.16.*

(1) *There exists a unique linear form*

$$t_{\mathbf{u}}\colon \mathcal{H}(W,\mathbf{u}) \to \mathbb{Z}[\mathbf{u},\mathbf{u}^{-1}]$$

with the following properties.

(a) *$t_{\mathbf{u}}$ is a central form on the algebra $\mathcal{H}(W,\mathbf{u})$, i.e., for all h, $h' \in \mathcal{H}(W,\mathbf{u})$, we have $t_{\mathbf{u}}(hh') = t_{\mathbf{u}}(h'h)$, and it endows $\mathcal{H}(W,\mathbf{u})$ with a structure of symmetric algebra over the ring $\mathbb{Z}[\mathbf{u},\mathbf{u}^{-1}]$, i.e., the map*

$$\widehat{t_{\mathbf{u}}}\colon \mathcal{H}(W,\mathbf{u}) \to \operatorname{Hom}(\mathcal{H}(W,\mathbf{u}),\mathbb{Z}[\mathbf{t},\mathbf{t}^{-1}])\ ,\ h \mapsto (h' \mapsto t_{\mathbf{u}}(hh'))$$

is an isomorphism of $\mathbb{Z}[\mathbf{u},\mathbf{u}^{-1}]$-modules between $\mathcal{H}(W,\mathbf{u})$ and its dual as a $\mathbb{Z}[\mathbf{u},\mathbf{u}^{-1}]$-module.

(b) *Through the specialization $u_{\mathcal{C},j} \mapsto \zeta_{e_{\mathcal{C}}}^j$, the form $t_{\mathbf{u}}$ becomes the canonical linear form on the group algebra.*

(c) *For all $b \in B$, we have*

$$t_{\mathbf{u}}(b^{-1})^{\vee} = \frac{t_{\mathbf{u}}(b\boldsymbol{\pi})}{t_{\mathbf{u}}(\boldsymbol{\pi})}\,.$$

(2) *The form $t_{\mathbf{u}}$ satisfies the following conditions.*

(a) *For $b \in B$, $t_{\mathbf{u}}(b)$ is invariant under the action of $\mathfrak{S}_W$.*

(b) *As an element of $\mathbb{Z}[\mathbf{u},\mathbf{u}^{-1}]$, $t_{\mathbf{u}}(b)$ is multi-homogeneous with degree $\ell_{\mathcal{C}}(b)$ in the indeterminates $u_{\mathcal{C},j}$ for $j = 0,1,\dots,e_{\mathcal{C}}-1$. In particular, we have*

$$t_{\mathbf{u}}(1) = 1 \quad \textit{and} \quad t_{\mathbf{u}}(\boldsymbol{\pi}) = (-1)^{N^{\vee}} \prod_{\substack{H\in\mathcal{A}\\ 0\le j\le e_H-1}} u_{H,j}\,.$$

(c) *If W' is a parabolic subgroup of W, the restriction of $t_{\mathbf{u}}$ to a parabolic sub-algebra $\mathcal{H}(W',W,\mathbf{u})$ is the corresponding specialization of $t_{\mathbf{u}'}(W')$ (we recall that $\mathcal{H}(W',W,\mathbf{u})$ is a specialization of $\mathcal{H}(W',\mathbf{u}')$).*

EXAMPLE. Let us examine the value of $t_{\mathbf{u}}$ in the case where W is cyclic (see [**BrMa2**]). We denote by e the order of W, we set $\mathbf{u} = \{u_0, u_1, \dots, u_{e-1}\}$, and we denote by $\mathbf{s}$ the "positive" generator of B.

- If $j > 0$, $t_{\mathbf{u}}(\mathbf{s}^j)$ is $(-1)^{e-1}$ times the sum of all monomials in $\mathbf{u}$ of degree j, where each indeterminate occurs with a strictly positive exponent.
- If $j \le 0$, $t_{\mathbf{u}}(\mathbf{s}^j)$ is the sum of all monomials in $\mathbf{u}$ of degree j (where each indeterminate occurs with a non-positive exponent).

Thus we have in particular

$$\begin{aligned} &t_{\mathbf{u}}(\mathbf{s}^j) = 0 \quad \text{for } 1 \le j < e\,,\\ &t_{\mathbf{u}}(\mathbf{s}^e) = (-1)^{e-1}\, u_0u_1\cdots u_{e-1}\,,\\ &t_{\mathbf{u}}(\mathbf{s}^{-1}) = \sum_{j=0}^{j=e-1} u_j^{-1}\,. \end{aligned}$$

Notice that, by (2)(c) in Theorem–Assumption above, it results from what precedes that, for W any complex reflection group, we have

$$t_{\mathbf{u}}(\mathbf{s}) = 0 \quad \text{whenever } \mathbf{s} \text{ is a distinguished braid reflection in } B\,.$$

Schur elements.

Since the ring $\mathbb{Z}_K[\mathbf{t},\mathbf{t}^{-1}]$ is integrally closed (see [**Bou2**], §3, Corollaire 3), the first assertion of the following proposition follows from Appendix 2, 8.14 below. Assertion (2) follows from the fact that the functions $\chi_{\mathbf{t}}$ (for $\chi \in \mathrm{Irr}(W)$) are linearly independent.

3.31. PROPOSITION. *Assume that $t_{\mathbf{u}}$ is a linear form on $\mathcal{H}(W,\mathbf{u})$ such that assertion (1)(a) of 3.30 holds, i.e., the map*

$$\widehat{t_{\mathbf{u}}}\colon \mathcal{H}(W,\mathbf{u}) \to \mathrm{Hom}(\mathcal{H}(W,\mathbf{u}), \mathbb{Z}[\mathbf{u},\mathbf{u}^{-1}])$$

is an isomorphism.

(1) *There exist elements $S_\chi(\mathbf{t}) \in \mathbb{Z}_K[\mathbf{t},\mathbf{t}^{-1}]$ for $\chi \in \mathrm{Irr}(W)$, such that*

$$t_{\mathbf{u}} = \sum_{\chi\in\mathrm{Irr}(W)} \frac{1}{S_\chi(\mathbf{t})}\,\chi_{\mathbf{t}}\,.$$

(2) *For all $g \in \mathrm{Gal}(K(\mathbf{t})/K(\mathbf{u}))$ we have $S_{g(\chi)}(\mathbf{t}) = g(S_\chi(\mathbf{t}))$.*

The element $S_\chi(\mathbf{t})$ is called *the Schur element* associated with χ.

It results from (3.20) that

$$S_{\det_{C,j}}(\mathbf{t}) \in \mathbb{Z}[\mathbf{u},\mathbf{u}^{-1}]\,. \tag{3.32}$$

The following palindromicity property of the Schur elements is the translation of the invariance property of $t_{\mathbf{u}}$ under the involution $x \mapsto x^\vee$.

3.33. PROPOSITION. *Assume W satisfies Theorem-Assumption 3.30. Whenever $\chi \in \mathrm{Irr}(W)$, its Schur element $S_\chi(\mathbf{t})$ satisfies the following property:*

$$S_\chi(\mathbf{t}^{-1})^* = \frac{t_{\mathbf{u}}(\boldsymbol{\pi})}{\omega_{\chi_{\mathbf{t}}}(\boldsymbol{\pi})} S_\chi(\mathbf{t})\,.$$

The following property of Schur elements, which in a sense "explains" their palindromicity, results from a case by case study of the groups W satisfying 3.16 (cf. [**BrMa**], [**GIM**], [**Ma2**], [**Ma3**], [**Ma5**]).

3.34. PROPOSITION. *Let $\chi \in \mathrm{Irr}(W)$. The Schur element $s_\chi(\mathbf{v})$ associated with the characters $\chi_{\mathbf{v}}$ of $K(\mathbf{v})\mathcal{H}$ is such that*

$$s_\chi(\mathbf{v}) = \xi_\chi \prod_{\Phi\in C_\chi} \Phi(\mathbf{v})^{n_\chi}\,,$$

where

- $\xi_\chi \in \mathbb{Z}_K$,
- *C_χ is a set of degree zero homogeneous elements of $\mathbb{Z}_K[\mathbf{v},\mathbf{v}^{-1}]$, dividing (in $\mathbb{Z}_K[\mathbf{v},\mathbf{v}^{-1}]$) a Laurent polynomial of the shape $M_1(\mathbf{v}) - M_2(\mathbf{v})$, where $M_1(\mathbf{v})$ and $M_2(\mathbf{v})$ are degree zero unitairy (Laurent) monomials,*
- $n_\chi \in \mathbb{N}$.

EXAMPLE. Consider the case where W is the Weyl group of type G_2. The generic Hecke algebra (an algebra over the ring $\mathbb{Z}[u_0, u_1, v_0, v_1, u_0^{-1}, u_1^{-1}, v_0^{-1}, v_1^{-1}]$) is generated by two elements $\mathbf{s}$ et $\mathbf{t}$ satisfying the relations

$$\mathbf{ststst} = \mathbf{tststs} \quad \text{and} \quad (\mathbf{s} - u_0)(\mathbf{s} - u_1) = (\mathbf{t} - v_0)(\mathbf{t} - v_1) = 0\,.$$

Let x_0, x_1, y_0, y_1 be indeterminates such that

$$x_0^2 = u_0\,,\ x_1^2 = -u_1\,,\ y_0^2 = v_0\,,\ y_1^2 = -v_1\,.$$

We set $\mathbf{x} := (x_0, x_1)\,, \mathbf{y} := (y_0, y_1)\,,$ and

$$\begin{cases} S(\mathbf{x},\mathbf{y}) := (x_1y_1)^{-6}(x_0^2 + x_1^2)(y_0^2 + y_1^2)(x_0^2y_0^2 + x_0x_1y_0y_1 + x_1^2y_1^2)(x_0^2y_0^2 - x_0x_1y_0y_1 + x_1^2y_1^2) \\ T(\mathbf{x},\mathbf{y}) := 2(x_0x_1y_0y_1)^{-2}(x_0^2y_0^2 - x_0x_1y_0y_1 + x_1^2y_1^2)(x_0^2y_1^2 + x_0x_1y_0y_1 + x_1^2y_0^2) \end{cases}$$

Then the Schur elements of the algebra $\mathcal{H}(W)$ are

$$S(x_0, x_1, y_0, y_1)\,, S(x_0, x_1, y_1, y_0)\,,\ S(x_1, x_0, y_0, y_1)\,,\ S(x_1, x_0, y_1, y_0)\,,$$
$$T(x_0, x_1, y_0, y_1)\,,\ T(x_1, x_0, y_0, y_1)\,.$$

Spetsial and cyclotomic specializations of Hecke algebras

Cyclotomic Hecke algebras.

Let us start with some definitions and notation which will be justified later on.

Let W be a complex reflection group, with field of definition K. Let d be an integer, a divisor of the order $|ZW|$ of the center ZW of W. Let ζ be a root of the unity in K, of order d. Let y be an indeterminate. We set

$$x := \zeta y^{|\boldsymbol{\mu}(K)|}\,.$$

3.35. DEFINITION. A *ζ–cyclotomic specialization* of the set of generic indeterminates $\mathbf{t}$ is a morphism of $\mathbb{Z}_K$–algebras

$$\phi\colon \mathbb{Z}_K[\mathbf{t}, \mathbf{t}^{-1}] \to \mathbb{Z}_K[y, y^{-1}]$$

with the following properties:

(a) $\phi\colon t_{\mathcal{C},j} \mapsto y^{n_{\mathcal{C},j}}$, with $n_{\mathcal{C},j} \in \mathbb{Z}$ for all $\mathcal{C}$ and all j .

(b) For all $\mathcal{C} \in \mathcal{A}/W$, the polynomial in t

$$\prod_{j=0}^{j=e_{\mathcal{C}}-1} \left(t - \zeta_{e_{\mathcal{C}}}^j y^{n_{\mathcal{C},j}}\right)$$

is invariant under the action of the Galois group $\mathrm{Gal}(K(y)/K(x))$, *i.e.*, belongs to $\mathbb{Z}_K[x, x^{-1}][t]$.

We shall write $m_{\mathcal{C},j} := n_{\mathcal{C},j}/|\boldsymbol{\mu}(K)|$ and then set $\phi\colon u_{\mathcal{C},j} \mapsto \zeta_{e_{\mathcal{C}}}^j(\zeta^{-1}x)^{m_{\mathcal{C},j}}$.

From now on, we assume that Theorem-Assumption 3.30 holds for W.

The corresponding *cyclotomic Hecke algebra* is the $\mathbb{Z}_K[x, x^{-1}]$–algebra $\mathcal{H}_\phi(W)$ obtained as the specialization of the algebra $\mathcal{H}(W, \mathbf{u})$ through the specialization ϕ, endowed with the symmetrizing form t_ϕ specialized from the form $t_{\mathbf{u}}$.

The algebra $\mathcal{H}_\phi(W)$ is an image of the group algebra $\mathbb{Z}_K[x, x^{-1}]B$ under a morphism which we still denote by $\phi\colon \mathbb{Z}_K[x, x^{-1}]B \to \mathcal{H}_\phi(W)$, and for $x = \zeta$ it specializes onto the group algebra $\mathbb{Z}_K W^{\mathrm{op}}$ (while its chosen symmetrizing form specializes onto the canonical symmetric form on $\mathbb{Z}_K W^{\mathrm{op}}$).

Schur elements.

Let y be an indeterminate. We call *K–cyclotomic polynomials* the monic minimal polynomials (over $K[y]$) of roots of unity. The K–cyclotomic polynomials are irreducible elements of $\mathbb{Z}_K[y]$. We denote by $\mathrm{Cycl}(K)$ the set of all K–cyclotomic polynomials.

The following property is an immediate consequence of 3.34 above.

3.36. PROPOSITION. *Let $\mathcal{H}_\phi(W)$ be a ζ–cyclotomic Hecke algebra, defined by $\phi\colon t_{C,j} \mapsto y^{n_{C,j}}$. The Schur element of any irreducible character $\chi_\phi \in K(y)\mathcal{H}_\phi(W)$ has the following shape :*

$$s_{\chi_\phi}(y) = \psi_\chi\, y^{a_\chi} \prod_{\Phi(y)\in \mathrm{Cycl}(K)_\chi} \Phi(y)^{m_\chi}$$

where

- $\psi_\chi \in \mathbb{Z}_K$,
- $a_\chi \in \mathbb{Z}$,
- $\mathrm{Cycl}(K)_\chi$ *is a set of K–cyclotomic polynomials, and* $m_\chi \in \mathbb{N}$.

Linear characters of the cyclotomic Hecke algebra.

Let $\mathcal{H}_\phi(W)$ be a ζ–cyclotomic Hecke algebra. Let

$$\theta\colon \mathcal{H}_\phi(W) \longrightarrow \mathbb{Z}_K[y, y^{-1}]$$

be a linear character of $\mathcal{H}_\phi(W)$ (recall that $y^{|\mu(K)|} = \zeta^{-1}x$). We denote by $\mathbf{j}_\theta := (j_C)_{C\in\mathcal{A}/W}$ the family of integers ($0 \le j_C < e_C$) such that $\theta(\mathbf{s}_C) = \zeta_{e_C}^{j_C}(\zeta^{-1}x)^{m_{C,j_C}}$.

We set

$$D_\theta := \sum_{C\in\mathcal{A}/W} m_{C,j_C} N_C e_C\,.$$

Let us denote by $\overline{\theta}$ the linear character of W obtained through the specialization $x \mapsto \zeta$. Thus we have

$$\overline{\theta}(s_C) = \zeta_{e_C}^{j_C}\,.$$

3.37. LEMMA. *Let $\mathbf{w}$ be an element of B such that $\mathbf{w}^d = \boldsymbol{\pi}$, and let w be its image in W. We have*

$$\theta(\mathbf{w}) = \overline{\theta}(w)(\zeta^{-1}x)^{D_\theta/d}\,.$$

[This is a particular case of 3.22.] □

Principal cyclotomic Hecke algebras.

3.38. DEFINITION. A ζ–cyclotomic algebra $\mathcal{H}_\phi(W)$ is said to be *principal* if there exists a linear character

$$\theta_0\colon \mathcal{H}_\phi(W) \longrightarrow \mathbb{Z}_K[x, x^{-1}] \quad , \quad \mathbf{s}_C \mapsto \zeta_{e_C}^{j_C}(\zeta^{-1}x)^{m_C}$$

such that

(p1) we have $m_C \ge m_{C,j}$ for all j ($0 \le j < e_C$),

(p2) whenever $\mathbf{w}$ is an element of B such that $\mathbf{w}^d = \boldsymbol{\pi}$, we have $\theta_0(\mathbf{w}) = x^{D_{\theta_0}/d}$.

The character θ_0 is then called a *principal character of* $\mathcal{H}_\phi(W)$.

EXAMPLES.

(1) The *spetsial Hecke algebra* $\mathcal{H}(W)$ is the principal cyclotomic Hecke algebra associated with the 1–cyclotomic specialization

$$\left\{\begin{array}{l} u_{\mathcal{C},0} \mapsto x\,, \\ u_{\mathcal{C},j} \mapsto \zeta_{e_\mathcal{C}}^j \quad \text{for } j>0\,. \end{array}\right. \tag{3.39}$$

In other words, the images of the elements $\mathbf{s}_H$ in the algebra $\mathcal{H}_x(W)$ satisfy the equations

$$(\mathbf{s}_H - x)(1+\mathbf{s}_H+\mathbf{s}_H^2+\cdots+\mathbf{s}_H^{e_H-1}) = 0\,.$$

In particular, one sees that if W is a Coxeter group, the spetsial Hecke algebra $\mathcal{H}(W)$ is the ordinary generic Hecke algebra over $\mathbb{Z}_K[x,x^{-1}]$.

(2) Let $W = G(4,2,2)$ and let $\zeta = i$. Recall (see for example [**BMR**]) that the corresponding braid group B is generated by three distinguished braid reflections $\mathbf{s}$, $\mathbf{s}'$, $\mathbf{s}''$, such that $\mathbf{s}\mathbf{s}'\mathbf{s}'' = \mathbf{s}'\mathbf{s}''\mathbf{s} = \mathbf{s}''\mathbf{s}\mathbf{s}'$, corresponding to reflections s, s', s'' of order 2 in W which satisfy the relations $s''s's = s'ss'' = ss''s'$.

The relations

$$\left\{\begin{array}{l} \mathbf{s}\mathbf{s}'\mathbf{s}'' = \mathbf{s}'\mathbf{s}''\mathbf{s} = \mathbf{s}''\mathbf{s}\mathbf{s}' \\ (\mathbf{s}-1)(\mathbf{s}-x^2) = (\mathbf{s}'-1)(\mathbf{s}'-x^2) = (\mathbf{s}''-1)(\mathbf{s}''-x^2) = 0 \end{array}\right.$$

define a principal cyclotomic Hecke algebra associated with W.

Some definitions and notation about palindromicity.

Let $P(x) \in \mathbb{C}(x)$ (or more generally suppose that P is a meromorphic function on $\mathbb{C}$). We say that $P(x)$ is *semi-palindromic* if there exist an integer $m \in \mathbb{Z}$ and a norm one complex number ξ such that

$$P(1/x)^* = \xi x^{-m} P(x)\,.$$

The following statements are obvious to check:

1. A constant is semi-palindromic.
2. If $P(x)$ and $Q(x)$ are semi-palindromic, so is $P(x)Q(x)$.
3. If ζ is a norm one complex number and if $P(x)$ is semi-palindromic, then the element $P^\zeta(x) := P(\zeta^{-1}x)$ is semi-palindromic.

For $P(x) \neq 0$, define the *valuation* $\mathrm{val}_x(P)$ (also denoted by a) as the order of 0 as a zero of $P(x)$ ($a<0$ if 0 is a pole of $P(x)$). Let c_a be the nonzero complex number defined by

$$c_a := \left(x^{-a}P(x)\right)_{|x=0}\,.$$

Define the *degree* $\deg_x(P)$ (also denoted by A) as the valuation of $P(1/x)$. Let c_A be the nonzero complex number defined by

$$c_A := \left(x^{A}P(1/x)\right)_{|x=0}\,.$$

Thus if $P(x) \in \mathbb{C}[x,x^{-1}]$ is a Laurent polynomial, we have

$$P(x) = c_a x^a + c_{a+1}x^{a+1} + \cdots + c_{A-1}c^{A-1} + c_A x^A$$

for some coefficients c_j. The following lemma is immediate.

3.40. Lemma. *Let $P(x)$ such that $P(1/x)^* = \xi x^{-m} P(x)$.*

(1) *We have $m = a + A$, and $c_A^* = \xi c_a$. Moreover,*

$$mP(x) = xP'(x) + \xi^* x^{m-1} P'(1/x)^* .$$

(2) *Let ζ be a norm one complex number such that $P(\zeta) \neq 0$.*

(a) *We have*

$$\xi = \zeta^m \frac{P(\zeta)^*}{P(\zeta)} \quad \text{and} \quad m = \zeta \frac{P'(\zeta)}{P(\zeta)} + \zeta^* \frac{P'(\zeta)^*}{P(\zeta)^*} = 2\Re\left(\zeta \frac{P'(\zeta)}{P(\zeta)}\right) .$$

(b) *If moreover $P(\zeta) \in \mathbb{R}$, we have $\xi = \zeta^m$. In particular if $\zeta = 1$, we have then*

$$\xi = 1 \quad \text{and} \quad m = \frac{P'(1) + P'(1)^*}{P(1)} .$$

More generally, assume that $P(y) \in \mathbb{C}(y)$, and set $x := y^n$ for some natural integer n. Then by abuse of notation, we define

$$\deg_x(P) := \frac{\deg_y(P)}{n} \quad \text{and} \quad \mathrm{val}_x(P) := \frac{\mathrm{val}_y(P)}{n} .$$

On spetsial cyclotomic Hecke algebras.

The results in this paragraph are taken from [**BMM2**], §6C. They are designed to be applied to characters of finite reductive groups (see below chap. 5 and 6).

We recall that we are now assuming Theorem-Assumption 3.30 holds.

3.41. Proposition. *Let us denote by $\mathcal{H}_x(W)$ the specialization of the generic Hecke algebra under (3.39) above. Then,*

(1) *the polynomial $P_W(\mathbf{u})$ specializes to a polynomial $P_W(x) \in \mathbb{Z}_K[x, x^{-1}]$ which is palindromic and satisfies the relation*

$$P_W(x) = x^{N^\vee} P_W(x^{-1}) ,$$

(2) *the absolute generic degree $\mathrm{Deg}_\chi(\mathbf{t})$ specializes to an element $\mathrm{Deg}_\chi(x) \in K(x^{1/|\mu(K)|})$ which satisfies the relation*

$$\mathrm{Deg}_\chi(x) = x^{\frac{N(\chi)+N(\chi^*)}{\chi(1)}} \mathrm{Deg}_\chi(x^{-1})^* ,$$

(3) *the character $\chi_\mathbf{t}$ specializes to a character χ_x which satisfies the relation*

$$\omega_{\chi_x}(\boldsymbol{\pi}) = x^{N+N^\vee - \frac{N(\chi)+N(\chi^*)}{\chi(1)}} .$$

3.42. Corollary. *The sum of the valuation a_χ and of the degree A_χ of the generic degree $\mathrm{Deg}_\chi(x)$ of χ is computable from its fake degree $\mathrm{Feg}_\chi(x)$: we have*

$$a_\chi + A_\chi = \frac{N(\chi) + N(\chi^*)}{\chi(1)} .$$

Indeed, this results from 3.41, (2) and 3.40, (1).

Palindromicity, fake degrees, generic degrees.

By [**Ma3**], 4.8, we know that the field $K(x^{1/|\boldsymbol{\mu}(K)|})$ contains the values of the absolutely irreducible characters of the spetsial cyclotomic Hecke algebra $\mathcal{H}_x(W)$. Following *loc.cit.*, we denote by δ the generator of $\mathrm{Gal}(K(x^{1/|\boldsymbol{\mu}(K)|})/K(x))$ defined by

$$\delta(x^{1/|\boldsymbol{\mu}(K)|}) = \exp\left(2\pi i/|\boldsymbol{\mu}(K)|\right) x^{1/|\boldsymbol{\mu}(K)|}\,.$$

We denote by ι the automorphism of order 2 of $K(x^{1/|\boldsymbol{\mu}(K)|})$ defined as the composition of the complex conjugation by the automorphism δ, *i.e.*,

$$\iota(\lambda) := \delta(\lambda^*) \quad \text{for } \lambda \in K(x^{1/|\boldsymbol{\mu}(K)|})\,.$$

We have an operation of $\mathrm{Gal}(K(x^{1/|\boldsymbol{\mu}(K)|})/K(x))$ on the set $\mathrm{Irr}(W)$ of irreducible characters of W (see also above §1) defined as follows:

$$g(\chi)_x = g(\chi_x)$$

(for $g \in \mathrm{Gal}(K(x^{1/|\boldsymbol{\mu}(K)|})/K(x))$ and $\chi \in \mathrm{Irr}(W)$).

The following property is proved in *loc.cit.*, (6.5), through a case by case analysis. Note that Opdam [**Op1**] has given a general proof for the case where W is a Coxeter group, and that his proof can be generalized to any complex reflection group W (see [**Op2**]) provided one assumes that presentations like in Appendix 1 hold for B (see above, comments before 3.18).

3.43. PROPOSITION. *Let W satisfy Theorem-Assumption 3.16. Then, for any $\chi \in \mathrm{Irr}(W)$, there exists an integer m such that*

$$\mathrm{Feg}_\chi(x) = x^m \mathrm{Feg}_{\iota(\chi)}(1/x)\,.$$

As in [**BMM2**], §6.D, one deduces easily from the preceding proposition a formula which was only known case by case for W a Weyl group.

Let us introduce the valuation b_χ and the degree B_χ of $\mathrm{Feg}_\chi(x)$.

3.44. COROLLARY. *We have*

$$\frac{(b_\chi + B_\chi) + (b_{\iota(\chi)} + B_{\iota(\chi)})}{2} = a_\chi + A_\chi\,.$$

Indeed, proposition 3.43 implies that the polynomial $P_\chi(x) := \mathrm{Feg}_\chi(x)\mathrm{Feg}_{\iota(\chi)}(x)$ satisfies the relation

$$P_\chi(x) = x^{2m} P_\chi(1/x)\,,$$

hence is palindromic. Then lemma 3.40, (2)(b) above shows that

$$2m = 2\frac{P'_\chi(1)}{P_\chi(1)}\,,$$

hence

$$m = \frac{N(\chi) + N(\iota(\chi))}{\chi(1)}\,.$$

By 3.24 above, we know that $N(\iota(\chi)) = N(\chi^*)$, which implies

$$m = \frac{N(\chi) + N(\chi^*)}{\chi(1)}\,.$$

Since the valuation and the degree of the polynomial $\mathrm{Feg}_\chi(x)\mathrm{Feg}_{\iota(\chi)}(x)$ are respectively $b_\chi + b_{\iota(\chi)}$ and $B_\chi + B_{\iota(\chi)}$, we have

$$\frac{(b_\chi + B_\chi) + (b_{\iota(\chi)} + B_{\iota(\chi)})}{2} = N^\vee - \sum_{\rho \in \mathrm{Ref}(W)} \frac{\chi(\rho)}{\chi(1)} = \frac{N(\chi) + N(\chi^*)}{\chi(1)}.$$

It results from 3.41 and 3.42 above that $\mathrm{Deg}_\chi(x)$ is semi–palindromic, and that, if a_χ and A_χ are respectively its valuation and degree in x, we have

$$a_\chi + A_\chi = \frac{N(\chi) + N(\chi^*)}{\chi(1)}.$$

Corollary 3.44 is now immediate.

On generic degrees of cyclotomic Hecke algebras.

Let ζ be a root of unity of order d, and let as above $\mathcal{H}_\phi(W)$ be a ζ–cyclotomic Hecke algebra defined by a cyclotomic specialization $\phi\colon t_{C,j} \mapsto y^{n_{C,j}}$ where $y^{|\mu(K)|} = \zeta^{-1}x$.

Let $\mathrm{Irr}(\mathcal{H}_\phi(W))$ denote the set of (absolutely) irreducible characters of the split semi–simple algebra $K(y)\mathcal{H}_\phi(W)$.

Let $P(y) \in K(y)$ be semi–palindromic.

Let $\psi \in \mathrm{Irr}(\mathcal{H}_\phi(W))$.

• We denote by $\overline{\psi}$ the character of W defined by ψ through the specialization $y \mapsto 1$. Thus in particular we have $\overline{\psi}(s_C) = (\psi(\mathbf{s}_C))_{y=1}$.

• The *P–generic degree* of ψ is the element $\mathrm{Deg}_\psi^{(P)}(y) \in K(y)$ defined as follows:

$$\mathrm{Deg}_\psi^{(P)}(y) := \frac{P(y)}{S_\psi(y)},$$

where we denote by $S_\psi(y)$ the Schur element of ψ relative to the form t_ϕ.

COMMENT. In the applications to finite reductive groups, we take $P(y) = \mathrm{Deg}(R_{\mathbf{L}}^{\mathbf{G}}(1))(x)\mathrm{Deg}(\lambda)(x)$.

• We denote by a_ψ and A_ψ respectively the (generalized) valuation and degree of $\mathrm{Deg}_\psi^{(P)}(y)$ in x.

Note that $S_\psi(y)$ is semi–palindromic, since by 3.33 it satisfies the equality

$$S_\chi(y^{-1})^* = \frac{t_\phi(\boldsymbol{\pi})}{\omega_\psi(\boldsymbol{\pi})} S_\chi(y). \tag{3.45}$$

It follows that $\mathrm{Deg}_\psi^{(P)}(y)$ is semi–palindromic, so satisfies the identity

$$\mathrm{Deg}_\psi^{(P)}(y^{-1})^* = \xi_\psi x^{-(a_\psi + A_\psi)} \mathrm{Deg}_\psi^{(P)}(y)$$

for some root of the unity ξ_ψ.

In what follows, we choose a linear character θ_0 of $\mathcal{H}_\phi(W)$, such that $\theta_0(\mathbf{s}_C) = \zeta_{e_C}^{j_C}(\zeta^{-1}x)^{m_C}$. We set

$$\begin{cases} D_0 := D_{\theta_0} = \displaystyle\sum_{C \in \mathcal{A}/W} m_C N_C e_C, \\ a_0 := a_{\theta_0} \quad \text{and} \quad A_0 := A_{\theta_0}. \end{cases}$$

The following results are Propositions 6.15 and 6.16 in [**BMM2**].

3.46. PROPOSITION. *Let* $\mathbf{w}$ *be an element of* B *such that* $\mathbf{w}^d = \boldsymbol{\pi}$.

(1) *We have*

$$\omega_\psi(\mathbf{w}) = \omega_{\overline{\psi}}(w)(\zeta^{-1}x)^{\frac{D_0-((a_\psi+A_\psi)-(a_0+A_0))}{d}}.$$

(2) *Assume that* $\mathcal{H}_\phi(W)$ *is principal, and that* θ_0 *is its principal character. Assume moreover that* $\mathrm{Deg}^{(P)}_{\theta_0}(y) = 1$. *We have*

$$\omega_\psi(\mathbf{w}) = \omega_{\overline{\psi}}(w)(\zeta^{-1}x)^{\frac{D_0-(a_\psi+A_\psi)}{d}}.$$

For $\mathcal{C} \in \mathcal{A}/W$ and $0 \leq j < e_\mathcal{C}$, we denote by $\theta_{\mathcal{C},j}$ the linear character of $\mathcal{H}_\phi(W)$ defined by

$$\theta_{\mathcal{C},j}(\mathbf{s}_{\mathcal{C}'}) = \begin{cases} \zeta_{e_\mathcal{C}}^j(\zeta^{-1}x)^{m_{\mathcal{C},j}} & \text{if } \mathcal{C}' = \mathcal{C}\,, \\ \theta_0(\mathbf{s}_{\mathcal{C}'}) & \text{if } \mathcal{C}' \neq \mathcal{C}\,. \end{cases}$$

We set

$$\begin{cases} \mathrm{Deg}^{(P)}_{\mathcal{C},j}(y) := \mathrm{Deg}^{(P)}_{\theta_{\mathcal{C},j}}(y)\,, \\ \omega_{\mathcal{C},j}(y) := \omega_{\theta_{\mathcal{C},j}}(y)\,, \\ a_{\mathcal{C},j} := a_{\theta_{\mathcal{C},j}} \quad \text{and} \quad A_{\mathcal{C},j} := A_{\theta_{\mathcal{C},j}}\,. \end{cases}$$

3.47. PROPOSITION.

(1)

$$(m_\mathcal{C} - m_{\mathcal{C},j})N_\mathcal{C}e_\mathcal{C} = (a_{\mathcal{C},j} + A_{\mathcal{C},j}) - (a_0 + A_0)\,.$$

(2) *In particular, if* $\mathcal{H}_\phi(W)$ *is principal, if* θ_0 *is the principal character and if* $\mathrm{Deg}^{(P)}_{\theta_0} = 1$, *we have*

$$(m_\mathcal{C} - m_{\mathcal{C},j})N_\mathcal{C}e_\mathcal{C} = (a_{\mathcal{C},j} + A_{\mathcal{C},j})\,.$$

CHAPTER IV

REFLECTION DATA

Definitions

From now on we assume K *is a subfield of the field* $\mathbb{Q}(\boldsymbol{\mu}_\infty)$ generated by all roots of unity. The complex conjugation $z \mapsto z^*$ induces an automorphism of K. We denote by $\mathbb{Z}_K$ the ring of integers of K.

4.1. Definitions.

1. A *reflection datum* $\mathbb{G}$ on K is a pair (V, Wf), where
- V is a finite dimensional vector space over K,
- W is a finite reflection group on V,
- f is an element of finite order of $\mathrm{GL}(V)$ which normalizes W.

Note that only the coset Wf (and not the automorphism f) is defined by $\mathbb{G}$.

The vector space V is called the space of the reflection datum and its dimension r is called *the rank of the reflection datum.* The group W is called the reflection group of the reflection datum. The image of Wf modulo W is denoted by $\overline{f}$ and called *the twist of the reflection datum.* Its order is denoted by $\delta(\mathbb{G})$: we have $f^m \in W$ if and only if $\delta(\mathbb{G})$ divides m.

A reflection datum is called *split* if $Wf = W$, *i.e.*, if $f \in W$.

2. A *sub–reflection datum* of $\mathbb{G} = (V, Wf)$ is a reflection datum of the shape $\mathbb{G}' = (V', W'(wf)_{|_{V'}})$, where V' is a subspace of V, W' is a reflection subgroup of $N_W(V')|_{V'}$ (the restriction to V' of the stabilizer $N_W(V')$ of V', isomorphic to $N_W(V')/C_W(V')$), and wf is an element of Wf which stabilizes V' and normalizes W'.

3. A *toric reflection datum (or torus)* is a reflection datum whose reflection group is trivial. A torus of $\mathbb{G} = (V, Wf)$ is a sub–reflection datum of the form $(V', (wf)_{|_{V'}})$, where V' is a subspace of V, and wf is an element of Wf which stabilizes V'.

4. If $\mathbb{T} = (V', (wf)_{|_{V'}})$ is a torus of $\mathbb{G}$, the centralizer of $\mathbb{T}$ in $\mathbb{G}$ is the reflection datum defined by $C_{\mathbb{G}}(\mathbb{T}) := (V, C_W(V')wf)$ (notice that $C_W(V')wf$ is the set of all elements in Wf which act like wf on V'). Such a reflection datum is also called a *Levi sub-reflection datum* of $\mathbb{G}$.

5. The *center of a reflection datum* $\mathbb{G} = (V, Wf)$ is the torus $Z(\mathbb{G}) := (V^W, f_{|_{(V^W)}})$ (notice that it does not depend on the choice of f in Wf).

6. We say that an extension K' of K *splits* $\mathbb{G}$ if it contains the m-th roots of unity, where m is the l.c.m. of the orders of the elements of Wf.

Remarks.

1. The reflection group of a reflection datum over $\mathbb{Q}$ (resp. over a subfield of $\mathbb{R}$) is a Weyl group (resp. a Coxeter group).

2. Any torus of $\mathbb{G}$ is contained in a *maximal* torus of $\mathbb{G}$, a reflection datum of the form (V, wf) where $w \in W$.

4.2. LEMMA.

(1) *The Levi sub-reflection data of* $\mathbb{G} = (V, Wf)$ *are the reflection data of the form* $\mathbb{L} = (V, W'wf)$, *where* W' *is a parabolic subgroup of* W *and* wf *is an element of* Wf *which normalizes* W'.

(2) *If* $\mathbb{L}$ *is a Levi sub-reflection datum of* $\mathbb{G}$, *then* $C_{\mathbb{G}}(Z(\mathbb{L})) = \mathbb{L}$.

Indeed, by the above definition of Levi sub–reflection datum (4.1, 4) it is enough to prove that if $\mathbb{L}$ is defined as in (1), then $C_{\mathbb{G}}(Z(\mathbb{L})) = \mathbb{L}$. By 4.1, 5, we have $Z(\mathbb{L}) = (V^{W'}, (wf)|_{(V^{W'})})$, hence by 4.1, 3, we have $C_{\mathbb{G}}(Z(\mathbb{L})) = (V, C_W(V^{W'})wf)$, and the result follows from Steinberg's theorem 1.2. □

Reflection data, generic groups and finite reductive groups

We start with a connected reductive algebraic group $\mathbf{G}$ over the algebraic closure of a finite field of characteristic $p > 0$. We assume that $\mathbf{G}$ is already defined over a finite field and let $F : \mathbf{G} \to \mathbf{G}$ be the corresponding Frobenius morphism. The group of fixed points $\mathbf{G}^F$ is then a finite reductive group. The choice of an F-stable maximal torus $\mathbf{T}$ of $\mathbf{G}$ and a Borel subgroup containing $\mathbf{T}$ gives rise to a root datum $(X, R, Y, R^\vee)$, consisting of the character and cocharacter groups X, Y of $\mathbf{T}$, the set of roots $R \subset X$ and the set of coroots $R^\vee \subset Y$. The Frobenius map F acts on $Y_{\mathbb{R}} := Y \otimes_{\mathbb{Z}} \mathbb{R}$ as qf where q is a power of p and f is an automorphism of finite order. Replacing the Borel subgroup by another one containing $\mathbf{T}$ changes f by an element of the Weyl group W of $\mathbf{G}$ with respect to $\mathbf{T}$. Hence f is uniquely determined as automorphism of $Y_{\mathbb{R}}$ up to elements of W.

We can thus naturally associate to $(\mathbf{G}, \mathbf{T}, F)$ the data $(X, R, Y, R^\vee, Wf)$. Here,

(i) X, Y are free $\mathbb{Z}$-modules of equal finite rank, endowed with a duality $X \times Y \to \mathbb{Z}$, $(x, y) \mapsto \langle x, y \rangle$,

(ii) $R \subset X$ and $R^\vee \subset Y$ are root systems with a bijection $R \to R^\vee$, $\alpha \mapsto \alpha^\vee$, such that $\langle \alpha, \alpha^\vee \rangle = 2$

(iii) W is the Weyl group of the root system $R^\vee$ in Y and f is an automorphism of Y of finite order stabilizing $R^\vee$.

A quintuple $(X, R, Y, R^\vee, Wf)$ satisfying these properties is called a *generic finite reductive group*.

Conversely, let's start from a generic group $\mathbb{G} = (X, R, Y, R^\vee, Wf)$. Then for any choice of a prime number p, $\mathbb{G}$ determines a pair $(\mathbf{G}, \mathbf{T})$ as above, up to inner automorphisms of $\mathbf{G}$ induced by $\mathbf{T}$. Moreover, the additional choice of a power q of p determines a triple $(\mathbf{G}, \mathbf{T}, F)$ as above. In this way the generic finite reductive group $\mathbb{G}$ gives rise to a whole series $\{\mathbb{G}(q) := \mathbf{G}^F \mid q \text{ a prime power}\}$ of finite reductive groups.

EXAMPLE. Let $\mathbf{G} = \mathrm{GL}_n(\overline{\mathbb{F}_q})$ be the group of invertible $n \times n$-matrices over the algebraic closure of the finite field $\mathbb{F}_q$ with the maximal torus $\mathbf{T}$ consisting of the diagonal matrices in $\mathbf{G}$. Then $\mathbf{T}$ is F-stable for the Frobenius map $F : \mathbf{G} \to \mathbf{G}$ which raises every matrix entry to its qth power, as well as for the product F^- of F with the transpose-inverse map on $\mathbf{G}$. In the first case, the group of F-fixed points is the general linear group over $\mathbb{F}_q$, while for the second Frobenius map F^- we obtain the general unitary group $\mathrm{U}_n(q)$. One easily checks that $(\mathbf{G}, \mathbf{T}, F)$ gives rise to a generic finite reductive group of the form

$$\mathrm{GL}_n = (\mathbb{Z}^n, R, \mathbb{Z}^n, R^\vee, \mathfrak{S}_n \cdot \mathrm{Id})$$

with $R = R^\vee = \{e_i - e_j \mid i \neq j\}$, where $\{e_1, \dots, e_n\}$ is the standard basis of $\mathbb{Z}^n$, while $(\mathbf{G}, \mathbf{T}, F^-)$ gives rise to $\mathrm{U}_n = (\mathbb{Z}^n, R, \mathbb{Z}^n, R^\vee, \mathfrak{S}_n(-\mathrm{Id}))$. In this sense we may think of $\mathrm{U}_n(q)$ as being $\mathrm{GL}_n(-q)$.

Now we relate the generic groups to reflection data.

For the simplicity of the exposition, we restrict ourselves to the case of generic groups over $\mathbb{Q}$ *(for the "very twisted" cases* ${}^2\mathrm{B}_2$, ${}^2\mathrm{F}_4$, ${}^2\mathrm{G}_2$, *see for example* [**BMM2**]*).*

Let $\mathbb{G} = (V, Wf)$ be a reflection datum on $\mathbb{Q}$, so that W is a Weyl group. Let $(R, R^\vee)$ be a root system for W (so in particular R is a finite subset of V, while $R^\vee$ is a finite subset of $V^\vee$). Let us denote by $Q(R)$ and $Q(R^\vee)$ the $\mathbb{Z}$-submodules of respectively $V^\vee$ and V generated respectively by R and $R^\vee$.

We denote by $V^\vee := \mathrm{Hom}(V, \mathbb{Q})$ the dual space of V, we still denote by W the image of W acting through the contragredient operation on the dual vector space $V^\vee$, and we set $\phi^\vee := {}^t\phi^{-1}$.

• Any choice of a pair of dual lattices X and Y in respectively $V^\vee$ and V such that

(a) X is $W.\phi^\vee$-stable and Y is Wf-stable,

(b) $Q(R) \subseteq X$ and $Q(R^\vee) \subseteq Y$

provides a generic group:

$$\mathbb{G}_{(X,Y)} := ((X, R, Y, R^\vee), Wf)\,.$$

• Reciprocally, any generic group $((X, R, Y, R^\vee), Wf)$ defines a reflection datum

$$\mathbb{G} := (\mathbb{Q} \otimes_{\mathbb{Z}} Y, Wf)\,,$$

and we have $\mathbb{G}_{(X,Y)} = ((X, R, Y, R^\vee), Wf)$.

Two generic groups $((X, R, Y, R^\vee), Wf)$, $((X_1, R_1, Y_1, R_1^\vee), W_1\phi_1)$ such that $\mathbb{Q} \otimes Y = \mathbb{Q} \otimes Y_1$ and $Wf = W_1\phi_1$ define the same reflection datum. Thus reflection data classify generic groups up to isogeny.

With the previous notation, the map $\mathbb{L}_{(X,Y)} \mapsto \mathbb{L}$ is then a bijection between Levi sub–reflection data of $\mathbb{G}$ and generic Levi subgroups of $\mathbb{G}_{(X,Y)}$ which respects all the invariants which will be introduced for reflection data and which were previously introduced (see [**BMM1**]) for generic groups (such as polynomial orders, signs, graded representation, class functions) or will be introduced later on (such as unipotent degrees). Moreover, it "commutes" with the usual constructions (adjoint, dual), and $\mathbb{L}$ is Φ_d–split if and only if $\mathbb{L}_{(X,Y)}$ is.

COMMENT. The motivation for the definition of generic groups was to formalize the properties of reductive algebraic groups over a finite field $\mathbb{F}_q$ which are independent of q.

The properties of reflection data that we will discuss in the remainder of this paper are motivated by the properties of unipotent class functions on reductive groups (which have been observed to depend only on the associated reflection datum). We will reflect on that by comments of the form "for $\mathbf{G}$ this means" and "for $\mathbf{G}^F$ this means" where by $\mathbf{G}$ we will mean *any* of the algebraic groups over $\overline{\mathbb{F}}_q$ with reflection datum $\mathbb{G}$ (in general, we will use the same letter as for the reflection datum, but in bold instead of blackboard bold font).

The order of a reflection datum

Let $\mathbb{G} = (V, Wf)$ be a reflection datum. We set $N^\vee(\mathbb{G}) := N^\vee(W)$ (or simply $N^\vee$), the number of pseudo–reflections in W, and we set $N(\mathbb{G}) := N(W)$ (or simply N), the number of reflecting hyperplanes of W.

Generalized degrees of a reflection datum.

Let K' be an extension of K which splits $\mathbb{G}$. We set $V' := K' \otimes_K V$.

Let S be the symmetric algebra of V. Similarly, let S' be the symmetric algebra of V'. Since f acts completely reducibly on the subspace of elements of $(S')^W$ with given degree, we can always choose a family $\{f_1, \ldots, f_r\}$ of basic homogeneous invariants which are all eigenvectors for the action of f (see [**St2**], 2.1), with eigenvalues respectively $\zeta_1, \ldots, \zeta_r$. Then the family of pairs $\{(d_1, \zeta_1), \ldots, (d_r, \zeta_r)\}$ depends only on $\mathbb{G}$ (cf. for example [**Sp**], 6.1), and we call them *the generalized degrees* of the reflection datum $\mathbb{G}$.

Generalized sign.

We recall that R_+ denotes the ideal of R consisting of elements without degree zero terms. The vector space $R_+/(R_+)^2$ has dimension r, and is endowed with an action of the image $\overline{f}$ of f modulo W. Any family of basic homogeneous invariants $\{f_1, \ldots, f_r\}$ as above provides a basis of $(S')^W_+/((S')^W_+)^2$ on which $\overline{f}$ is diagonal. The *generalized sign* (note that in general it is not a sign !) of the reflection datum $\mathbb{G}$ is by definition:

$$\varepsilon_{\mathbb{G}} := (-1)^r \zeta_1 \cdots \zeta_r \,.$$

One can prove (cf. [**Sp**], 6.5) that if f admits a fixed point in $V - \bigcup_{H \in \mathcal{A}} H$, then $\{\zeta_1, \ldots, \zeta_r\}$ is the spectrum of f (in its action on V'). In particular, we have then $\varepsilon_{\mathbb{G}} := (-1)^r \mathrm{det}_V(f)$.

It follows that

4.3. If $\mathbb{T} = (V, wf)$ is a maximal torus of $\mathbb{G}$, we have $\varepsilon_{\mathbb{T}} = (-1)^r \mathrm{det}_V(wf)$. Moreover, if f admits a fixed point in $V - \bigcup_{H \in \mathcal{A}} H$, then $\varepsilon_{\mathbb{T}} = \varepsilon_{\mathbb{G}} \mathrm{det}_V(w)$.

REMARK. Contrary to the case of real reflection data, the group $\langle Wf \rangle$ (subgroup of $\mathrm{GL}(V)$ generated by Wf) is not always a semi–direct product. For example, for $W = G(4,2,4)$, there exists an element f of order 4 in $N_{\mathrm{GL}(V)}(W)$ such that $\langle Wf \rangle = G(4,1,4)$, which is a non split extension of W (see proposition 4.12 below).

The graded regular representation.

We set $R\mathbb{G} := S_W = S/(S.R_+)$ the coinvariant algebra viewed as endowed with the natural action of the group $W\langle f \rangle)$. Recall that this finite dimensional graded algebra is isomorphic, as a KW–module, to the regular representation of W. We call $R\mathbb{G}$ the *graded regular representation* of the reflection datum $\mathbb{G}$. We denote by $R^n\mathbb{G}$ the subspace of elements of degree n of $R\mathbb{G}$. Then we have (see 1.19):

$$R\mathbb{G} = \sum_{n=0}^{N^\vee(\mathbb{G})} R^n\mathbb{G} \,.$$

The polynomial order.

The *polynomial order of* $\mathbb{G}$ is the polynomial (in $K[x]$) denoted by $|\mathbb{G}|$ and defined by the formula

$$(4.4) \qquad |\mathbb{G}| := \frac{\varepsilon_{\mathbb{G}} x^{N(\mathbb{G})}}{\frac{1}{|W|}\sum_{w\in W}\frac{1}{\det_V(1-xwf)^*}} .$$

where $\det^*$ denotes the complex conjugate of the determinant (Note that in this case, we have $\det_V^*(1-xwf) = \det_V(1-x(wf)^{-1})$).

COMMENT. In the case of reductive groups, $|\mathbb{G}|(q)$ is the order of $\mathbf{G}^F$. See e.g. [**BrMa1**], th. 2.2.

4.5. PROPOSITION. *We have*

$$|\mathbb{G}| = \varepsilon_{\mathbb{G}} x^{N(\mathbb{G})} \prod_{j=1}^{j=r} (1-\zeta_j^* x^{d_j}) = x^{N(\mathbb{G})} \prod_{j=1}^{j=r} (x^{d_j} - \zeta_j) \quad ,$$

and in particular $|\mathbb{G}| \in \mathbb{Z}_K[x]$.

Indeed, by Molien's formula (making free use of the notion of "graded character" of a graded module as in [**BrMa1**]) we have

$$\frac{1}{\det_V(1-xwf)} = \mathrm{tr}(wf\,; S)\,,$$

Thus

$$\begin{aligned} \frac{1}{|W|}\sum_{w\in W}\frac{1}{\det_V(1-xwf)} &= \frac{1}{|W|}\sum_{w\in W}\mathrm{tr}(wf;S) \\ &= \frac{1}{|W|}\sum_{w\in W}\mathrm{tr}(fw;S) = \mathrm{tr}(f;R)\,. \end{aligned}$$

It results from the definition of generalized degrees that, as an $\langle f\rangle$-module, R is isomorphic to $K[f_1]\otimes\ldots\otimes K[f_r]$ where the action of f is given by $f\cdot f_j = \zeta_j f_j$. We see that the first equality of the proposition results from this, and the second one results from the first one and from 1.10. □

Recall that the *cyclotomic polynomials over* K are the minimal polynomials over K of roots of unity. We see that an irreducible divisor of $|\mathbb{G}|$ in $K[x]$ is either x or a cyclotomic polynomial over K. More precisely,

4.6. COROLLARY.

(1) *We have* $|\mathbb{G}| = x^{N(\mathbb{G})}\prod_\Phi \Phi(x)^{a(\Phi)}$, *where* Φ *runs over the set of cyclotomic polynomials over* K, *and where* $a(\Phi)$ *is the number of indices* j *such that* $\zeta^{d_j} = \zeta_j$ *(where* ζ *is a chosen root of* Φ*).*
(2) *The degree of the polynomial* $|\mathbb{G}|$ *is* $(N(\mathbb{G}) + N^\vee(\mathbb{G}) + r)$.
(3) *We have* $|\mathbb{G}|(1/x) = \varepsilon_{\mathbb{G}} x^{-(2N(\mathbb{G})+N^\vee(\mathbb{G})+r)}|\mathbb{G}|(x)^*$.

Various constructions with reflection data

Product of reflection data.

Let $\mathbb{G} = (V, Wf)$ and $\mathbb{G}' = (V', W'f')$ be two reflection data. Their product is the reflection datum

$$\mathbb{G} \times \mathbb{G}' := (V \times V', (W \times W')(f \times f')) .$$

COMMENT. This corresponds to the product of reductive groups.

Extending scalars.

Let $\mathbb{G} = (V, Wf)$ be a reflection datum on the subfield K of $\mathbb{Q}(\boldsymbol{\mu}_\infty)$, and let K' be an extension of K contained in $\mathbb{Q}(\boldsymbol{\mu}_\infty)$. The reflection datum $K'\mathbb{G}$ is then defined as the image of $\mathbb{G}$ under the functor $K' \otimes \cdot$:

$$K'\mathbb{G} := (K' \otimes V, 1 \otimes (Wf)) .$$

COMMENT. This construction has no analogue for reductive groups.

Lifting scalars.

Let $\mathbb{G} = (V, Wf)$ be a reflection datum and let $a \in \mathbb{N}$.

(4.7) We define the reflection datum $\mathbb{G}^{(a)} = (V^{(a)}, W^{(a)} f^{(a)})$ by the following rules:

(ls.1) $V^{(a)} := V \times \cdots \times V$ *(a times), and* $W^{(a)} := W \times \cdots \times W$ *(a times),*

(ls.2) $f^{(a)}$ *is the product of f (acting diagonally on $V \times \cdots \times V$) by the a-cycle which permutes cyclically the factors V of $V^{(a)}$.*

We have

$$|\mathbb{G}^{(a)}|(x) = |\mathbb{G}|(x^a) .$$

COMMENT. This corresponds to an extension of scalars from $\mathbb{F}_q$ to $\mathbb{F}_{q^a}$ for reductive groups.

Changing x into ζx and generalized Ennola duality.

Let $\mathbb{G} = (V, Wf)$ be a reflection datum. Let $\zeta \in K$ be a root of unity. We define the reflection datum $\mathbb{G}^\zeta$ by

$$\mathbb{G}^\zeta := (V, W\zeta f) . \tag{4.8}$$

From the definition of the polynomial order, we get

$$|\mathbb{G}^\zeta|(x) = \varepsilon_{\mathbb{G}^\zeta} \varepsilon_{\mathbb{G}}^{-1} \zeta^{N(\mathbb{G})} |\mathbb{G}|(\zeta^{-1} x)$$

and, since

$$\varepsilon_{\mathbb{G}^\zeta} = \varepsilon_{\mathbb{G}} \zeta^{d_1 + d_2 + \cdots + d_r} ,$$

we get (by 1.10)

$$|\mathbb{G}^\zeta|(x) = \zeta^{(N + N^\vee + r)} |\mathbb{G}|(\zeta^{-1} x) .$$

If $\zeta \mathrm{Id} \in W$, then $\mathbb{G}^\zeta = \mathbb{G}$, and all the invariants of $\mathbb{G}$ and of $\mathbb{G}^\zeta$ coincide.

COMMENT. A prototype for this construction is the *Ennola* duality between the linear and unitary groups, which corresponds to changing x into $-x$ in the reflection datum associated to linear groups.

The dual, the "semi–simple quotient" and adjoint reflection data.

• We recall that we denote by $V^\vee$ the dual space of V. We still denote by W the image in $\mathrm{GL}(V^\vee)$ of the group W acting through the contragredient representation $w \mapsto {}^t w^{-1}$ and we set $f^\vee := {}^t f^{-1}$. Then the *dual reflection datum* of $\mathbb{G}$ is

$$\mathbb{G}^\vee := (V^\vee, W f^\vee)\,.$$

COMMENT. This corresponds to the Langlands dual for algebraic groups. However, since Weyl groups are real reflection groups, the reflection data associated to an algebraic group and its Langlands dual are isomorphic.

• The "semi–simple quotient" of $\mathbb{G}$ is "the quotient of $\mathbb{G}$ by its center $Z(\mathbb{G})$", namely

$$\mathbb{G}_{\mathrm{ss}} := (V/V^W, Wf)\,,$$

where we identify W and f with their images in the linear group of V/V^W.

REMARK. One may also define the *adjoint reflection datum* of $\mathbb{G}$ as

$$\mathbb{G}_{\mathrm{ad}} := (((V^\vee)^W)^\perp, Wf)\,,$$

where we identify W, f with their images in the linear group of $((V^\vee)^W)^\perp$. Denoting by pr_W the projector of V defined by $\mathrm{pr}_W := \dfrac{1}{|W|}\sum_{w\in W} w$, it is readily checked that the map $1-\mathrm{pr}_W$ induces a W–isomorphism from V/V^W onto $((V^\vee)^W)^\perp$, which shows that $\mathbb{G}_{\mathrm{ss}}$ and $\mathbb{G}_{\mathrm{ad}}$ are isomorphic.

COMMENT. As the chosen names reflect, this corresponds to the quotient by the radical (resp. to the adjoint group) for reductive groups.

Intersection of a sub-reflection datum of maximal rank with a Levi sub-reflection datum.

Let $\mathbb{L} = (V, W_{\mathbb{L}} v f)$ be a Levi sub-reflection datum of $\mathbb{G}$ and let $\mathbb{M} = (V, W_{\mathbb{M}} w f)$ be a sub-reflection datum of maximal rank of $\mathbb{G}$.

(4.9) We say that $\mathbb{L} \cap \mathbb{M}$ *is defined if*

$$W_{\mathbb{L}} v f \cap W_{\mathbb{M}} w f \neq \emptyset\,.$$

In that case, choosing an element $u \in W_{\mathbb{L}} v \cap W_{\mathbb{M}} w$, *we define*

$$\mathbb{L} \cap \mathbb{M} := (V, (W_{\mathbb{L}} \cap W_{\mathbb{M}}) u f)\,.$$

Note that $W_{\mathbb{L}} v f = W_{\mathbb{L}} u f$ and $W_{\mathbb{M}} w f = W_{\mathbb{M}} u f$, hence $W_{\mathbb{L}} v f \cap W_{\mathbb{M}} w f = (W_{\mathbb{L}} \cap W_{\mathbb{M}}) u f$. Notice also that $W_{\mathbb{L}} \cap W_{\mathbb{M}}$ is a parabolic subgroup of $W_{\mathbb{M}}$. It is easy to see that $\mathbb{L} \cap \mathbb{M}$ is a well–defined (*i.e.*, independent of the choice of u) Levi sub-reflection datum of $\mathbb{M}$, and a sub-reflection datum of maximal rank of $\mathbb{L}$.

By the preceding definition, it is clear that

(4.10) $\mathbb{L} \cap \mathbb{M}$ *is defined if and only if there exists a maximal sub-torus* $\mathbb{T}$ *of* $\mathbb{G}$ *contained in both* $\mathbb{L}$ *and* $\mathbb{M}$*. In this case,* $\mathbb{T}$ *is contained in* $\mathbb{L} \cap \mathbb{M}$*.*

For $w \in W$ and $\mathbb{L}$ a sub–reflection datum of maximal rank of $\mathbb{G}$, we denote by ${}^w\mathbb{L} := w\mathbb{L}w^{-1}$ the conjugate of $\mathbb{L}$ under w.

4.11. Definition. *For* $\mathbb{L}$ *a Levi sub-reflection datum and* $\mathbb{M}$ *a sub-reflection datum of maximal rank of* $\mathbb{G}$*, we denote by* $\mathcal{T}_{W_{\mathbb{G}}}(\mathbb{L}, \mathbb{M})$ *the set of all* $w \in W_{\mathbb{G}}$ *such that* ${}^w\mathbb{L} \cap \mathbb{M}$ *is defined.*

Classification of reflection data

Using the classification of finite irreducible complex reflection groups by Shephard and Todd (cf. above 1.5) it is possible to give a complete description of all reflection data as follows.

Let $\mathbb{G} = (V, Wf)$ be a reflection datum. Let $V_1 \leq V$ be a $W\langle f\rangle$–invariant subspace of V and V_2 a $W\langle f\rangle$-invariant complement. Any reflection in W either acts trivially on V_1 or on V_2. Thus, as W is generated by reflections, the decomposition $V = V_1 \times V_2$ induces a f-invariant direct product decomposition $W = W_1 \times W_2$, such that W_i acts trivially on V_{3-i} $(i = 1, 2)$. This shows that $\mathbb{G}$ is a direct product

$$\mathbb{G} = \mathbb{G}_1 \times \mathbb{G}_2 , \qquad \text{with } \mathbb{G}_i = (V_i, W_i f|_{V_i}) \text{ for } i = 1, 2 .$$

Now assume that $W\langle f\rangle$ acts irreducibly on V. Let $V = V_1 \times \ldots \times V_a$ be the decomposition of V into W-irreducible subspaces. As above this induces a direct product decomposition $W = W_1 \times \ldots \times W_a$ of W into reflection subgroups W_i such that W_i acts trivially on all V_j for $j \neq i$. Moreover, the (V_i, W_i) are permuted transitively by f. Thus $\mathbb{G}$ is the lifting of scalars

$$\mathbb{G} = (V_1, W_1\psi)^{(a)}, \qquad \text{with } \psi = f^a|_{V_1}$$

of the reflection datum $(V_1, W_1\psi)$ where W_1 acts irreducibly (hence absolutely irreducibly) on V_1.

It hence remains to classify those reflection data $\mathbb{G}$ where V is an (absolutely) irreducible W-module. For this, it is useful to first determine normal embeddings of irreducible reflection groups in the same dimension.

4.12 Proposition. *Let* W *be an irreducible finite complex reflection group on the complex vector space* V *and* W' *a proper irreducible (normal) subgroup of* W *generated by a union of reflection classes of* W*, maximal in* W *with respect to these properties. Then* (W, W') *are in the following list (where we adopt the notation of*

[**ShTo**] *for irreducible reflection groups):*

W	W'
$G(dep,e,r)$	$G(dep,ep,r)$, p *prime*
$G(2de,2e,2)$	$G(de,e,2)$
G_5	G_4
G_6	$G(4,2,2)$, G_4
G_7	G_6, G_5
G_8	$G(4,2,2)$
G_9	G_{13}, G_8
G_{10}	G_7, G_8
G_{11}	G_{15}, G_9, G_{10}

W	W'
G_{13}	$G(4,2,2)$, G_{12}
G_{14}	G_{12}, G_5
G_{15}	G_{13}, G_{14}, G_7
G_{17}	G_{22}, G_{16}
G_{18}	G_{20}, G_{16}
G_{19}	G_{21}, G_{17}, G_{18}
G_{21}	G_{22}, G_{20}
G_{26}	$G(3,3,3)$, G_{25}
G_{28}	$G(2,2,4)$

The proof is a routine case-by-case check, using for example the tables on p. 395 and p. 412 of [**Co**].

The following proposition is 3.13 in [**BMM2**].

4.13 Proposition. *Let W be an irreducible complex reflection group on the n-dimensional complex vector space V and f an automorphism of finite order of V normalizing W. Then up to multiples of the identity, either $\overline{f}=1$ or we are in one of the following cases :*

- *$W=G(de,e,r)$, with $e>1$, and $\overline{f}$ of order dividing e comes from the embedding $G(de,e,r)<G(de,1,r)$,*
- *$W=G(4,2,2)$ and f of order 3 comes from the embedding $G(4,2,2)<G_6$,*
- *$W=G(3,3,3)$ and f of order 4 comes from the embedding $G(3,3,3)<G_{26}$,*
- *$W=G(2,2,4)$ and f of order 3 comes from the embedding $W(D_4)<W(F_4)$,*
- *$W=G_5$ and f of order 2 comes from the embedding $G_5<G_{14}$,*
- *$W=G_7$ and f of order 2 comes from the embedding $G_7<G_{10}$,*
- *$W=G_{28}$ and f realizes the graph-automorphism of $W(F_4)$,*

Here $W(F_4)$, $W(D_4)$, denote the Weyl groups of type F_4, D_4 respectively.

Note that most of these can be visualized by graph–automorphisms of the corresponding diagram (see appendix 1).

The polynomial orders of the twisted reflection data corresponding to the above automorphisms f can easily be calculated from Proposition 4.5 in terms of the generalized degrees. For the infinite families ${}^tG(de,e,r)$, $t|e$, we have

$$|{}^tG(de,e,r)|=x^{N(G(de,e,r))}(x^{de}-1)(x^{2de}-1)\dots(x^{(r-1)de}-1)(x^{rd}-\zeta)\,,$$

where ζ is a primitive t–th root of unity, while for the exceptional cases we obtain :

$$|{}^3G(4,2,2)|=x^6(x^4-1)(x^4-\zeta_3^2)$$
$$|{}^4G(3,3,3)|=x^9(x^6-1)(x^6+1)$$
$$|{}^3G(2,2,4)|=x^{12}(x^2-1)(x^6-1)(x^8+x^4+1)$$
$$|{}^2G_5|=x^8(x^6-1)(x^{12}+1)$$
$$|{}^2G_7|=x^{14}(x^{12}-1)(x^{12}+1)$$
$$|{}^2G_{28}|=x^{24}(x^2-1)(x^6+1)(x^8-1)(x^{12}+1)$$

(notice that ${}^3G(2,2,4) = {}^3D_4$ and ${}^2G_{28} = {}^2F_4$).

Uniform class functions on a reflection datum

Generalities, induction and restriction.

In this section, we assume that K *splits the reflection datum* $\mathbb{G} = (V, Wf)$.

Let $\mathrm{CF}_{\mathrm{uf}}(\mathbb{G})$ be the $\mathbb{Z}_K$–module of all W–invariant functions on the coset Wf (for the natural action of W on Wf by conjugation) with values in $\mathbb{Z}_K$, called *uniform class functions on* $\mathbb{G}$. For $\alpha \in \mathrm{CF}_{\mathrm{uf}}(\mathbb{G})$, we denote by α^* its complex conjugate.

If α and $\alpha' \in \mathrm{CF}_{\mathrm{uf}}(\mathbb{G})$, we set $\langle \alpha, \alpha' \rangle_{\mathbb{G}} := \frac{1}{|W|} \sum_{w \in W} \alpha(wf)\alpha'(wf)^*$.

Notation.

- If $\mathbb{Z}_K \to \mathcal{O}$ is a ring morphism, we denote by $\mathrm{CF}_{\mathrm{uf}}(\mathbb{G}, \mathcal{O})$ the $\mathcal{O}$–module of W–invariant functions on Wf with values in $\mathcal{O}$, which we call the module of *uniform class functions on* $\mathbb{G}$ with values in $\mathcal{O}$. We have $\mathrm{CF}_{\mathrm{uf}}(\mathbb{G}, \mathcal{O}) = \mathcal{O} \otimes_{\mathbb{Z}_K} \mathrm{CF}_{\mathrm{uf}}(\mathbb{G})$.
- For $wf \in Wf$, we denote by $\mathrm{ch}^{\mathbb{G}}_{wf}$ (or simply ch_{wf}) the characteristic function of the orbit of wf under W. The family $\left(\mathrm{ch}^{\mathbb{G}}_{wf}\right)$ (where wf runs over a complete set of representatives of the orbits of W on Wf) is a basis of $\mathrm{CF}_{\mathrm{uf}}(\mathbb{G})$.
- For $wf \in Wf$, we set

$$R^{\mathbb{G}}_{wf} := |C_W(wf)| \mathrm{ch}^{\mathbb{G}}_{wf}$$

(or simply R_{wf}). The $\mathbb{Z}_K$–module generated by the functions $R^{\mathbb{G}}_{wf}$ is denoted by ${\mathrm{CF}_{\mathrm{uf}}}^{\mathrm{pr}}(\mathbb{G})$. For $\alpha \in \mathrm{CF}_{\mathrm{uf}}(\mathbb{G})$ and $wf \in Wf$ we have $\langle \alpha, R_{wf} \rangle_{\mathbb{G}} = \alpha(wf)$, so $\mathrm{CF}_{\mathrm{uf}}(\mathbb{G})$ is the $\mathbb{Z}_K$-dual of ${\mathrm{CF}_{\mathrm{uf}}}^{\mathrm{pr}}(\mathbb{G})$.

[The exponent "pr" stands for "projective", by analogy with the vocabulary of modular representation theory of finite groups]

COMMENT. In the case of reductive groups, we have $K = \mathbb{Q}$. Let $\mathrm{Uch}(\mathbf{G}^F)$ be the set of unipotent characters of $\mathbf{G}^F$: then the map which associates to $R^{\mathbb{G}}_{wf}$ the Deligne–Lusztig character $R^{\mathbf{G}}_{\mathbf{T}_{wF}}(\mathrm{Id})$ defines an isometric embedding (for the scalar products $\langle \alpha, \alpha' \rangle_{\mathbb{G}}$ and $\langle \alpha, \alpha' \rangle_{\mathbf{G}^F}$) from $\mathrm{CF}_{\mathrm{uf}}(\mathbb{G})$ onto the sub-$\mathbb{Z}$-module of $\mathbb{Q}\mathrm{Uch}(\mathbf{G}^F)$ of the $\mathbb{Q}$-linear combinations of Deligne-Lusztig characters (i.e. "unipotent uniform functions") which have an integral scalar product with the Deligne-Lusztig characters.

- Let $\langle Wf \rangle$ be the subgroup of $\mathrm{GL}(V)$ generated by Wf. We recall that we denote by $\overline{f}$ the image of f in $\langle Wf \rangle / W$ — thus $\langle Wf \rangle / W$ is cyclic and generated by $\overline{f}$.

For $\psi \in \mathrm{Irr}(\langle Wf \rangle)$, we denote by $R^{\mathbb{G}}_{\psi}$ (or simply R_ψ) the restriction of ψ to the coset Wf. We have

$$R^{\mathbb{G}}_{\psi} = \frac{1}{|W|} \sum_{w \in W} \psi(wf) R^{\mathbb{G}}_{wf} ,$$

and we call such a function a *uniform almost character* of $\mathbb{G}$.

Let $\mathrm{Irr}(W)^{\overline{f}}$ denote the set of $\overline{f}$–stable irreducible characters of W. For $\chi \in \mathrm{Irr}(W)^{\overline{f}}$, we denote by $E_{\mathbb{G}}(\chi)$ (or simply $E(\chi)$) the set of restrictions to Wf of the extensions of χ to characters of $\langle Wf \rangle$. Since K contains the $\delta(\mathbb{G})$–th roots of unity (where $\delta(\mathbb{G})$ denotes the order of the twist $\overline{f}$), the group $\langle\overline{f}\rangle$ acts regularly on $E(\chi)$.

Each element of $E_{\mathbb{G}}(\chi)$ has norm 1, the sets $E_{\mathbb{G}}(\chi)$ for $\chi \in \mathrm{Irr}(W)^{\overline{f}}$ are mutually orthogonal, and we have

$$\mathrm{CF}_{\mathrm{uf}}(\mathbb{G}, K) = \overset{\perp}{\bigoplus_{\chi\in\mathrm{Irr}(W)^{\overline{f}}}} KE_{\mathbb{G}}(\chi)\,,$$

where we set $KE_{\mathbb{G}}(\chi) := KR_{\psi}^{\mathbb{G}}$ for some (any) $\psi \in E_{\mathbb{G}}(\chi)$.

Induction and restriction.

Let $\mathbb{L} = (V, W_{\mathbb{L}}wf)$ be a sub-reflection datum of maximal rank of $\mathbb{G}$, and let $\alpha \in \mathrm{CF}_{\mathrm{uf}}(\mathbb{G})$ and $\beta \in \mathrm{CF}_{\mathrm{uf}}(\mathbb{L})$. We denote

- by $\mathrm{Res}_{\mathbb{L}}^{\mathbb{G}}\alpha$ the restriction of α to the coset $W_{\mathbb{L}}wf$,
- by $\mathrm{Ind}_{\mathbb{L}}^{\mathbb{G}}\beta$ the uniform class function on $\mathbb{G}$ defined by

$$\mathrm{Ind}_{\mathbb{L}}^{\mathbb{G}}\beta(uf) := \frac{1}{|W_{\mathbb{L}}|}\sum_{v\in W}\tilde{\beta}(vufv^{-1}) \quad \text{for } uf \in W_{\mathbb{G}}f, \tag{4.14}$$

where $\tilde{\beta}(xf) = \beta(xf)$ if $x \in W_{\mathbb{L}}w$, and $\tilde{\beta}(xf) = 0$ if $x \notin W_{\mathbb{L}}w$. In other words, we have

$$\mathrm{Ind}_{\mathbb{L}}^{\mathbb{G}}\beta(uf) = \sum_{v\in W_{\mathbb{G}}/W_{\mathbb{L}},\, {}^{v}(uf)\in W_{\mathbb{L}}wf} \beta({}^{v}(uf))\,. \tag{4.15}$$

For any reflection datum $\mathbb{G}$ we denote by $1^{\mathbb{G}}$ the constant function on Wf with value 1. For $w \in W$, let us denote by $\mathbb{T}_{wf}$ the maximal torus of $\mathbb{G}$ defined by $\mathbb{T}_{wf} := (V, wf)$. It follows from the definitions that

$$R_{wf}^{\mathbb{G}} = \mathrm{Ind}_{\mathbb{T}_{wf}}^{\mathbb{G}} 1^{\mathbb{T}_{wf}}\,. \tag{4.16}$$

For $\alpha \in \mathrm{CF}_{\mathrm{uf}}(\mathbb{G})$, $\beta \in \mathrm{CF}_{\mathrm{uf}}(\mathbb{L})$ we have the *Frobenius reciprocity*:

$$\langle \alpha, \mathrm{Ind}_{\mathbb{L}}^{\mathbb{G}}\beta\rangle_{\mathbb{G}} = \langle \mathrm{Res}_{\mathbb{L}}^{\mathbb{G}}\alpha, \beta\rangle_{\mathbb{L}}\,. \tag{4.17}$$

COMMENT. In the case of reductive groups, assume that $\mathbb{L}$ is a Levi sub-reflection datum. Then $\mathrm{Ind}_{\mathbb{L}}^{\mathbb{G}}$ corresponds to Lusztig induction from $\mathbf{L}$ to $\mathbf{G}$ (this results from definition 4.14 applied to a Deligne–Lusztig character which, using the transitivity of Lusztig induction, agrees with Lusztig induction). Similarly, the Lusztig restriction of a uniform function is uniform by [**DeLu**], theorem 7, so by (4.17) $\mathrm{Res}_{\mathbb{L}}^{\mathbb{G}}$ corresponds to Lusztig restriction.

When $\mathbb{L}$ corresponds to a reductive subgroup of maximal rank which is *not* a Levi subgroup, then $\mathrm{Ind}_{\mathbb{L}}^{\mathbb{G}}$ corresponds to a generalization of Lusztig induction to that setting. However, this generalization does not necessarily map $\mathbb{Z}\mathrm{Uch}(\mathbf{L}^F)$ to $\mathbb{Z}\mathrm{Uch}(\mathbf{G}^F)$. For instance, when $\mathbf{L}$ is of type A_2 corresponding to the long roots of $\mathbf{G}$ of type G_2, then the image of $\mathbb{Z}\mathrm{Uch}(\mathbf{L}^F)$ by this generalized induction is only in $\mathbb{Z}[1/3]\mathrm{Uch}(\mathbf{G}^F)$.

The Mackey formula.

Let now $\mathbb{L}$ be a Levi sub-reflection datum and let $\mathbb{M}$ be a sub-reflection datum of maximal rank of $\mathbb{G}$.

It is clear that $W_{\mathbb{L}}$ acts on $\mathcal{T}_{W_{\mathbb{G}}}(\mathbb{L},\mathbb{M})$ (see definition 4.11 above) from the right, while $W_{\mathbb{M}}$ acts on $\mathcal{T}_{W_{\mathbb{G}}}(\mathbb{L},\mathbb{M})$ from the left. If we let w run over a chosen double coset $W_{\mathbb{M}}vW_{\mathbb{L}}$ for some $v \in \mathcal{T}_{W_{\mathbb{G}}}(\mathbb{L},\mathbb{M})$, we see that ${}^w\mathbb{L} \cap \mathbb{M}$ is defined up to $W_{{}^w\mathbb{L}}$–conjugation as a sub-reflection datum of ${}^w\mathbb{L}$ and up to $W_{\mathbb{M}}$–conjugation as a sub-reflection datum of $\mathbb{M}$, which proves that the operations $\mathrm{Ind}_{\mathbb{M}\cap{}^w\mathbb{L}}^{\mathbb{M}}$ and $\mathrm{Res}_{\mathbb{M}\cap{}^w\mathbb{L}}^{{}^w\mathbb{L}}$ depend only on the double coset of w. This gives sense to the following formula where $\mathrm{ad}(w)$ denotes the operator of conjugation by w

$$\mathrm{Res}_{\mathbb{M}}^{\mathbb{G}} \cdot \mathrm{Ind}_{\mathbb{L}}^{\mathbb{G}} = \sum_{w \in W_{\mathbb{M}} \backslash \mathcal{T}_{W_{\mathbb{G}}}(\mathbb{L},\mathbb{M})/W_{\mathbb{L}}} \mathrm{Ind}_{\mathbb{M}\cap{}^w\mathbb{L}}^{\mathbb{M}} \cdot \mathrm{Res}_{\mathbb{M}\cap{}^w\mathbb{L}}^{{}^w\mathbb{L}} \cdot \mathrm{ad}(w)\,, \tag{4.18}$$

whose proof is a straightforward calculation as in the case of ordinary induction and restriction (note that $\mathcal{T}_{W_{\mathbb{G}}}(\mathbb{L},\mathbb{M})$ may be empty).

COMMENT. In the case of reductive groups, assuming that both $\mathbb{L}$ and $\mathbb{M}$ are Levi sub-reflection data, the Mackey formula corresponds to the Mackey formula for Lusztig induction and restriction (projected on uniform function).

Uniform class functions on $\mathbb{G}^\zeta$.

The map $\sigma_{\mathbb{G}}^\zeta \colon \mathrm{CF}_{\mathrm{uf}}(\mathbb{G}) \to \mathrm{CF}_{\mathrm{uf}}(\mathbb{G}^\zeta)$, given by $\sigma_{\mathbb{G}}^\zeta\alpha(w\zeta f) = \alpha(wf)$, is an isometry. The map $\mathbb{L} \mapsto \mathbb{L}^\zeta$ is a $W_{\mathbb{G}}$–equivariant bijection from the set of all sub-reflection data of maximal rank of $\mathbb{G}$ (resp. of all Levi sub–reflection data of $\mathbb{G}$) onto the set of all sub-reflection data of maximal rank of $\mathbb{G}^\zeta$ (resp. of all Levi sub–reflection data of $\mathbb{G}^\zeta$). It is clear that

$$\sigma_{\mathbb{G}}^\zeta \cdot \mathrm{Ind}_{\mathbb{L}}^{\mathbb{G}} = \mathrm{Ind}_{\mathbb{L}^\zeta}^{\mathbb{G}^\zeta} \cdot \sigma_{\mathbb{L}}^\zeta \quad , \quad \sigma_{\mathbb{L}}^\zeta \cdot \mathrm{Res}_{\mathbb{L}}^{\mathbb{G}} = \mathrm{Res}_{\mathbb{L}^\zeta}^{\mathbb{G}^\zeta} \cdot \sigma_{\mathbb{G}}^\zeta\,. \tag{4.19}$$

Uniform class functions on $\mathbb{G}^{(a)}$.

Let $a \in \mathbb{N}$. We have (see 4.7) $W_{\mathbb{G}^{(a)}} = (W_{\mathbb{G}})^a$, and the map

$$(w_1, w_2, \dots, w_a)f^{(a)} \mapsto w_1w_2\cdots w_af$$

defines a bijection between the set of classes of $W_{\mathbb{G}^{(a)}}f^{(a)}$ under $W_{\mathbb{G}^{(a)}}$–conjugacy and the set of classes of $W_{\mathbb{G}}f$ under $W_{\mathbb{G}}$–conjugacy. Thus it induces an isometry

$$\sigma_{\mathbb{G}}^{(a)} \colon \mathrm{CF}_{\mathrm{uf}}(\mathbb{G}^{(a)}) \xrightarrow{\sim} \mathrm{CF}_{\mathrm{uf}}(\mathbb{G})\,.$$

4.20. PROPOSITION. *We have*

$$\sigma_{\mathbb{G}}^{(a)} \cdot \mathrm{Ind}_{\mathbb{L}^{(a)}}^{\mathbb{G}^{(a)}} = \mathrm{Ind}_{\mathbb{L}}^{\mathbb{G}} \cdot \sigma_{\mathbb{L}}^{(a)}$$
$$\mathrm{Res}_{\mathbb{L}}^{\mathbb{G}} \cdot \sigma_{\mathbb{G}}^{(a)} = \sigma_{\mathbb{L}}^{(a)} \cdot \mathrm{Res}_{\mathbb{L}^{(a)}}^{\mathbb{G}^{(a)}}\,.$$

Degrees.

• We denote by $\mathrm{tr}_{R\mathbb{G}}$ the uniform class function on $\mathbb{G}$ (with values in the polynomial ring $\mathbb{Z}_K[x]$) defined by the character of the "graded regular representation" $R\mathbb{G}$ (see above). Thus the value of the function $\mathrm{tr}_{R\mathbb{G}}$ on wf is

$$\mathrm{tr}_{R\mathbb{G}}(wf) := \sum_{n=0}^{N^\vee(\mathbb{G})} \mathrm{tr}(wf; R^n\mathbb{G})x^n\,.$$

We call $\mathrm{tr}_{R\mathbb{G}}$ *the regular character* of $\mathbb{G}$.

• We define the *degree*, a linear function

$$\mathrm{Deg}_{\mathbb{G}}\colon \mathrm{CF}_{\mathrm{uf}}(\mathbb{G}) \to K[x]\,,$$

as follows: for $\alpha \in \mathrm{CF}_{\mathrm{uf}}(\mathbb{G})$, we set

$$\mathrm{Deg}_{\mathbb{G}}(\alpha) := \langle \alpha, \mathrm{tr}_{R\mathbb{G}}\rangle_{\mathbb{G}} = \sum_{n=0}^{N^{\vee}(\mathbb{G})} \left(\frac{1}{|W|} \sum_{w\in W} \alpha(wf)\mathrm{tr}(wf; R^n\mathbb{G})^* \right) x^n\,. \tag{4.21}$$

We shall often omit the subscript $\mathbb{G}$ (writing then $\mathrm{Deg}(\alpha)$) when the context allows it.

Notice that

$$\mathrm{Deg}(R_{wf}^{\mathbb{G}}) = \mathrm{tr}_{R\mathbb{G}}(wf)\,, \tag{4.22}$$

and so in particular that

$$\mathrm{Deg}(R_{wf}^{\mathbb{G}}) \in \mathbb{Z}_K[x]\,. \tag{4.23}$$

Degrees of almost characters.

Let E be a $K\langle Wf\rangle$–module. Let χ_E be the restriction of the character of E (a uniform class function on $\langle Wf\rangle$) to Wf. Then the degree of R_{χ_E} is the "graded multiplicity" of E in the graded regular representation $R\mathbb{G}$:

$$\mathrm{Deg}_{\mathbb{G}}(R_{\chi_E}) = \mathrm{tr}(f\,;\, \mathrm{Hom}_{KW}(R\mathbb{G}, E))\,. \tag{4.24}$$

Notice that

$$\mathrm{Deg}_{\mathbb{G}}(R_{\chi_E}) \in \mathbb{Z}[\zeta_{\delta(\mathbb{G})}][x]\,, \tag{4.25}$$

(we recall that $\delta(\mathbb{G})$ is the order of the twist $\overline{f}$ of $\mathbb{G}$).

Remark. Let $\chi \in \mathrm{Irr}(W)$. Then χ is a uniform function on the split reflection datum $\mathbb{G}_0 := (V, W)$, and we have (see above §1, 1.20)

$$\mathrm{Deg}_{\mathbb{G}_0}(R_\chi) = \mathrm{Feg}_\chi(x)\,.$$

Let $\chi \in \mathrm{Irr}(W)^{\overline{f}}$. If $\psi \in \mathrm{Irr}(\langle Wf\rangle)$ runs over the set of extensions of χ to $\langle Wf\rangle$, and since K contains the $\delta(\mathbb{G})$–th roots of unity, the set of degrees $\mathrm{Deg}(R_\psi)$ is an orbit of $\boldsymbol{\mu}_{\delta(\mathbb{G})}(K)$ on $\mathbb{Z}_K[x]$. The element $\mathrm{Deg}_{\mathbb{G}}(R_\psi)^* \cdot R_\psi$ depends only on χ and is the orthogonal projection of $\mathrm{tr}_{R\mathbb{G}}$ onto $K[x]E_{\mathbb{G}}(\chi)$. We set

$$\mathrm{Reg}_\chi^{\mathbb{G}} := \mathrm{Deg}_{\mathbb{G}}(R_\psi)^* \cdot R_\psi\,, \tag{4.26}$$

and in other words, we have

$$\mathrm{tr}_{R\mathbb{G}} = \sum_{\chi\in\mathrm{Irr}(W)^{\overline{f}}} \mathrm{Reg}_\chi^{\mathbb{G}}\,. \tag{4.27}$$

The proof of the following lemma is immediate.

4.28. Lemma. *We have*

$$\mathrm{tr}_{R\mathbb{G}} = \frac{1}{|W|}\sum_{w\in W} \mathrm{Deg}_{\mathbb{G}}(R^{\mathbb{G}}_{wf})^* R^{\mathbb{G}}_{wf}\,,$$

and in particular

$$\mathrm{Deg}_{\mathbb{G}}(\mathrm{tr}_{R\mathbb{G}}) = \frac{|\mathbb{G}||\mathbb{G}|^*}{x^{2N(\mathbb{G})}}\frac{1}{|W|}\sum_{w\in W}\frac{1}{\det_V(1-xwf)\det_V(1-xwf)^*}\,.$$

Polynomial order and degrees.

From the isomorphism of $\langle Wf\rangle$–modules :

$$S \simeq R\mathbb{G}\otimes_K R\,,$$

we deduce

$$\frac{1}{|W|}\sum_{w\in W}\frac{\alpha(wf)}{\det_V(1-xwf)^*} = \mathrm{Deg}(\alpha)\frac{1}{|W|}\sum_{w\in W}\frac{1}{\det_V(1-xwf)^*}\,, \tag{4.29}$$

or, in other words

$$\langle\alpha,\mathrm{tr}_S\rangle_{\mathbb{G}} = \langle\alpha,\mathrm{tr}_{R\mathbb{G}}\rangle_{\mathbb{G}}\langle 1^{\mathbb{G}},\mathrm{tr}_S\rangle_{\mathbb{G}}\,. \tag{4.30}$$

Let us set $S_{\mathbb{G}}(\alpha) := \langle\alpha,\mathrm{tr}_S\rangle_{\mathbb{G}}$. Then (4.30) becomes:

$$S_{\mathbb{G}}(\alpha) = \mathrm{Deg}_{\mathbb{G}}(\alpha)S_{\mathbb{G}}(1^{\mathbb{G}})\,. \tag{4.31}$$

By (4.4) we get

$$|\mathbb{G}| = \varepsilon_{\mathbb{G}}\frac{x^{N(\mathbb{G})}}{S_{\mathbb{G}}(1^{\mathbb{G}})}\,,\quad\text{hence}$$

$$S_{\mathbb{G}}(\alpha)|\mathbb{G}| = \varepsilon_{\mathbb{G}}x^{N(\mathbb{G})}\mathrm{Deg}_{\mathbb{G}}(\alpha)\,. \tag{4.32}$$

• For a sub-reflection datum $\mathbb{L}$ of maximal rank of $\mathbb{G}$, by the Frobenius reciprocity (4.17), we have

$$\mathrm{Deg}(\mathrm{Ind}_{\mathbb{L}}^{\mathbb{G}}1^{\mathbb{L}}) = \langle 1^{\mathbb{L}},\mathrm{Res}_{\mathbb{L}}^{\mathbb{G}}\mathrm{tr}_{R\mathbb{G}}\rangle_{\mathbb{L}} = \sum_{n=0}^{N^\vee(\mathbb{G})}\mathrm{tr}(wf;(R^n\mathbb{G})^{W_{\mathbb{L}}})^*x^n \tag{4.33}$$

where $W_{\mathbb{L}}wf$ is the coset associated to $\mathbb{L}$ and $(R^n\mathbb{G})^{W_{\mathbb{L}}}$ are the $W_{\mathbb{L}}$–invariants in $R^n\mathbb{G}$.

4.34. Proposition. *We have*

$$|\mathbb{G}|/|\mathbb{L}| = \varepsilon_{\mathbb{G}}\varepsilon_{\mathbb{L}}^{-1}x^{N(\mathbb{G})-N(\mathbb{L})}\mathrm{Deg}(\mathrm{Ind}_{\mathbb{L}}^{\mathbb{G}}1^{\mathbb{L}})\,,$$

and in particular $|\mathbb{L}|$ *divides* $|\mathbb{G}|$ *(in* $\mathbb{Z}_K[x]$*).*

• Let us recall that every element $wf\in Wf$ defines a maximal torus (or, equivalently, a minimal Levi sub-reflection datum) $\mathbb{T}_{wf} := (V,wf)$. By (4.16) and (4.22) we have

$$\mathrm{tr}_{R\mathbb{G}}(wf) = \mathrm{Deg}(\mathrm{Ind}_{\mathbb{T}_{wf}}^{\mathbb{G}}1^{\mathbb{T}_{wf}})^*\,.$$

So

$$|\mathbb{G}|/|\mathbb{T}_{wf}| = \varepsilon_{\mathbb{G}}\varepsilon_{\mathbb{T}_{wf}}^{-1}x^{N(\mathbb{G})}\mathrm{tr}_{R\mathbb{G}}(wf)^*\,. \tag{4.35}$$

It follows from 4.34 and (4.35) that

(4.36) $$\mathrm{Res}_{\mathbb{L}}^{\mathbb{G}}\mathrm{tr}_{R\mathbb{G}}^{*} = \mathrm{Deg}_{\mathbb{G}}(\mathrm{Ind}_{\mathbb{L}}^{\mathbb{G}}1^{\mathbb{L}})\mathrm{tr}_{R\mathbb{L}}^{*}\,.$$

(4.37) For $\beta \in \mathrm{CF}_{\mathrm{uf}}(\mathbb{L})$, *we have* $\mathrm{Deg}_{\mathbb{G}}(\mathrm{Ind}_{\mathbb{L}}^{\mathbb{G}}\beta) = \mathrm{Deg}_{\mathbb{G}}(\mathrm{Ind}_{\mathbb{L}}^{\mathbb{G}}1^{\mathbb{L}})\mathrm{Deg}_{\mathbb{L}}(\beta)$.

Indeed, $\mathrm{Deg}_{\mathbb{G}}(\mathrm{Ind}_{\mathbb{L}}^{\mathbb{G}}\beta) = \langle \beta, \mathrm{Res}_{\mathbb{L}}^{\mathbb{G}}\mathrm{tr}_{R\mathbb{G}}\rangle_{\mathbb{L}} = \mathrm{Deg}_{\mathbb{G}}(\mathrm{Ind}_{\mathbb{L}}^{\mathbb{G}}1^{\mathbb{L}})\langle \beta, \mathrm{tr}_{R\mathbb{L}}\rangle_{\mathbb{L}}$.

COMMENT. In the case of reductive groups, it follows from (4.22) and (4.35) that $\mathrm{Deg}(R_{wf})(q)$ is the degree of the Deligne–Lusztig character $R_{\mathbf{T}_{wf}}^{\mathbf{G}}$. Since the regular representation of $\mathbf{G}(q)$ is uniform, it follows that $\mathrm{tr}_{R\mathbf{G}}$ corresponds to a (graded by x) version of the unipotent part of the regular representation of $\mathbf{G}(q)$, and that Deg corresponds indeed to the (generic) degree for unipotent uniform functions on $\mathbf{G}(q)$.

Changing x to $1/x$.

As a particular uniform class function on $\mathbb{G}$, we can consider the function $\det_V$ restricted to $W_{\mathbb{G}}f$, which we still denote by $\det_V$. Notice that this restriction might also be denoted by $R_{\det_V}^{\mathbb{G}}$, since it is the almost character associated to the character of $\langle Wf\rangle$ defined by $\det_V$.

The following two results are 4.25 and 4.26 in [**BMM2**].

4.38. PROPOSITION. *Let α be a uniform class function on $\mathbb{G}$. We have*

$$S_{\mathbb{G}}(\alpha\det_V^*)(x) = (-1)^r x^{-r} S_{\mathbb{G}}(\alpha^*)(1/x)^*\,,$$
$$\mathrm{Deg}_{\mathbb{G}}(\alpha\det_V^*) = (-1)^r \varepsilon_{\mathbb{G}}^* x^{N^\vee(\mathbb{G})}\mathrm{Deg}_{\mathbb{G}}(\alpha^*)(1/x)^*\,.$$

4.39. COROLLARY. *We have*

$$\mathrm{Deg}_{\mathbb{G}}(\det_V^*) = (-1)^r \varepsilon_{\mathbb{G}}^* x^{N^\vee(\mathbb{G})}\,,$$
$$\mathrm{Deg}_{\mathbb{G}}(\det_V) = (-1)^r \varepsilon_{\mathbb{G}} x^{N(\mathbb{G})}\,.$$

Changing x to ξx.

Let E be a $K\langle Wf\rangle$-module, and let χ be the restriction of the character of E to Wf.

Let $\xi \in \boldsymbol{\mu}(K)$ such that $\xi\mathrm{Id} \in Z(\langle Wf\rangle)$.

Then we have

(4.40) $$\mathrm{Deg}(\chi)(x) = \omega_\chi(\xi)\mathrm{Deg}(\chi)(\xi x)\,.$$

Indeed, by formula 4.29, we have

$$\frac{1}{|W|}\sum_{w\in W}\frac{\chi(wf)}{\det_V(1-xwf)^*} = \mathrm{Deg}(\alpha)(x)\frac{1}{|W|}\sum_{w\in W}\frac{1}{\det_V(1-xwf)^*}\,.$$

The preceding formula may be rewritten

$$\frac{1}{|W|}\sum_{w\in W}\frac{\chi(w\xi f)}{\det_V(1-xw\xi f)^*} = \mathrm{Deg}(\alpha)(x)\frac{1}{|W|}\sum_{w\in W}\frac{1}{\det_V(1-xw\xi f)^*}\,,$$

hence

$$\frac{\omega_\chi(\xi)}{|W|}\sum_{w\in W}\frac{\chi(wf)}{\det_V(1-xw\xi f)^*} = \mathrm{Deg}(\alpha)(x)\frac{1}{|W|}\sum_{w\in W}\frac{1}{\det_V(1-xw\xi f)^*}\,,$$

from which one deduces formula 4.40.

Φ–Sylow theory and Φ–split Levi sub-reflection data

The Sylow theorems.

Let $\mathbb{T} = (V, wf)$ be a maximal torus. It is easily checked that in this case the polynomial order $|\mathbb{T}|$ is $\det_V(x - wf)$.

Let $\Phi(x) \in K[x]$ be a cyclotomic polynomial. *A Φ–reflection datum* is a torus whose polynomial order is a power of Φ.

The following theorem is 5.1 in [**BMM2**].

4.41. THEOREM. *Let $\mathbb{G}$ be a reflection datum over K and let Φ be a cyclotomic polynomial on K.*

(1) *If Φ divides $|\mathbb{G}|$, there exist non trivial Φ-sub-reflection data of $\mathbb{G}$.*
(2) *Let $\mathbb{S}$ be a maximal Φ-sub-reflection datum of $\mathbb{G}$. Then $|\mathbb{S}| = \Phi^{a(\Phi)}$, contribution of Φ to $|\mathbb{G}|$.*
(3) *Two maximal Φ-sub-reflection data of $\mathbb{G}$ are conjugate under $W_{\mathbb{G}}$.*
(4) *Let $\mathbb{S}$ be a maximal Φ-sub-reflection datum of $\mathbb{G}$. We set $\mathbb{L} := C_{\mathbb{G}}(\mathbb{S})$ and $W_{\mathbb{G}}(\mathbb{L}) := N_{W_{\mathbb{G}}}(\mathbb{L})/W_{\mathbb{L}}$. Then*

$$|\mathbb{G}|/(|W_{\mathbb{G}}(\mathbb{L})||\mathbb{L}|) \equiv 1 \mod \Phi .$$

The maximal Φ–sub-reflection data of $\mathbb{G}$ are called the Sylow Φ–sub-reflection data.

EXAMPLE. Let $\mathbb{G} = (V, Wf)$ be the reflection datum attached to the series of Steinberg triality groups ${}^3D_4(q)$. Thus W is the Weyl group of type D_4 and f is a non-trivial automorphism of W of order 3 induced by the inclusion of W into the Weyl group of type F_4. The polynomial order of $\mathbb{G}$ is

$$|\mathbb{G}| = x^{12}(x^2-1)(x^8+x^4+1)(x^6-1) = x^{12}\Phi_1^2\Phi_2^2\Phi_3^2\Phi_6^2\Phi_{12} .$$

According to the Sylow theorems there exist (maximal) tori of $\mathbb{G}$ of orders respectively $\Phi_3^2, \Phi_6^2, \Phi_{12}$, namely the Sylow d-tori for $d = 3, 6, 12$. Furthermore, there exist Sylow tori of orders Φ_1^2, Φ_2^2, but these tori are not maximal.

Let $\mathbb{G}$ be a generic group and q a prime power. The Sylow Theorem 4.41 also translates to an assertion about Sylow ℓ-subgroups of $\mathbb{G}(q)$ for large primes ℓ not dividing q (see [**BMM2**], Cor. 3.13) :

4.42. PROPOSITION. *Let ℓ be a prime dividing $|\mathbb{G}(q)|$ but not dividing $q|W\langle f\rangle|$.*

(1) *There exists a unique d such that ℓ divides $\Phi_d(q)$ and Φ_d divides $|\mathbb{G}|$.*
(2) *Any Sylow ℓ-subgroup of $\mathbb{G}(q)$ is contained in the group $\mathbb{S}(q)$ for some Sylow d-torus $\mathbb{S}$ of $\mathbb{G}$.*
(3) *The Sylow ℓ-subgroups are isomorphic to a direct product of $a(d)$ cyclic groups of order ℓ^a, where $\Phi_d^{a(d)}$ is the precise power of Φ_d dividing $|\mathbb{G}|$ and where ℓ^a is the precise power of ℓ dividing $\Phi_d(q)$.*

Let us state and comment on the following fundamental supplementary property.

4.43. THEOREM. *Let $\mathbb{S}$ be a maximal Φ-sub-reflection datum of $\mathbb{G}$ and let $\mathbb{L} := C_{\mathbb{G}}(\mathbb{S})$. If $\mathbb{L} = (V, W_{\mathbb{L}}wf)$, we set $V(\mathbb{L}, \Phi) := \ker \Phi(wf) \cap V^{W_{\mathbb{L}}}$, viewed as a vector space over the field $K[x]/\Phi(x)$ through its natural structure of $K[wf]$-module. Then the pair $(V(\mathbb{L}, \Phi), W_{\mathbb{G}}(\mathbb{L}))$ is a reflection group.*

REMARK. This is a slight reformulation of the main result of [**LeSp**]. It had first been proved by Springer [**Sp**] in the particular case where $\mathbb{L}$ is a torus, *i.e.*, when wf is a regular element (see next paragraph below) and it had been checked case by case for many groups (see [**BMM1**]).

It should be noticed that, at least for W *spetsial* (see [**BMM2**]), 4.43 is a particular case of a more general result involving normalizers of "cuspidal Φ–pairs" (see below 5.7).

4.44. PROPOSITION. *Let $\mathbb{S}$ be a maximal Φ–sub-reflection datum of $\mathbb{G}$, and set $\mathbb{L} = C_{\mathbb{G}}(\mathbb{S})$. Let α be a class function on $\mathbb{G}$. We have:*

$$\mathrm{Deg}_{\mathbb{G}}(\alpha) \equiv \mathrm{Deg}_{\mathbb{L}}(\mathrm{Res}_{\mathbb{L}}^{\mathbb{G}}(\alpha)) \mod \Phi .$$

In particular, we have

$$\begin{aligned} \varepsilon_{\mathbb{G}} x^{N(\mathbb{G})} &\equiv \varepsilon_{\mathbb{L}} x^{N(\mathbb{L})} \mod \Phi , \\ \varepsilon_{\mathbb{G}}^* x^{N^\vee(\mathbb{G})} &\equiv \varepsilon_{\mathbb{L}}^* x^{N^\vee(\mathbb{L})} \mod \Phi , \\ x^{N(\mathbb{G})+N^\vee(\mathbb{G})} &\equiv x^{N(\mathbb{L})+N^\vee(\mathbb{L})} \mod \Phi . \end{aligned}$$

Φ–split Levi sub-reflection data.

We call *Φ–split Levi sub-reflection data* the centralizers in $\mathbb{G}$ of the Φ–sub-reflection data of $\mathbb{G}$. In particular, the centralizers of the Sylow Φ–sub-reflection data are the minimal Φ–split Levi sub-reflection data. It then follows that the preceding congruence holds for every Φ–split Levi sub-reflection datum of $\mathbb{G}$, from which we deduce by 4.34:

(4.45) For any Φ–split Levi sub-reflection datum $\mathbb{L}$ of $\mathbb{G}$, we have

$$|\mathbb{G}|/|\mathbb{L}| \equiv \mathrm{Deg}(\mathrm{Ind}_{\mathbb{L}}^{\mathbb{G}} 1^{\mathbb{L}}) \mod \Phi .$$

Given a Levi sub-reflection datum $(V, W_{\mathbb{L}} wf)$, we define its image in $\mathbb{G}_{\mathrm{ss}}$ (resp. its image in $\mathbb{G}_{\mathrm{ad}}$) to be $(V/V^W, W_{\mathbb{L}} wf)$ (resp. $\big(((V^\vee)^W)^\perp, W_{\mathbb{L}} wf\big)$).

4.46. LEMMA. *A Levi sub-reflection datum is Φ-split if and only if its image in $\mathbb{G}_{\mathrm{ss}}$ (resp. in $\mathbb{G}_{\mathrm{ad}}$) is Φ–split.*

Mackey formula for Φ–split sub-reflection data.

The Mackey formula becomes particularly simple when we restrict ourselves to pairs of Φ–split Levi sub-reflection data (see [**BMM2**], 5.7).

4.47. PROPOSITION. *Let $\mathbb{M}_1$, $\mathbb{M}_2$ be Φ–split Levi sub-reflection data of $\mathbb{G}$.*

(1) *$\mathbb{M}_1 \cap \mathbb{M}_2$ is defined if and only if $\mathbb{M}_1$ and $\mathbb{M}_2$ contain a common Sylow Φ–sub-reflection datum of $\mathbb{G}$. In particular $\mathcal{T}_{W_{\mathbb{G}}}(\mathbb{M}_1, \mathbb{M}_2) \neq \emptyset$.*
(2) *If $\mathbb{M}_1 \cap \mathbb{M}_2$ is defined, then it is a Φ–split Levi sub-reflection datum of $\mathbb{G}$.*
(3) *Let $\mathbb{M}_1$ and $\mathbb{M}_2$ be two Φ–split Levi sub-reflection data containing the minimal Φ–split Levi sub-reflection datum $\mathbb{L}$. Then we have*

$$\mathcal{T}_{W_{\mathbb{G}}}(\mathbb{M}_1, \mathbb{M}_2) = W_{\mathbb{M}_2} N_{W_{\mathbb{G}}}(\mathbb{L}) W_{\mathbb{M}_1} .$$

In particular, we have

$$\mathrm{Res}_{\mathbb{M}_2}^{\mathbb{G}} \cdot \mathrm{Ind}_{\mathbb{M}_1}^{\mathbb{G}} = \sum_{w \in W_{\mathbb{M}_2}(\mathbb{L}) \backslash W_{\mathbb{G}}(\mathbb{L}) / W_{\mathbb{M}_1}(\mathbb{L})} \mathrm{Ind}_{\mathbb{M}_2 \cap {}^w\mathbb{M}_1}^{\mathbb{M}_2} \cdot \mathrm{Res}_{\mathbb{M}_2 \cap {}^w\mathbb{M}_1}^{{}^w\mathbb{M}_1} \cdot \mathrm{ad}(w) ,$$

where, for a Levi sub-reflection datum $\mathbb{L}$ of $\mathbb{G}$, we put $W_{\mathbb{G}}(\mathbb{L}) = N_{W_{\mathbb{G}}}(\mathbb{L})/W_{\mathbb{L}}$.

Regular elements and Sylow theorems.

This section is taken from [**BMM2**], §5B.

Let $\mathbb{G} = (V, Wf)$ be a reflection datum on K. Let K' be an extension of K which splits $\mathbb{G}$ and $V' := K' \otimes V$.

The following definition extend (cf. [**Sp**]) the definition of regular elements given in §1 above.

DEFINITION.

• We say that an element wf in the coset Wf is regular if it has a regular eigenvector.

• Moreover, if wf has a regular eigenvector for the eigenvalue ζ, we say that wf is ζ–regular.

• Note that if wf is ζ–regular, it is also ζ'–regular for any root ζ' of the minimal polynomial Φ of ζ over K. We then say that wf is Φ–regular.

For $wf \in Wf$ and ζ a root of unity in K', we denote by $V'(wf, \zeta)$ the eigenspace of wf corresponding to the eigenvalue ζ.

4.48. PROPOSITION. *Let ζ be a root of unity in K' and let Φ be its minimal polynomial over K. Let $\mathbb{G}$ be a reflection datum on K, and let $\Phi^{a(\Phi)}$ be the largest power of Φ which divides $|\mathbb{G}|$. For $wf \in Wf$, we set $\mathbb{S}(wf, \Phi) := (\ker(\Phi(wf)), (wf)_{|\ker(\Phi(wf))})$ (a Φ–sub-reflection datum of $\mathbb{G}$). Let $\mathbb{L}(wf, \Phi)$ denote its centralizer. The following assertions are equivalent.*

(i) *wf is ζ–regular,*

(ii) *$C_W(V'(wf, \zeta)) = \{1\}$,*

(iii) *$\mathbb{L}(wf, \Phi)$ is a torus,*

(iv) *for $w'f \in Wf$, $\Phi^{a(\Phi)}$ divides the characteristic polynomial of $w'f$ if and only if $w'f$ is W–conjugate to wf.*

4.49. LEMMA. *If a reflection datum $\mathbb{G} = (V, Wf)$ is not a torus (i.e., if $W \neq 1$), it has two maximal tori with different order (in other words, there are two elements wf and $w'f$ in Wf with different characteristic polynomials).*

Indeed, let $|\mathbb{G}| = x^{N(\mathbb{G})} \prod_\zeta (x - \zeta)^{a(\zeta)}$ be the factorization of the polynomial order $|\mathbb{G}|$ over $K'[x]$. By Sylow theorems (4.41) applied to the reflection datum defined over K', for each root of unity ζ there is $wf \in Wf$ whose characteristic polynomial is divisible by $(x - \zeta)^{a(\zeta)}$. Hence it suffices to check that, if W is not trivial, we have $\sum_\zeta a(\zeta) > \dim(V)$. Indeed, we have $\sum_\zeta a(\zeta) = \deg |\mathbb{G}| - N(\mathbb{G})$, *i.e.*, by 4.6, (2), $\sum_\zeta a(\zeta) = N^\vee(\mathbb{G}) + r$. □

We say that a cyclotomic polynomial Φ on K is *regular* for $\mathbb{G}$ if there is a Φ–regular element in Wf.

4.50. COROLLARY. *Let Φ be a cyclotomic polynomial on K. Let $\mathbb{G}$ be a reflection datum on K, and let $\Phi^{a(\Phi)}$ be the largest power of Φ which divides $|\mathbb{G}|$. The following assertions are equivalent:*

(i) *Φ is regular for $\mathbb{G}$,*

(ii) *the centralizer of a Sylow Φ–sub-reflection datum of $\mathbb{G}$ is a torus,*

(iii) *there is only one conjugacy class under W of maximal tori of $\mathbb{G}$ whose order is divisible by $\Phi^{a(\Phi)}$.*

It is an easy consequence of 4.48 and the fact (see the proof of 4.41 above) that any Sylow Φ–sub-reflection datum is of the form $\mathbb{S}(wf,\Phi)$.

Degrees and regular elements.

Let α be a class function on $\mathbb{G}$. Let wf be Φ–regular, so that the associated maximal toric sub-reflection datum $\mathbb{T}_{wf}$ of $\mathbb{G}$ is a Sylow Φ–sub-reflection datum by 4.48. Then the congruence given in 4.44 becomes

$$\mathrm{Deg}(R_\alpha) \equiv \alpha(wf) \mod \Phi\,,$$

which can be reformulated into the following proposition.

4.51. PROPOSITION. *If wf is ζ–regular for some root of unity ζ, for any $\alpha \in \mathrm{CF}_{\mathrm{uf}}(\mathbb{G})$ we have*

$$\mathrm{Deg}(R_\alpha)(\zeta) = \alpha(wf)\,.$$

CHAPTER V

GENRALIZED HARISH–CHANDRA THEORIES

Generic unipotent characters

In the previous chapters we have introduced the concept of reflection datum to capture much of the generic nature of the subgroup structure and of uniform functions of finite reductive groups. We now turn to the description of the generic behaviour of irreducible characters.

From now on, we consider a reflection datum $\mathbb{G} = (V, Wf)$ which is associated with a finite reductive group $(\mathbf{G}, F)$.

An irreducible character $\gamma \in \mathrm{Irr}(\mathbf{G}^F)$ is called *unipotent* if there exists an F-stable maximal torus $\mathbf{S} \leq \mathbf{G}$ such that γ is a constituent of the Deligne–Lusztig induced character $R_{\mathbf{S}}^{\mathbf{G}}(1)$. Let $\mathrm{Uch}(\mathbf{G}^F)$ denote the set of unipotent characters of $\mathbf{G}^F$.

For any F-stable Levi subgroup $\mathbf{L} \leq \mathbf{G}$ Deligne and Lusztig (cf. for example [**Lu6**]) defined functors

$$R_{\mathbf{L}}^{\mathbf{G}} : \mathbb{Z}\mathrm{Irr}(\mathbf{L}^F) \to \mathbb{Z}\mathrm{Irr}(\mathbf{G}^F), \quad {}^*R_{\mathbf{L}}^{\mathbf{G}} : \mathbb{Z}\mathrm{Irr}(\mathbf{G}^F) \to \mathbb{Z}\mathrm{Irr}(\mathbf{L}^F),$$

between the character groups of $\mathbf{L}^F$ and $\mathbf{G}^F$, adjoint to each other with respect to the usual scalar product of characters.

More precisely, the definition of these functors also involves the choice of a parabolic subgroup $\mathbf{P}$ containing $\mathbf{L}$, but in our situation it turns out that they are in fact independent of this choice (see below §).

The results of Lusztig on unipotent characters (cf. for example [**Lu3**]), completed by some results of Shoji (see [**BMM1**], Ths. 1.26 and 1.33) may be rephrased as follows.

5.1. Theorem. *There exists a set* $\mathrm{Uch}(\mathbb{G})$ *and a map*

$$\mathrm{Deg} : \mathrm{Uch}(\mathbb{G}) \to \mathbb{Q}[x], \quad \gamma \mapsto \mathrm{Deg}(\gamma),$$

such that for any choice of p and q (and hence of $\mathbf{G}$ *and F) there is a bijection*

$$\psi_q^{\mathbb{G}} : \mathrm{Uch}(\mathbb{G}) \xrightarrow{\sim} \mathrm{Uch}(\mathbf{G}^F)$$

such that

(1) *The degree of* $\psi_q^{\mathbb{G}}(\gamma)$ *is* $\psi_q^{\mathbb{G}}(\gamma)(1) = \mathrm{Deg}(\gamma)(q)$.

(2) *For any generic Levi subgroup* $\mathbb{L}$ *of* $\mathbb{G}$ *there exist linear maps*

$$R_{\mathbb{L}}^{\mathbb{G}} : \mathbb{Z}\mathrm{Uch}(\mathbb{L}) \to \mathbb{Z}\mathrm{Uch}(\mathbb{G}), \quad {}^*R_{\mathbb{L}}^{\mathbb{G}} : \mathbb{Z}\mathrm{Uch}(\mathbb{G}) \to \mathbb{Z}\mathrm{Uch}(\mathbb{L}),$$

satisfying $\psi_q^{\mathbb{G}} R_{\mathbb{L}}^{\mathbb{G}} = R_{\mathbf{L}}^{\mathbf{G}} \psi_q^{\mathbb{L}}$ *for all q (we extend* $\psi_q^{\mathbb{G}}$ *linearly to* $\mathbb{Z}\mathrm{Uch}(\mathbb{G})$*).*

The set Uch($\mathbb{G}$) *is called the set of (generic) unipotent characters of* $\mathbb{G}$.

In this sense, the sets of unipotent characters together with the collection of functors $R_{\mathbf{L}}^{\mathbf{G}}$ and ${}^*R_{\mathbf{L}}^{\mathbf{G}}$ are generic for a series of finite reductive groups. of Lie type.

EXAMPLE. In the case of the finite reductive group $\mathbf{G}^F = \mathrm{GL}_n(q)$ the unipotent characters are just the constituents of the permutation character $\mathrm{Ind}_{\mathbf{B}^F}^{\mathbf{G}^F}$ on the F-fixed points of an F-stable Borel subgroup $\mathbf{B}$ of $\mathbf{G}$. The endomorphism algebra of this permutation module is the Hecke algebra of type A_{n-1} with parameter q. Its irreducible characters are in bijection with the irreducible characters of the symmetric group $\mathfrak{S}_n$, the Weyl group of $\mathbf{G}^F$. Since the latter are naturally indexed by partitions $\alpha \vdash n$, the same is true for the unipotent characters of $\mathbf{G}^F$. Hence in this case we have $\mathrm{Uch}(\mathrm{GL}_n) = \{\gamma_\alpha \mid \alpha \vdash n\}$. The function $\mathrm{Deg} : \mathrm{Uch}(\mathrm{GL}_n) \to \mathbb{Q}[x]$ can be described as follows: for a partition $\alpha = (\alpha_1 \leq \ldots \leq \alpha_m)$ of n let $\beta_i := \alpha_i + i - 1$, $1 \leq i \leq m$. Then

$$\mathrm{Deg}(\gamma_\alpha) = \frac{(x-1)\ldots(x^n-1)\prod_{j>i}(x^{\beta_j} - x^{\beta_i})}{x^{\binom{m-1}{2}+\binom{m-2}{2}+\ldots}\prod_i\prod_{j=1}^{\beta_i}(x^j-1)}.$$

(This always is a polynomial.) Furthermore, there exists a bijection

$$\mathrm{Uch}(\mathrm{GL}_n) \to \mathrm{Uch}(\mathrm{U}_n), \quad \gamma \mapsto \gamma^-,$$

such that $\mathrm{Deg}(\gamma^-)(x) = \pm\mathrm{Deg}(\gamma)(-x)$. This is another consequence of the Ennola duality between GL_n and U_n.

Φ-Harish-Chandra theories

The idea that a generalized Harish-Chandra theory should exist for the unipotent characters of a finite reductive group was implicit in many instances of Lusztig's papers, and occurred also in the papers by Fong–Srinivasan [**FoSr**j] ($j = 1,2,3$) and Schewe [**Sch**] in particular cases. The general case was settled in [**BMM1**].

Φ-cuspidal characters.

Let $\mathbb{G}$ be a reflection datum and Φ a K–cyclotomic polynomial such that Φ divides $|\mathbb{G}|$. We first have to introduce a generalization of cuspidal characters.

DEFINITION. A generic unipotent character $\gamma \in \mathrm{Uch}(\mathbb{G})$ is called Φ-*cuspidal* if ${}^*R_{\mathbb{L}}^{\mathbb{G}}(\gamma) = 0$ for all Φ-split Levi subgroups $\mathbb{L}$ properly contained in $\mathbb{G}$.

Thus in the case $\Phi = x - 1$ the Φ-cuspidal characters of $\mathbb{G}$ are those whose image under $\psi_q^{\mathbb{G}}$ is a cuspidal unipotent character as defined in the ordinary Harish–Chandra theory. We have the following alternative characterization of Φ-cuspidality (see [**BMM1**], Prop. 2.9):

5.2. PROPOSITION. *A unipotent character* $\gamma \in \mathrm{Uch}(\mathbb{G})$ *is* Φ-*cuspidal if and only if*

$$\mathrm{Deg}(\gamma)_\Phi = |\mathbb{G}_{\mathrm{ss}}|_\Phi .$$

Here, we write f_Φ for the Φ-part of $f \in \mathbb{Q}[x]$. The semisimple quotient $\mathbb{G}_{\mathrm{ss}}$ of $\mathbb{G}$ was defined above (chap. iv).

EXAMPLE. We continue the example of GL_n (see above). We assume here that $\Phi = \Phi_d$, the d-th cyclotomic polynomial. Let $\alpha = (\alpha_1 \leq \ldots \leq \alpha_m)$ be a partition of n. A *d-hook of α* is a pair $h = (\nu, \beta)$ of integers $0 \leq \nu < \beta$ such that β occurs among the $\{\alpha_i + i - 1 \mid 1 \leq i \leq m\}$ but ν does not. The *length* of the hook h is $l(h) = \beta - \nu$ and h is also called an $l(h)$-hook. A moments thought shows that up to a power of x the degree given above of the unipotent character γ_α of the generic group of type GL_n is of the form

$$\frac{(x-1)\ldots(x^n-1)}{\prod_{h \text{ hook of } \alpha}(x^{l(h)}-1)}.$$

Note that the existence of an ad-hook of α for some $a \geq 1$ implies the existence of a d-hook. Thus, by Proposition 5.2 the unipotent character γ_α is Φ_d-cuspidal if and only if α has no d-hook. Such partitions α are also called *d-cores.* In particular, GL_n has a 1-cuspidal unipotent character if and only if $n = 0$, since the empty partition is the only 1-core. This shows that indeed all unipotent characters of GL_n occur in $R_{\mathbb{S}}^{\mathbb{G}}(1)$ for the maximally split torus $\mathbb{S}$.

Φ-Harish-Chandra series.

A pair $(\mathbb{L}, \lambda)$ consisting of a Φ-split Levi subgroup $\mathbb{L}$ of $\mathbb{G}$ and a unipotent character $\lambda \in \mathrm{Uch}(\mathbb{L})$ is called *Φ-split.* It is called *Φ-cuspidal* if moreover λ is Φ-cuspidal. We introduce the following relation on the set of Φ-split pairs:

DEFINITION. Let $(\mathbb{M}_1, \mu_1)$ and $(\mathbb{M}_2, \mu_2)$ be Φ-split in $\mathbb{G}$. Then we say that $(\mathbb{M}_1, \mu_1) \leq_\Phi (\mathbb{M}_2, \mu_2)$ if $\mathbb{M}_1$ is a Φ-split Levi subgroup of $\mathbb{M}_2$ and μ_2 occurs in $R_{\mathbb{M}_1}^{\mathbb{M}_2}(\mu_1)$.

For a Φ-cuspidal pair $(\mathbb{L}, \lambda)$ of $\mathbb{G}$ we write

$$\mathrm{Uch}(\mathbb{G}, (\mathbb{L}, \lambda)) := \{\gamma \in \mathrm{Uch}(\mathbb{G}) \mid (\mathbb{L}, \lambda) \leq_\Phi (\mathbb{G}, \gamma)\}$$

for the set of unipotent characters of $\mathbb{G}$ lying above $(\mathbb{L}, \lambda)$. We call $\mathrm{Uch}(\mathbb{G}, (\mathbb{L}, \lambda))$ the *Φ-Harish-Chandra series above* $(\mathbb{L}, \lambda)$ because of the following fundamental result, which shows that for any Φ we obtain a generalized Harish-Chandra theory.

5.3. THEOREM. *Let $\mathbb{G}$ be a reflection datum and $d \geq 1$ such that Φ divides* $|\mathbb{G}|$.

(1) (Disjointness) *The sets* $\mathrm{Uch}(\mathbb{G}, (\mathbb{L}, \lambda))$ *(where $(\mathbb{L}, \lambda)$ runs over a system of representatives of the $W_\mathbb{G}$-conjugacy classes of Φ-cuspidal pairs) form a partition of* $\mathrm{Uch}(\mathbb{G})$.
(2) (Transitivity) *Let $(\mathbb{L}, \lambda)$ be Φ-cuspidal and $(\mathbb{M}, \mu)$ be Φ-split. We assume that $(\mathbb{L}, \lambda) \leq_\Phi (\mathbb{M}, \mu)$ and $(\mathbb{M}, \mu) \leq_\Phi (\mathbb{G}, \gamma)$. Then $(\mathbb{L}, \lambda) \leq_\Phi (\mathbb{G}, \gamma)$.*

The statement of Theorem 5.3 has been checked case by case (except for the case $d = 1$ where a conceptual proof is known). It is a consequence of the more precise Theorem 5.6 about the decomposition of twisted induction of unipotent characters from Φ-split Levi subgroups.

EXAMPLE. We continue the example of GL_n, where we saw that the Φ_d-cuspidal characters of GL_n are indexed by partitions of n which are d-cores. Let $h = (\nu, \beta)$ be a d-hook of the partition $\alpha = (\alpha_1 \leq \ldots \leq \alpha_m)$, and let j be the index with $\alpha_j + j - 1 = \beta$. Then $\alpha' = (\alpha'_1 \leq \ldots \leq \alpha'_m)$ is called the partition obtained from α by *removing the d-hook h* if the set $\{\alpha'_i + i - 1\}$ coincides with the set

$$\{\alpha_i + i - 1 \mid i \neq j\} \cup \{\alpha_j + j - 1 - (\beta - \nu)\}.$$

The d-core obtained from α by successively removing all possible d-hooks is called the *d-core of α*. It can be shown that the Φ_d-Harish-Chandra series of GL_n above the Φ_d-cuspidal character indexed by the d-core α consists of the unipotent characters indexed by partitions of n whose d-core is α.

Generic blocks

One importance of Φ-Harish-Chandra series lies in their connection with ℓ-blocks of finite groups of Lie type for primes ℓ not dividing q. Before continuing the exposition of the generic theory we therefore briefly explain this application.

We fix $\mathbb{G}$ and a choice of a prime power q, hence a pair $(\mathbf{G}, F)$. A prime ℓ not dividing q is called *large for* $\mathbb{G}$ if ℓ does not divide the order of $W\langle f\rangle$. If ℓ is large then by Proposition4.42(a) there exists a unique Φ such that ℓ divides $\Phi(q)$ and Φ divides $|\mathbb{G}|$.

The following result is proved in [**BMM1**], Th. 5.24.

5.4. THEOREM. *Let ℓ be large for $\mathbb{G}$, and assume that ℓ divides $\Phi(q)$ and Φ divides $|\mathbb{G}|$. Then the partition of* $\mathrm{Uch}(\mathbf{G}^F)$ *into ℓ-blocks coincides with the image under $\psi_q^{\mathbb{G}}$ of the partition of* $\mathrm{Uch}(\mathbb{G})$ *into Φ-Harish-Chandra series.*

Thus the preceding theorem shows that distribution of unipotent characters of $\mathbf{G}^F$ into ℓ-blocks is generic.

Let us write $b_\ell(\mathbf{L}, \lambda)$ for the unipotent ℓ-block of $\mathbf{G}^F$ indexed by $(\mathbb{L}, \lambda)$. The unipotent blocks and their defect groups can be described more precisely.

DEFINITION. For $\gamma \in \mathrm{Uch}(\mathbb{G})$ let $S_\Phi(\gamma)$ denote the set of Φ-tori of $\mathbb{G}$ contained in a maximal torus $\mathbb{S}$ of $\mathbb{G}$ such that ${}^*R_{\mathbb{S}}^{\mathbb{G}}(\gamma) \neq 0$. The maximal elements of $S_\Phi(\gamma)$ are called the *Φ-defect tori of γ*.

The Φ-defect tori of a unipotent character can be characterized as follows (see [**BMM1**], Th. 4.8).

PROPOSITION. *Let $(\mathbb{L}, \lambda)$ be a Φ-cuspidal pair and $\gamma \in \mathrm{Uch}(\mathbb{G}, (\mathbb{L}, \lambda))$. Then the Φ-defect tori of γ are conjugate to* $\mathrm{Rad}(\mathbb{L})_\Phi$.

Here $\mathrm{Rad}(\mathbb{L})_\Phi$ denotes the Sylow Φ-torus of the torus $\mathrm{Rad}(\mathbb{L})$. For an F-stable Levi subgroup $\mathbf{L}$ of $\mathbf{G}$ denote by $\mathrm{Ab}_\ell\mathrm{Irr}(\mathbf{L}^F)$ the group of characters (over a splitting field of characteristic 0) of ℓ-power order of the abelian group $\mathbf{L}^F/[\mathbf{L}^F, \mathbf{L}^F]$. Then [**BMM1**], Th. 5.24, gives the following sharpening of Theorem 5.4:

5.5. THEOREM. *Let ℓ be large for $\mathbb{G}$, and assume that ℓ divides $\Phi(q)$ and Φ divides $|\mathbb{G}|$. Let $(\mathbb{L}, \lambda)$ be a Φ-cuspidal pair of $\mathbb{G}$.*

(1) *The ℓ-block $b_\ell(\mathbf{L}, \lambda)$ of $\mathbf{G}^F$ consists of the irreducible constituents of the virtual characters $R_{\mathbf{L}}^{\mathbf{G}}(\theta\lambda)$, where $\theta \in \mathrm{Ab}_\ell\mathrm{Irr}(\mathbf{L}^F)$.*
(2) *The defect groups of $b_\ell(\mathbf{L}, \lambda)$ are the Sylow ℓ-subgroups of the groups of F-fixed points of the Φ-defect tori of unipotent characters in $b_\ell(\mathbf{L}, \lambda)$.*

The structure of the Sylow ℓ-subgroup of the group of F-fixed points of a Φ-torus was described in Proposition 4.42(c).

EXAMPLE. Theorem 5.5 and the description of Φ-Harish-Chandra series for GL_n in examples above provide the following description (first proved by Fong and Srinivasan) of unipotent ℓ-blocks of $\mathrm{GL}_n(q)$ for large primes ℓ dividing $\Phi_d(q)$. Two unipotent characters γ_α, $\gamma_{\alpha'}$ lie in the same ℓ-block if and only if the d-cores of α and α' coincide.

Relative Weyl groups

The Φ-Harish-Chandra theories presented above seem to be just the shadow of a much deeper theory describing the decomposition of the functor of twisted induction. Much of this is still conjectural (see below chap. VI).

The decomposition of $R_{\mathbb{L}}^{\mathbb{G}}$.

The only known proof of Theorem 5.3 in the case $\Phi \neq x - 1$ consists in determining the decomposition of the Deligne-Lusztig induced of Φ-cuspidal unipotent characters. To state this result we need to introduce an important invariant of a Φ-Harish-Chandra series.

Let $(\mathbb{L}, \lambda)$ be a Φ-cuspidal pair in $\mathbb{G}$. Let $(\mathbf{G}, F)$ be a finite reductive group associated to $\mathbb{G}$, and let $\mathbf{L}$ be an F-stable Levi subgroup of $\mathbf{G}$ corresponding to $\mathbb{L}$. Then the relative Weyl group $W_{\mathbf{G}^F}(\mathbf{L}) = N_{\mathbf{G}^F}(\mathbf{L})/\mathbf{L}^F$ acts on $\mathrm{Irr}(\mathbf{L}^F)$. By results of Lusztig, this leaves the subset $\mathrm{Uch}(\mathbf{L}^F)$ of unipotent characters invariant. Moreover, the action on $\mathrm{Uch}(\mathbf{L}^F)$ is generic in the sense that it is possible to define an action of $W_{\mathbb{G}}(\mathbb{L}) = N_{W_{\mathbb{G}}}(W_{\mathbb{L}})/W_{\mathbb{L}}$ on $\mathrm{Uch}(\mathbb{L})$ which, under all $\psi_q^{\mathbb{G}}$, specializes to the action of $W_{\mathbf{G}^F}(\mathbf{L})$ on $\mathrm{Uch}(\mathbf{L}^F)$ (see [**BMM1**]). This gives sense to the definition

$$W_{\mathbb{G}}(\mathbb{L}, \lambda) := N_{W_{\mathbb{G}}}(W_{\mathbb{L}}, \lambda)/W_{\mathbb{L}}$$

of the *relative Weyl group of* $(\mathbb{L}, \lambda)$ *in* $\mathbb{G}$.

The following result was proved in [**BMM1**] (Th. 3.2) by using results of Asai on the decomposition of $R_{\mathbf{L}}^{\mathbf{G}}$ in the case of classical groups, and by explicit determination of these decompositions in the case of exceptional groups.

5.6. THEOREM. *For each Φ there exists a collection of isometries*

$$I_{(\mathbb{L},\lambda)}^{\mathbb{M}} : \mathbb{Z}\mathrm{Irr}(W_{\mathbb{M}}(\mathbb{L}, \lambda)) \to \mathbb{Z}\mathrm{Uch}(\mathbb{M}, (\mathbb{L}, \lambda)),$$

(where $\mathbb{M}$ runs over the Φ-split Levi subgroups of $\mathbb{G}$, and $(\mathbb{L}, \lambda)$ over the set of Φ-cuspidal pairs of $\mathbb{M}$) such that for all $\mathbb{M}$ and all $(\mathbb{L}, \lambda)$ we have

$$R_{\mathbb{M}}^{\mathbb{G}} I_{(\mathbb{L},\lambda)}^{\mathbb{M}} = I_{(\mathbb{L},\lambda)}^{\mathbb{G}} \mathrm{Ind}_{W_{\mathbb{M}}(\mathbb{L},\lambda)}^{W_{\mathbb{G}}(\mathbb{L},\lambda)}.$$

An isometry I from $\mathbb{Z}\mathrm{Irr}(W_{\mathbb{G}}(\mathbb{L}, \lambda))$ to $\mathbb{Z}\mathrm{Uch}(\mathbb{G}, (\mathbb{L}, \lambda))$ is nothing else but a bijection

$$\mathrm{Irr}(W_{\mathbb{G}}(\mathbb{L}, \lambda)) \xrightarrow{\sim} \mathrm{Uch}(\mathbb{G}, (\mathbb{L}, \lambda)), \quad \chi \mapsto \gamma_{\mathrm{ch}},$$

together with a collection of signs

$$\{\epsilon(\gamma) \mid \gamma \in \mathrm{Uch}(\mathbb{G}, (\mathbb{L}, \lambda))\},$$

such that

$$I(\chi) = \epsilon(\gamma_\chi)\, \gamma_\chi .$$

Thus, Theorem 5.6 states that up to an adjustment by suitable signs, twisted induction from Φ-split Levi subgroups is nothing but ordinary induction in relative

Weyl groups. As a consequence of [**BMM1**], Th. 5.24, we moreover have the following congruence of character degrees

$$\epsilon(\gamma_\chi)\,\mathrm{Deg}(\gamma_\chi) \equiv \chi(1) \pmod{\Phi} \quad \text{in } \mathbb{Q}[x]$$

in the situation of Theorem 5.6.

EXAMPLE. We continue the example of reflection datum of type ${}^3D_4$. The relative Weyl groups $W_{\mathbb{G}}(\mathbb{S}_\Phi)$ for the Sylow Φ-tori $\mathbb{S}_\Phi$ with $d \in \{3,6\}$ turn out to be isomorphic to $\mathrm{SL}_2(3)$. Since $\mathrm{SL}_2(3)$ has 7 irreducible characters Theorem 5.6 implies that $\mathrm{Uch}(\mathbb{G}, (\mathbb{S}_\Phi, 1))$ has cardinality 7 for $d \in \{3,6\}$. Moreover, the vector of degrees of the irreducible characters of $\mathrm{SL}_2(3)$ is $(1,1,1,2,2,2,3)$, so $R_{\mathbb{S}_\Phi}^{\mathbb{G}}(1)$ contains three constituents with multiplicity ±1, three with multiplicity ±2 and one with multiplicity ±3.

Relative Weyl groups are pseudo-reflection groups.

Theorem 5.6 turns attention towards the relative Weyl groups of Φ-cuspidal pairs $(\mathbb{L}, \lambda)$. Lusztig proved that in the case $d = 1$ the relative Weyl groups are Coxeter groups.

If $\mathbb{L} = (V, W_{\mathbb{L}}wf)$, we denote by $Z(\mathbb{L})_\Phi$ the Sylow Φ–sub-reflection datum of the center of $\mathbb{L}$, defined by

$$Z(\mathbb{L})_\Phi = (\ker \Phi(wf) \cap V^{W_{\mathbb{L}}}, wf)\,.$$

In particular, we denote by $V(\mathbb{L}, \Phi)$ its vector space $\ker \Phi(wf) \cap V^{W_{\mathbb{L}}}$, viewed as a vector space over the field $K[x]/\Phi(x)$ through its natural structure of $K[wf]$–module.

Let us set

$$K_{\mathbb{G}}(\mathbb{L}, \lambda) := K[x]/\Phi(x)\,.$$

In a case-by-case analysis the following surprising fact can be verified (see [**BrMa2**]). It is a generalization, in this context, of the main result of [**LeSp2**], which concerns the case where $\mathbb{L}$ is a minimal split Levi reflection datum and λ is the trivial character.

5.7. THEOREM.
(1) *The pair* $(V(\mathbb{L}, \Phi), W_{\mathbb{G}}(\mathbb{L}, \lambda))$ *is a* $K_{\mathbb{G}}(\mathbb{L}, \lambda)$*–complex reflection group.*
(2) *The* $W_{\mathbb{G}}(\mathbb{L}, \lambda)$ *is irreducible on* $V(\mathbb{L}, \Phi)$ *if* W *is irreducible on* V.

EXAMPLE. Let $\mathbb{G}$ be a generic group of type E_7 and let $d = 4$. A Sylow 4-torus $\mathbb{S}$ of $\mathbb{G}$ has order Φ_4^2 and its centralizer is a 4-split Levi subgroup $\mathbb{L} = C_{\mathbb{G}}(\mathbb{S})$ with semisimple part $(\mathrm{PGL}_2)^3$. Since $\mathbb{L}$ is minimal 4-split all its unipotent characters are 4-cuspidal. The relative Weyl group $W_{\mathbb{G}}(\mathbb{L})$ is a two-dimensional complex reflection group of order 96, denoted G_8 by Shephard and Todd. It has two orbits of length 3 and two fixed points on the set $\mathrm{Uch}(\mathbb{L})$. In particular, the relative Weyl group $W_{\mathbb{G}}(\mathbb{L}, \lambda)$ for a (4-cuspidal) character λ in one of the orbits of length 3 is strictly smaller than $W_{\mathbb{G}}(\mathbb{L})$. It turns out to be the imprimitive complex reflection group denoted $G(4,1,2)$ of order 32.

The relative Weyl groups occurring in reflection data of exceptional type are collected in [**BMM1**], Tables 1 and 3, while those for classical types are described in [**BMM1**], Sec. 3, see also [**BrMa2**], 3B.

CHAPTER VI

THE ABELIAN DEFECT GROUP CONJECTURE FOR FINITE REDUCTIVE GROUPS AND SOME CONSEQUENCES

The origin of the following conjectures is a general conjecture about "abstract" finite groups (see [**Bro1**]), which concerns the structure (up to equivalence) of the derived bounded category of the ℓ–adic algebra of a block of any finite group with an abelian defect group. This "abstract" conjecture inspired more precise guesses in the particular case of finite reductive groups, which first appeared in [**Bro1**] (last paragraph). These conjectures have also been partly stated at the conference held in honor of Charlie Curtis at the University of Oregon in September 1991.

Notation.

Let $\mathbf{G}$ be a connected reductive algebraic group over an algebraic closure $\overline{\mathbb{F}}_p$ of the prime field with p elements. Let q be a power of p and let $\mathbb{F}_q$ be the subfield of cardinal q of $\overline{\mathbb{F}}_p$. We assume that $\mathbf{G}$ is endowed with a Frobenius endomorphism F which defines a rational structure on $\mathbb{F}_q$. We denote by $2N$ the number of roots of $\mathbf{G}$.

Let $\mathbf{P}$ be a parabolic subgroup of $\mathbf{G}$, with unipotent radical $\mathbf{U}$, and with F–stable Levi complement $\mathbf{L}$.

We denote by $\mathrm{Y}(\mathbf{U})$ the associated Deligne–Lusztig variety defined (cf. for example [**Lu6**]) by

$$\mathrm{Y}(\mathbf{U}) := \{g(\mathbf{U}\cap F(\mathbf{U})) \in \mathbf{G}/\mathbf{U}\cap F(\mathbf{U})\,;\, g^{-1}F(g) \in F(\mathbf{U})\}\,.$$

Notice that $\mathbf{G}^F$ acts on $\mathrm{Y}(\mathbf{U})$ by left multiplication while $\mathbf{L}^F$ acts on $\mathrm{Y}(\mathbf{U})$ by right multiplication.

It is known (cf. for example [**Lu6**]) that $\mathrm{Y}(\mathbf{U})$ is an $\mathbf{L}^F$–torsor on a variety $\mathrm{X}(\mathbf{U})$, which is smooth of pure dimension equal to $\dim(\mathbf{U}/\mathbf{U}\cap F(\mathbf{U}))$, and which is affine (at least if q is large enough). In particular $\mathrm{X}(\mathbf{U})$ is endowed with a left action of $\mathbf{G}^F$. If $\mathcal{O}$ is a commutative ring, the image of the constant sheaf $\mathcal{O}$ on $\mathrm{Y}(\mathbf{U})$ through the finite morphism $\pi\colon \mathrm{Y}(\mathbf{U}) \to \mathrm{X}(\mathbf{U})$ is a locally constant constant sheaf $\pi_*(\mathcal{O})$ on $\mathrm{X}(\mathbf{U})$. We denote this sheaf by $\mathcal{F}_{\mathcal{O}\mathbf{L}^F}$.

In the particular case where $\mathbf{B} = \mathbf{TU}$ is a Borel subgroup such that $\mathbf{B}$ and $F(\mathbf{B})$ are in relative position w for some $w \in W$ (cf. [**DeLu**], 1.2), we set $\mathrm{X}_w := \mathrm{X}(\mathbf{U})$.

Let ℓ be a prime number which does not divide q and let $\mathcal{O}$ be the ring of integers of a finite extension of the field $\mathbb{Q}_\ell$ of ℓ–adic numbers. For any $\mathbf{G}^F$–equivariant torsion free $\mathcal{O}$–sheaf $\mathcal{F}$ on $\mathrm{X}(\mathbf{U})$, we denote by $\mathcal{H}_\mathcal{O}(\mathrm{X}(\mathbf{U}),\mathcal{F})$ the algebra of endomorphisms of the "ℓ–adic cohomology" complex $\mathrm{R}\Gamma_c(\mathrm{X}(\mathbf{U}),\mathcal{F})$ viewed as an element of the derived bounded category $\mathcal{D}^b(\mathcal{O}\mathbf{G}^F)$ of the category of finitely generated $\mathcal{O}\mathbf{G}^F$–modules.

We set

$$\mathrm{R}\Gamma_c(\mathrm{Y}(\mathbf{U})) := \mathrm{R}\Gamma_c(\mathrm{X}(\mathbf{U}), \mathcal{F}_{\mathcal{O}\mathbf{L}^F}) \quad \text{and} \quad \mathcal{H}_{\mathcal{O}}(\mathrm{Y}(\mathbf{U})) := \mathcal{H}_{\mathcal{O}}(\mathrm{X}(\mathbf{U}), \mathcal{F}_{\mathcal{O}\mathbf{L}^F}) .$$

Note that the algebra $\mathcal{H}_{\mathcal{O}}(\mathrm{Y}(\mathbf{U}))$ contains the group algebra $\mathcal{O}\mathbf{L}^F$ as a subalgebra.

For K an extension of $\mathcal{O}$, we set $\mathcal{H}_{\mathrm{K}}(\mathrm{X}(\mathbf{U}), \mathcal{F}) := \mathrm{K} \otimes_{\mathcal{O}} \mathcal{H}_{\mathcal{O}}(\mathrm{X}(\mathbf{U}), \mathcal{F})$.

Finally, if $\mathbf{\Gamma} = (\cdots \to \mathbf{\Gamma}^{n-1} \to \mathbf{\Gamma}^n \to \mathbf{\Gamma}^{n+1} \to \cdots)$ is a complex of $\mathcal{O}$–modules, we denote by $\mathbf{\Gamma}^\vee$ the "ordinary" $\mathcal{O}$–dual of $\mathbf{\Gamma}$: we have $\mathbf{\Gamma}^{\vee m} := \mathrm{Hom}_{\mathcal{O}}(\mathbf{\Gamma}^{-m}, \mathcal{O})$, and the differentials are defined by $\mathcal{O}$–transposition.

Conjectures about ℓ–blocks

The data.

(H1) Let $\ell \neq p$ be a prime number which does not divide $|Z(\mathbf{G})/Z^o(\mathbf{G})|$ nor $|Z(\mathbf{G}^*)/Z^o(\mathbf{G}^*)|$, and which is good for $\mathbf{G}$.

(H2) Let $\mathcal{O}$ be the ring of integers of a finite unramified extension K of the field of ℓ–adic numbers $\mathbb{Q}_\ell$, with residue field k, such that the finite group algebra $\mathrm{k}\mathbf{G}^F$ is split.

(H3) Let e be a primitive central idempotent of $\mathcal{O}\mathbf{G}^F$ (an "ℓ–block" of $\mathbf{G}^F$) with abelian defect group D. Let $\mathbf{L} := C_{\mathbf{G}}(D)$. Let f be a block of $\mathbf{L}^F$ such that (D, f) is an e–subpair of $\mathbf{G}^F$.

It results from (H1) that the group $\mathbf{L}$ is a rational Levi subgroup of $\mathbf{G}$.

We have $N_{\mathbf{G}^F}(D, f) = N_{\mathbf{G}^F}(\mathbf{L}, f)$, and we set $W_{\mathbf{G}^F}(\mathbf{L}, f) := N_{\mathbf{G}^F}(\mathbf{L}, f)/\mathbf{L}^F$. Note that by known properties of maximal subpairs (see for example [**AlBr**]), D is a Sylow ℓ–subgroup of $Z(\mathbf{L})^F$, and ℓ does not divide $|W_{\mathbf{G}^F}(\mathbf{L}, f)|$.

ℓ–Conjectures.

There exist

• a parabolic subgroup of $\mathbf{G}$ with unipotent radical $\mathbf{U}$ and Levi complement $\mathbf{L}$,

• a finite complex $\mathbf{\Upsilon} = (\cdots \to \mathbf{\Upsilon}^{n-1} \to \mathbf{\Upsilon}^n \to \mathbf{\Upsilon}^{n+1} \to \cdots)$ of $(\mathcal{O}\mathbf{G}^F, \mathcal{O}\mathbf{L}^F)$–bimodules, which are finitely generated projective as $\mathcal{O}\mathbf{G}^F$–modules as well as $\mathcal{O}\mathbf{L}^F$–modules,

with the following properties.

(ℓ-C1) Viewed as an object of the derived bounded category of the category of $(\mathcal{O}\mathbf{G}^F, \mathcal{O}\mathbf{L}^F)$–bimodules, $\mathbf{\Upsilon}$ is isomorphic to $\mathrm{R}\Gamma_c(\mathrm{Y}(\mathbf{U}))$. In particular, for each n, the n-th homology group of $\mathbf{\Upsilon}$ is isomorphic, as an $(\mathcal{O}\mathbf{G}^F, \mathcal{O}\mathbf{L}^F)$–bimodule, to $\mathcal{O} \otimes_{\mathbb{Z}_\ell} \mathrm{H}^n_c(\mathrm{Y}(\mathbf{U}), \mathbb{Z}_\ell)$.

(ℓ-C2) The idempotent e acts as the identity on the complex $\mathbf{\Upsilon}.f$,

(ℓ-C3) Let $\Delta(D)$ denote the diagonal embedding of D in $\mathbf{G}^F \times \mathbf{L}^F$. For each n, the $\mathcal{O}[\mathbf{G}^F \times \mathbf{L}^F]$–module $\mathbf{\Upsilon}^n.f$ is relatively $\Delta(D)$–projective, and its restriction to $\mathcal{O}\Delta(D)$ is a permutation module for $\Delta(D)$.

(ℓ-C4) • the structure of complex of $(\mathcal{O}\mathbf{G}^F e, \mathcal{O}\mathbf{L}^F f)$–bimodules of $\mathbf{\Upsilon}.f$ extends to a structure of complex of $(\mathcal{O}\mathbf{G}^F e, f\mathcal{H}_{\mathcal{O}}(\mathrm{Y}(\mathbf{U}))f)$–bimodules, all of which are projective as right $f\mathcal{H}_{\mathcal{O}}(\mathrm{Y}(\mathbf{U}))f$–modules,

• the complexes $(\mathbf{\Upsilon}.f \otimes_{f\mathcal{H}_{\mathcal{O}}(\mathrm{Y}(\mathbf{U}))f} f.\mathbf{\Upsilon}^\vee)$ and $\mathcal{O}\mathbf{G}^F e$ are homotopy equivalent as complexes of $(\mathcal{O}\mathbf{G}^F e, \mathcal{O}\mathbf{G}^F e)$–bimodules,

• the complexes $(f.\mathbf{\Upsilon}^\vee \otimes_{\mathcal{O}\mathbf{G}^F e} \mathbf{\Upsilon}.f)$ and $f\mathcal{H}_{\mathcal{O}}(\mathrm{Y}(\mathbf{U}))f$ are homotopy equivalent as complexes of $(f\mathcal{H}_{\mathcal{O}}(\mathrm{Y}(\mathbf{U}))f, f\mathcal{H}_{\mathcal{O}}(\mathrm{Y}(\mathbf{U}))f)$–bimodules.

(ℓ-C5) The algebra $f\mathcal{H}_{\mathcal{O}}(\mathrm{Y}(\mathbf{U}))f$ is isomorphic to the block algebra $\mathcal{O}N_{\mathbf{G}^F}(D, f)f$.

Let us make several remarks and draw some consequences of the preceding conjectures.

1. It can be proved, using some results of Jeremy Rickard [**Ri**], that there exists a complex $\boldsymbol{\Upsilon}$ of $(\mathcal{O}\mathbf{G}^F, \mathcal{O}\mathbf{L}^F)$–bimodules such that ($\ell$-C1) is satisfied, and such that for every integer n, the $\mathcal{O}[\mathbf{G}^F \times \mathbf{L}^F]$–module $\boldsymbol{\Upsilon}^n$ is relatively $\Delta(S)$–projective, and its restriction to $\mathcal{O}\Delta(S)$ is a permutation module for $\Delta(S)$, where $\Delta(S)$ is the diagonal embedding of a Sylow ℓ–subgroup S of $\mathbf{L}^F$ in $\mathbf{G}^F \times \mathbf{L}^F$. Then ($\ell$-C3) follows from ($\ell$-C2).

2. It follows from (ℓ-C4) that the categories $\mathcal{D}^b(\mathcal{O}\mathbf{G}^F e)$ and $\mathcal{D}^b(f\mathcal{H}_{\mathcal{O}}(\mathrm{Y}(\mathbf{U}))f)$ are equivalent, whence by (ℓ-C5) that

(ℓ-C6) the derived categories $\mathcal{D}^b(\mathcal{O}\mathbf{G}^F e)$ and $\mathcal{D}^b(\mathcal{O}N_{\mathbf{G}^F}(D,f)f)$ are equivalent.

This last equivalence is a particular case of a general conjecture about "abstract" finite groups stated in [**Bro1**].

3. For any chain map endomorphism α of $\boldsymbol{\Upsilon}$, we set

$$\mathrm{tr}_{\boldsymbol{\Upsilon}}(\alpha) := \sum_n (-1)^n \mathrm{tr}_{\boldsymbol{\Upsilon}^n}(\alpha)\,,$$

where $\mathrm{tr}_{\boldsymbol{\Upsilon}^n}(\alpha)$ denotes the trace of α as an endomorphism of the free $\mathcal{O}$–module $\boldsymbol{\Upsilon}^n$. Then it is easy to see that, since $f\mathcal{H}_{\mathcal{O}}(\mathbf{U})f$ is equal to $\mathrm{End}_{\mathcal{O}\mathbf{G}^F e}(\boldsymbol{\Upsilon})$,

(ℓ-C7) the linear form $\mathrm{tr}_{\boldsymbol{\Upsilon}}$ gives $f\mathcal{H}(\mathrm{Y}_{\mathcal{O}}(\mathbf{U}))f$ a structure of symmetric $\mathcal{O}$–algebra.

4. Set $\mathrm{H}^n_c(\mathrm{Y}(\mathbf{U}),\mathrm{K}) := \mathrm{K} \otimes_{\mathbb{Z}_\ell} \mathrm{H}^n_c(\mathrm{Y}(\mathbf{U}),\mathbb{Z}_\ell)$. We have

$$\mathrm{R}\Gamma_c(\mathrm{Y}(\mathbf{U}),\mathrm{K}) = \bigoplus_n \mathrm{H}^n_c(\mathrm{Y}(\mathbf{U}),\mathrm{K})[-n]\ ,$$

and it follows from (ℓ-C4) that

(ℓ-C8)
- The algebra $f\mathcal{H}_{\mathrm{K}}(\mathrm{Y}(\mathbf{U}))f$ is semi–simple,
- the $\mathrm{K}\mathbf{G}^F e$–modules $\mathrm{H}^n_c(\mathrm{Y}(\mathbf{U}),\mathrm{K})$ are all disjoint,
- $f\mathcal{H}_{\mathrm{K}}(\mathrm{Y}(\mathbf{U}))f$ is the algebra of endomorphisms of the finite graded $\mathrm{K}\mathbf{G}^F e$–module $\mathrm{R}\Gamma_c(\mathrm{Y}(\mathbf{U}),\mathrm{K})$.

5. Let S be a subgroup of D. We set $\mathbf{G}_S := C_{\mathbf{G}}(S)$, and $\mathbf{U}_S := \mathbf{U} \cap \mathbf{G}_S$. Let e_S be the block of $\mathbf{G}^F(S) = C_{\mathbf{G}^F}(S)$ such that $(S,e_S) \subseteq (D,f)$. Then the datum $(\mathbf{G}_S, e_S, \mathbf{L}^F, f, \mathbf{U}_S)$ owns the same properties as the datum $(\mathbf{G}, e, \mathbf{L}^F, f, \mathbf{U})$ and there exists a complex $\boldsymbol{\Upsilon}_S$ such that properties (ℓ-C$_i$) ($i = 1, \ldots, 5$) hold (with appropriate substitutions).

It is likely that $\mathrm{k}\otimes_{\mathcal{O}} \boldsymbol{\Upsilon}_S$ is the image of $\boldsymbol{\Upsilon}$ (viewed as a complex of $\mathcal{O}[\mathbf{G}^F \times \mathbf{L}^F]$–modules) through the Brauer morphism Br_S (see [**Ri**]).

The preceding considerations will probably be the starting point to explain the correspondance between e and f known as "isotypie" (cf. [**Bro1**] and [**BMM1**]).

Conjectures about Deligne–Lusztig induction

Assume now that ℓ is "large", and more precisely that

ℓ does not divide the product of the order of the Weyl group W of $\mathbf{G}$ by the order of the outer automorphism of W induced by F.

We still denote by $\mathcal{O}$ the ring of integers of a finite unramified extension K of the field of ℓ–adic numbers $\mathbb{Q}_\ell$, with residue field k, such that the finite group algebra $\mathrm{k}\mathbf{G}^F$ is split.

We denote by $e_\ell^{\mathbf{G}^F}$ the central idempotent of $\mathcal{O}\mathbf{G}^F$ associated with the subset $\mathcal{E}_\ell(\mathbf{G}^F, 1)$ of the set of irreducible characters of $\mathbf{G}^F$ (see [**BrMi1**]). We call "unipotent blocks of $\mathcal{O}\mathbf{G}^F$" the primitive central idempotents e such that $ee_\ell^{\mathbf{G}^F} \neq 0$.

Then the following hold.

• (see [**BrMa1**], 3.13) There exists a unique integer d such that the d–th cyclotomic polynomial divides the polynomial order of $\mathbf{G}^F$ and ℓ divides $\Phi_d(q)$. The ℓ–subgroups of $\mathbf{G}^F$ ar all contained, up to $\mathbf{G}^F$–conjugation, in a Sylow Φ_d–subgroup of $\mathbf{G}$.

• (see [**BMM1**]) If e is a unipotent block with maximal subpair (D, f) (where D is a defect group of e, and f is a block of $\mathcal{O}C_{\mathbf{G}^F}(D)$), then $\mathbf{L} := C_{\mathbf{G}}(D)$ is a d–split Levi subgroup of $\mathbf{G}$ (*i.e.*, $\mathbf{L} = C_{\mathbf{G}}(Z^o(\mathbf{L})_d)$), D is a Sylow ℓ–subgroup of $Z(\mathbf{L}^F)$, and the canonical character of f is a d–cuspidal unipotent character λ of $\mathbf{L}^F$. Moreover, the set of irreducible characters of $\mathcal{O}\mathbf{G}^F e$ coincides with the set irreducible constituents of $R_{\mathbf{L}}^{\mathbf{G}}(\tau\lambda)$, where τ runs over the set of characters of $\mathbf{L}^F/[\mathbf{L}, \mathbf{L}]^F$ whose order is a power of ℓ.

Restricting ourselves to unipotent characters, we see that the "ℓ–conjectures" have the following particular consequences.

More notation and data.

Let d be an integer such that Φ_d divides the polynomial order of $\mathbf{G}^F$. From now on, we assume that ℓ divides $\Phi_d(q)$.

Let $(\mathbf{L}, \lambda)$ be a d–cuspidal pair of $\mathbf{G}^F$. The character λ is a character with ℓ–defect zero of $\mathbf{L}^F/Z(\mathbf{L}^F)$. It follows that there exists a unique $\mathcal{O}$–free $\mathcal{O}\mathbf{L}^F$–module M_λ with character λ (M_λ is a projective $\mathcal{O}[\mathbf{L}^F/Z(\mathbf{L}^F)]$–module). The natural morphism $\mathcal{O}\mathbf{L}^F \to \mathrm{End}_{\mathcal{O}}(M_\lambda)$ is onto, and defines a torsion–free $\mathcal{O}$–sheaf on $\mathrm{X}(\mathbf{U})$ which we denote by $\mathcal{F}_\lambda$.

Let $\mathrm{Uch}(\mathbf{G}^F, (\mathbf{L}, \lambda))$ be the set of all (unipotent) irreducible characters γ of $\mathbf{G}^F$ such that $\left(R_{\mathbf{L}}^{\mathbf{G}}(\lambda), \gamma\right)_{\mathbf{G}^F} \neq 0$. For each $\gamma \in \mathrm{Uch}(\mathbf{G}^F, (\mathbf{L}, \lambda))$ we denote by $e_\gamma^{\mathbf{G}^F}$ the primitive central idempotent of $\mathrm{K}\mathbf{G}^F$ corresponding to γ.

We set

$$e_{(\mathbf{L},\lambda)}^{\mathbf{G}^F} := \sum_{\gamma \in \mathrm{Uch}(\mathbf{G}^F, (\mathbf{L},\lambda))} e_\gamma^{\mathbf{G}^F} .$$

d–**Conjectures.**

There exist

• a parabolic subgroup of $\mathbf{G}$ with unipotent radical $\mathbf{U}$ and Levi complement $\mathbf{L}$,

• a finite complex $\Xi = (\cdots \to \Xi^{n-1} \to \Xi^n \to \Xi^{n+1} \to \cdots)$ of finitely generated projective $\mathcal{O}\mathbf{G}^F$–modules,

with the following properties.

(d-C1) Viewed as an object of the derived bounded category of the category of $\mathcal{O}\mathbf{G}^F$–modules, Ξ is isomorphic to $\mathrm{R}\Gamma_c(\mathrm{X}(\mathbf{U}), \mathcal{F}_\lambda)$. In particular, for each n, the n-th homology group of Ξ is isomorphic, as an $\mathcal{O}\mathbf{G}^F$–module, to $\mathrm{H}_c^n(\mathrm{X}(\mathbf{U}), \mathcal{F}_\lambda)$.

(d-C2) • the structure of complex of $\mathcal{O}\mathbf{G}^F$–modules of $\Xi.f$ extends to a structure of complex of $(\mathcal{O}\mathbf{G}^F, \mathcal{H}_{\mathcal{O}}(\mathrm{X}(\mathbf{U}), \mathcal{F}_\lambda))$–bimodules, all of which are projective as right $\mathcal{H}_{\mathcal{O}}(\mathrm{X}(\mathbf{U}), \mathcal{F}_\lambda)$–modules,

• the complexes $(\Xi^\vee \otimes_{\mathcal{O}\mathbf{G}^F} \Xi)$ and $\mathcal{H}_{\mathcal{O}}(\mathrm{X}(\mathbf{U}), \mathcal{F}_\lambda)$ are homotopy equivalent as complexes of $(\mathcal{H}_{\mathcal{O}}(\mathrm{X}(\mathbf{U}), \mathcal{F}_\lambda), \mathcal{H}_{\mathcal{O}}(\mathrm{X}(\mathbf{U}), \mathcal{F}_\lambda))$–bimodules.

(d-C3) The algebra $\mathcal{H}_{\mathrm{K}}(\mathrm{X}(\mathbf{U}), \mathcal{F}_\lambda)$ is isomorphic to the algebra $\mathrm{K}W_{\mathbf{G}^F}(\mathbf{L}, \lambda)$ (recall that we set $W_{\mathbf{G}^F}(\mathbf{L}, \lambda) := N_{\mathbf{G}^F}(\mathbf{L}, \lambda)/\mathbf{L}^F$).

The following properties are consequences of the preceding conjectures.

(d-C4) We have

$$\mathrm{R}\Gamma_c(\mathrm{X}(\mathbf{U}, \mathcal{F}_\lambda; \mathrm{K}) = \bigoplus_n \mathrm{H}^n_c(\mathbf{U}, \mathcal{F}_\lambda; \mathrm{K})[-n] \,.$$

The cohomology groups $\mathrm{H}^n_c(\mathrm{X}(\mathbf{U}), \mathcal{F}_\lambda; \mathrm{K})$ are all disjoint as $\mathrm{K}\mathbf{G}^F$–modules, and the algebra $\mathcal{H}_{\mathrm{K}}(\mathrm{X}(\mathbf{U}), \mathcal{F}_\lambda) := \mathrm{K} \otimes_{\mathcal{O}} \mathcal{H}(\mathrm{X}(\mathbf{U}), \mathcal{F}_\lambda)$ is equal to the algebra of $\mathbf{G}^F$–endomorphisms of the graded module $\mathrm{R}\Gamma_c(\mathrm{X}(\mathbf{U}, \mathcal{F}_\lambda; \mathrm{K})$.

(d-C5) The preceding graded module induces an equivalence between the categories of graded modules over the algebras $\mathrm{K}\mathbf{G}^F e^{\mathbf{G}^F}_{(\mathbf{L},\lambda)}$ and $\mathrm{K}W_{\mathbf{G}^F}(\mathbf{L}, \lambda)$ respectively. In particular (see [**BMM1**], fundamental theorem) there exists an isometry

$$(*) \qquad I^{\mathbf{G}^F}_{(\mathbf{L},\lambda)} : \mathbb{Z}\mathrm{Irr}(W_{\mathbf{G}^F}(\mathbf{L}, \lambda)) \to \mathbb{Z}\mathrm{Uch}(\mathbf{G}^F, (\mathbf{L}, \lambda))$$

such that for all $\mathbf{G}$, we have

$$R^{\mathbf{G}}_{\mathbf{L}}(\lambda) = I^{\mathbf{G}^F}_{(\mathbf{L},\lambda)} \cdot \mathrm{Ind}_{\{1\}}^{W_{\mathbf{G}^F}(\mathbf{L},\lambda)}(1) \,.$$

Cyclotomic Hecke algebras conjectures

From now on, in order to settle things ready to be put in a more general setting, we denote by $\mathbb{G} = (V, Wf)$ a K–reflection datum. We assume that W acts irreducibly on V.

We assume that, for all choice of a prime number p, $\mathbb{G}$ defines to a reductive algebraic group $\mathbf{G}$ over an algebraic closure of a finite field together with a p–endomorphism F such that the fixed points group $\mathbf{G}^F$ is finite.

Let d be an integer, $d \geq 1$, and let Φ be an irreducible monic element of $\mathbb{Z}_K[x]$ which divides $x^d - 1$. Let ζ be a root of Φ in $\mathbb{C}$. In order to avoid trivialities, we assume that Φ divides the polynomial order $|\mathbb{G}|$ of $\mathbb{G}$.

Notation.

Here we use notation introduced in chap. 3 above in the section devoted to cyclotomic Hecke algebras.

We denote by $\mathrm{Uch}_\Phi(\mathbb{G})$ the set of all Φ–cuspidal unipotent characters of $\mathbb{G}$.

For a Φ–cuspidal pair $(\mathbb{L}, \lambda)$ of $\mathbb{G}$, where $\mathbb{L} = (V, W_{\mathbb{L}}wf)$ we recall that we set $W_{\mathbb{G}}(\mathbb{L}, \lambda) := N_W(\mathbb{L}, \lambda)/W_{\mathbb{L}}$ and $K_{\mathbb{G}}(\mathbb{L}, \lambda) := K[x]/\Phi(x)$, that we denote by $Z(\mathbb{L})_\Phi$ the Sylow Φ–sub-reflection datum of the center of $\mathbb{L}$, and by $V(\mathbb{L}, \Phi)$ its vector space $\ker \Phi(wf) \cap V^{W_{\mathbb{L}}}$, viewed as a vector space over the field $K[x]/\Phi(x)$ through its natural structure of $K[wf]$–module.

We recall that $W_{\mathbb{G}}(\mathbb{L}, \lambda)$ is a complex reflection group in its action on $V(\mathbb{L}, \Phi)$.

We denote by $\mathcal{A}_{\mathbb{G}}(\mathbb{L}, \lambda)$ the set of reflecting hyperplanes of $W_{\mathbb{G}}(\mathbb{L}, \lambda)$ in its action on $V(\mathbb{L}, \Phi)$.

By the Φ–Harish–Chandra theories (see above), we have a partition

$$\mathrm{Uch}(\mathbb{G}) = \coprod_{[(\mathbb{L},\lambda)]} \mathrm{Uch}(\mathbb{G}; (\mathbb{L}, \lambda))$$

indexed by the orbits of W on the set of all Φ–cuspidal pairs of $\mathbb{G}$, and a pair of maps

$$\begin{cases} \mathrm{Irr}(W_{\mathbb{G}}(\mathbb{L},\lambda)) \longrightarrow \mathrm{Uch}(\mathbb{G};(\mathbb{L},\lambda)) \,,\; \chi \mapsto \gamma_\chi \\ \mathrm{Irr}(W_{\mathbb{G}}(\mathbb{L},\lambda)) \longrightarrow \{\pm 1\} \,,\; \chi \mapsto \varepsilon_\chi \end{cases}$$

satisfying the following properties :

For each $\gamma \in \mathrm{Uch}(\mathbb{G})$, let us denote by a_γ and A_γ respectively the valuation and the degree (in x) of $\mathrm{Deg}_\gamma(x)$.

For each orbit $\mathcal{C}$ of $W_{\mathbb{G}}(\mathbb{L},\lambda)$ on its set $\mathcal{A}_{\mathbb{G}}(\mathbb{L},\lambda)$ of reflecting hyperplanes, let $\det_{\mathcal{C}}^j$ be the j-th power of the character $\det_{\mathcal{C}}$ of $W_{\mathbb{G}}(\mathbb{L},\lambda)$ such that

$$\det_{\mathcal{C}}(s_H) = \begin{cases} \det_{V(\mathbb{L},\Phi)}(s_H) & \text{if } H \in \mathcal{C}\,, \\ 1 \quad \text{if } H \notin \mathcal{C}\,. \end{cases}$$

Let us denote by $\gamma_{\mathcal{C},j}$ the element of $\mathrm{Uch}(\mathbb{G})$ which corresponds to $\det_{\mathcal{C}}^j$, and by $a_{\mathcal{C},j}$ and $A_{\mathcal{C},j}$ respectively the valuation and the degree (in x) of $\mathrm{Deg}(\gamma_{\mathcal{C},j})$.

There exist

- an element $\gamma_0 \in \mathrm{Uch}(\mathbb{G};\, (\mathbb{L},\lambda)))$ such that, for $a_0 := a_{\gamma_0}$ and $A_0 := A_{\gamma_0}$, we have (for all $\gamma \in \mathrm{Uch}(\mathbb{G};\, (\mathbb{L},\lambda)))$:

$$(a_0 + A_0) \le (a_\gamma + A_\gamma)\,,$$

- and (for each $\mathcal{C} \in \mathcal{A}_{\mathbb{G}}(\mathbb{L},\lambda))$ an integer $m_{\mathcal{C}} > 0$ allowing to define the rational numbers $m_{\mathcal{C},j}$ by the equalities

$$(m_{\mathcal{C}} - m_{\mathcal{C},j})N_{\mathcal{C}}e_{\mathcal{C}} = (a_{\mathcal{C},j} + A_{\mathcal{C},j}) - (a_0 + A_0)\,,$$

such that the cyclotomic Hecke algebra $\mathcal{H}(\mathbb{G};(\mathbb{L},\lambda))$ defined by the specialization

$$\mathcal{S}_{(\mathbb{G};(\mathbb{L},\lambda))} : u_{\mathcal{C},j} \mapsto \zeta_{e_{\mathcal{C}}}^j (\zeta^{-1}x)^{m_{\mathcal{C},j}}$$

satisfies the properties (U.5.3.j) below.

Let us first choose a suitable extension $K_{\mathbb{G}}(\mathbb{L},\lambda)(y)$ of the rational fraction field $K_{\mathbb{G}}(\mathbb{L},\lambda)(x)$ so that it induces a bijection

$$\mathrm{Irr}(W_{\mathbb{G}}(\mathbb{L},\lambda)) \xrightarrow{\sim} \mathrm{Irr}(\mathcal{H}(\mathbb{G};(\mathbb{L},\lambda))) \quad , \quad \chi \mapsto \chi_x\,.$$

It follows that we have a bijection

$$\mathrm{Irr}(\mathcal{H}(\mathbb{G};(\mathbb{L},\lambda))) \xrightarrow{\sim} \mathrm{Uch}(\mathbb{G};\, (\mathbb{L},\lambda)) \quad , \quad \psi \mapsto \gamma_\psi\,.$$

Let us call θ_0 the element of $\mathrm{Irr}(\mathcal{H}(\mathbb{G};(\mathbb{L},\lambda)))$ which corresponds to γ_0.

(U5.3.1) The character θ_0 is a linear character, and for all $\mathcal{C} \in \mathcal{A}_{\mathbb{G}}(\mathbb{L},\lambda)$, there exists and integer $j_{\mathcal{C}}$ such that we have

$$\theta_0(s_{\mathcal{C}}) = \zeta_{e_{\mathcal{C}}}^{j_{\mathcal{C}}}(\zeta^{-1}x)^{m_{\mathcal{C}}}\,.$$

Assume moreover that $\mathbb{L}$ is minimal (as a Φ–split Levi sub–reflection datum of $\mathbb{G}$) and that $\lambda = 1^{\mathbb{L}}$. Then

- the algebra $\mathcal{H}(\mathbb{G};(\mathbb{L},\lambda))$ is principal and θ_0 is a principal character,
- we have $\gamma_{\theta_0} = 1^{\mathbb{G}}$.

Assume moreover that the polynomial Φ is regular for $\mathbb{G}$. Then $\mathbb{L}$ is a (maximal) torus $\mathbb{T}$. We set

$$\mathcal{H}_{\mathbb{G}}(\mathbb{T}) := \mathcal{H}(\mathbb{G};(\mathbb{L},1))\,,\; W_{\mathbb{G}}(\mathbb{T}) := W_{\mathbb{G}}(\mathbb{L},1)\,,\; \mathcal{A}_{\mathbb{G}}(\mathbb{T}) := \mathcal{A}_{\mathbb{G}}(\mathbb{L},1)\,.$$

Then we have

$$N_{\mathbb{G}} + N_{\mathbb{G}}^{\vee} = \sum_{\mathcal{C}\in\mathcal{A}_{\mathbb{G}}(\mathbb{T})} N_{\mathcal{C}} e_{\mathcal{C}} m_{\mathcal{C}} .$$

(U5.3.2) For $\chi \in \mathrm{Irr}(W_{\mathbb{G}}(\mathbb{L},\lambda))$, let us denote by $S_\chi(x)$ the Schur element of χ_x, so that we have $t_\phi = \sum_{\chi\in\mathrm{Irr}(W_{\mathbb{G}}(\mathbb{L},\lambda))} \frac{1}{S_\chi(x)}\chi_x$, where t_ϕ is the specialisation of the canonical symmetrizing form $t_{\mathbf{u}}$ (see chap. III above). Then for all $\chi \in \mathrm{Irr}(W_{\mathbb{G}}(\mathbb{L},\lambda))$, we have

$$\varepsilon_\chi \mathrm{Deg}_{\mathbb{G}}(\gamma_\chi)(x) = \mathrm{Deg}_{\mathbb{L}}(\lambda)\mathrm{Deg}_{\mathbb{G}}(\mathrm{Ind}_{\mathbb{L}}^{\mathbb{G}} 1^{\mathbb{L}})\frac{1}{S_\chi(x)} .$$

Recall that $\mathrm{Deg}_{\mathbb{G}}(\mathrm{Ind}_{\mathbb{L}}^{\mathbb{G}} 1^{\mathbb{L}}) = \dfrac{|\mathbb{G}|/(\varepsilon_{\mathbb{G}} x^{N(\mathbb{G})})}{|\mathbb{L}|/(\varepsilon_{\mathbb{L}} x^{N(\mathbb{L})})}$.

(U5.3.3) Assume that ζ_d is a root of $\Phi(x)$.

The element $(wf)^{\delta_{\mathbb{G}}}$ belongs to the center of $W_{\mathbb{G}}(\mathbb{L},\lambda)$. Assume that it has order d, and that the center $Z(W_{\mathbb{G}}(\mathbb{L},\lambda))$ has order z.

Let $B_{\mathbb{G}}(\mathbb{L},\lambda)$ be a braid group associated with $W_{\mathbb{G}}(\mathbb{L},\lambda)$ (defined up to the choice of a base point), and let $P_{\mathbb{G}}(\mathbb{L},\lambda)$ be the corresponding pure braid group. We denote by $\boldsymbol{\pi}$ the positive generator of the center $Z(P_{\mathbb{G}}(\mathbb{L},\lambda))$ of the pure braid group, and by $\boldsymbol{\beta}$ the positive generator of the center $Z(B_{\mathbb{G}}(\mathbb{L},\lambda))$ of the braid group, so that we have $\boldsymbol{\beta}^z = \boldsymbol{\pi}$.

- There exists a choice of the base point such that $(wf)^{\delta_{\mathbb{G}}}$ is the image of the element $\boldsymbol{\beta}^{z/d}$.
- For all $\chi \in \mathrm{Irr}(W_{\mathbb{G}}(\mathbb{L},\lambda))$, we have

$$\frac{\varphi_{\mathbb{G}}(\gamma_\chi)}{\varphi_{\mathbb{L}}(\lambda)} x^{\frac{D_0-((a_{\gamma_\chi}+A_{\gamma_\chi})-(a_0+A_0))}{d}} = \omega_{\chi_x}(\boldsymbol{\beta}^{z/d}) ,$$

where we set (see above chap. III, notation before 3.46)

$$D_0 := \sum_{\mathcal{C}\in\mathcal{A}_{\mathbb{G}}(\mathbb{L},\lambda)} m_{\mathcal{C}} N_{\mathcal{C}} e_{\mathcal{C}} .$$

The following property is now a consequence of 3.46 above.

6.1. PROPOSITION. *Assume as above that ζ_d is a root of $\Phi(x)$. For $\chi \in \mathrm{Irr}(W_{\mathbb{G}}(\mathbb{L},\lambda))$, we have*

$$\varphi_{\mathbb{G}}(\gamma_\chi) = \varphi_{\mathbb{L}}(\lambda)\exp\left(-2i\pi\frac{D_0-((a_{\gamma_\chi}+A_{\gamma_\chi})-(a_0+A_0))}{d^2}\right)\omega_\chi((wf)^\delta) .$$

(U5.3.4) Let $\mathbb{M}$ be a sub-reflection datum of $\mathbb{G}$ containing $\mathbb{L}$. Then the specialisation $\mathcal{S}_{(\mathbb{M};(\mathbb{L},\lambda))}$ is the restriction of the specialisation $\mathcal{S}_{(\mathbb{G};(\mathbb{L},\lambda))}$: whenever s is a distinguished pseudo–reflection in $W_{\mathbb{M}}(\mathbb{L},\lambda)$ and $0 \le j \le e_s - 1$, we have $\mathcal{S}_{(\mathbb{M};(\mathbb{L},\lambda))}(u_{s,j}) = \mathcal{S}_{(\mathbb{G};(\mathbb{L},\lambda))}(u_{s,j})$.

In particular, the Hecke algebra $\mathcal{H}(\mathbb{M};(\mathbb{L},\lambda))$ is naturally identified with a parabolic subalgebra of $\mathcal{H}(\mathbb{G};(\mathbb{L},\lambda))$.

Connection with the actual finite groups.

For any suitable element ξ, we denote by $\mathcal{H}(\mathbf{G}^F, (\mathbf{L}, \lambda); \xi)$ the $\mathbb{Z}_K[\xi, \xi^{-1}]$–algebra defined by the specialization $x \mapsto \xi$.

Let q be a power of a prime number p and let $(\mathbf{G}, F)$ ($\mathbf{G}$ a connected reductive algebraic group over $\overline{\mathbb{F}}_p$, F a p–endomorphism of $\mathbf{G}$ such that $\mathbf{G}^F$ is finite) be defined by $\mathbb{G}$ and q.

Let ℓ be a prime number which does not divide $|W\langle f\rangle|$ and which divides $\Phi(q)$.

Let $\mathcal{O}$ be a suitable extension of $\mathbb{Z}_\ell[\zeta, q, q^{-1}]$. We use notation introduced in d–Conjectures above — with suitable extensions of scalars.

(HC) the morphism $\mathbb{Z}_K[x, x^{-1}] \to \mathcal{O}$ defined by $x \mapsto q$ induces an $\mathcal{O}$–algebra isomorphism

$$\mathcal{O} \otimes \mathcal{H}(\mathbb{G}, (\mathbb{L}, \lambda); q) \xrightarrow{\sim} \mathcal{H}_{\mathcal{O}}(\mathrm{X}(\mathbf{U}), \mathcal{F}_\lambda)$$

which send the canonical symmetrizing form t_x onto tr_Ξ.

More precise conjectures for roots of $\boldsymbol{\pi}$

We shall present here more precise conjectures for the case where the cuspidal pair $(\mathbb{L}, \lambda)$ considered above is of the shape $(\mathbb{T}, 1)$, where $\mathbb{T}$ is a (maximal) torus.

For the simplicity of the exposition, we restrict ourselves to the split case (*i.e.*, where $\mathbb{G} = (V, W)$).

In this case, the variety $X(\mathbf{U})$ is isomorphic to the Deligne–Lusztig variety X_w for an element w in the conjugacy class of W defined by $\mathbb{T}$. The complex denoted above by $\mathrm{R}\Gamma_c(X(\mathbf{U}), \mathcal{F}_\lambda)$ is nothing but $\mathrm{R}\Gamma_c(X_w, \mathbb{Z}_\ell)$.

The "good choice" of $\mathbf{U}$ mentioned in the above conjectures amounts to a good choice of w in its conjugacy class. We shall see that this choice has to do with the roots of $\boldsymbol{\pi}$ in the corresponding braid monoid (see above 2.23 for the occurrence of d–th roots of $\boldsymbol{\pi}$ in B), and that we have indeed some actions of braids on the corresponding Deligne–Lusztig varieties.

The variety of Borel subgroups and the braid group.

Prerequisites.

Let $\mathbf{G}$ be a reductive connected algebraic group over an algebraically closed field. As in [**DeLu**], we consider *its* Weyl group W endowed with *its* fundamental chamber and hence with *its* set of distinguished generators S.

We denote by $\mathbf{B}$ the associated braid group, well defined by the choice of a base point in the fundamental chamber. We recall that there exists a set $\mathbf{S}$ and a bijection $S \xrightarrow{\sim} \mathbf{S}$ such that

- if $\mathbf{s} \in \mathbf{S}$ corresponds to $s \in S$, then $\mathbf{s}$ is an s–distinguished braid reflection (see 2.13 above),
- $\mathbf{S}$, together with the braid relations on S, defines a presentation of $\mathbf{B}$.

We denote by $\mathbf{B}^+$ the sub–monoid of $\mathbf{B}$ generated by $\mathbf{S}$.

We have an injective map $W \hookrightarrow \mathbf{B}^+$ defined as follows : for $s_1 s_2 \cdots s_n$ a reduced decomposition of $w \in W$, we set

$$\mathbf{w} := \mathbf{s}_1 \mathbf{s}_2 \cdots \mathbf{s}_n \,.$$

We denote by $\mathbf{B}^+_{\mathrm{red}}$ the image of W under the preceding map.

Varieties of Borel subgroups and a theorem of Deligne.

Let $\mathcal{B}$ the variety of Borel subgroups of $\mathbf{G}$, a smooth projective homogeneous space for $\mathbf{G}$. Then (see for example [**DeLu**]), we have a natural bijection

$$(*) \qquad \mathcal{B} \times \mathcal{B}/\mathbf{G} \xrightarrow{\sim} W\,.$$

Moreover, for each choice of $B \in \mathcal{B}$, we have isomorphisms

$$\mathcal{B} \xrightarrow{\sim} \mathbf{G}/B \quad \text{and} \quad W \xrightarrow{\sim} B\backslash\mathbf{G}/B\,.$$

Let (B, B') be a pair of Borel subgroups. We say that B and B' are in relative position w (for $w \in W$) and we write $B \overset{w}{\to} B'$ if the orbit of (B, B') under $\mathbf{G}$ corresponds to the element w through the bijection (*) above.

We denote bu $\mathcal{O}(w)$ the variety of such pairs endowed with the left and right projections

$$\lambda\colon \mathcal{O}(w) \twoheadrightarrow \mathcal{B}\,,\ (B, B') \mapsto B \quad \text{and} \quad \rho\colon \mathcal{O}(w) \twoheadrightarrow \mathcal{B}\,,\ (B, B') \mapsto B'\,.$$

We define the composition

$$\mathcal{O}(w)\cdot\mathcal{O}(w') := \mathcal{O}(w) \underset{\mathcal{B}}{\times} \mathcal{O}(w')\,.$$

Whenever lengths of w and w' are additive, *i.e.*, $\ell(ww') = \ell(w) + \ell(w')$, Bruhat decomposition gives us an isomorphism

$$\mathcal{O}(ww') \xrightarrow{\sim} \mathcal{O}(w)\cdot\mathcal{O}(w')\,.$$

The main result of [**De3**] shows that one can extend the map $w \mapsto \mathcal{O}(w)$ (or, in other words, the map $\mathbf{w} \mapsto \mathcal{O}(\mathbf{w})$) to a construction which, to any $b \in \mathbf{B}^+$, associates a scheme $\mathcal{O}(b)$ on $\mathcal{B} \times \mathcal{B}$, defined up to a *unique* isomorphism, such that

$$\mathcal{O}(bb') = \mathcal{O}(b)\cdot\mathcal{O}(b') \quad \text{whenever } b, b' \in \mathbf{B}^+\,.$$

If $b = \mathbf{w}_1\mathbf{w}_2\cdots\mathbf{w}_n$, where $\mathbf{w}_j \in \mathbf{B}^+_{\text{red}}$ for $1 \le j \le n$, then we may think of $\mathcal{O}(b)$ as defined by

$$\mathcal{O}(b) = \{(B_0, B_1, \ldots, B_n) \mid (B_j \in \mathcal{B} \text{ for } 0 \le j \le n)(B_{j-1} \overset{w_j}{\to} B_j \text{ for } 1 \le j \le n)\}\,.$$

Deligne–Lusztig generalized varieties and actions of braids.

From now on, we follow closely [**BrMi2**], §2.

We consider again the case where $\mathbf{G}$ is a connected reductive algebraic group defined over an algebraic closure of the prime field with p elements, endowed with a p-endomorphism $F\colon \mathbf{G} \longrightarrow \mathbf{G}$. In order to simplify the exposition, we assume that $(\mathbf{G}, F)$ is *split*, *i.e.*, that the automorphism of W induced by F is the identity.

We let F act on the right on $\mathcal{B}$, by setting

$$B \cdot F := F(B)\,.$$

The following definition has been inspired by the variety described by Lusztig in [**Lu0**], page 25.

DEFINITION. Let $b \in \mathbf{B}^+$. The Deligne–Lusztig variety associated with b is the part of $\mathcal{O}(b)$ lying over the graph of F :

$$X_b^{(F)} := \{x \in \mathcal{O}(b) \mid \lambda(x) = \rho(x) \cdot F\}.$$

When there is no ambiguity, we set $X_b = X_b^{(F)}$.

In other words, if $b = \mathbf{w}_1\mathbf{w}_2 \cdots \mathbf{w}_n$, we have

$$X_b^{(F)} = \{(B_0, B_1, \ldots, B_n) \mid (B_{j-1} \overset{w_j}{\to} B_j \text{ for } 1 \leq j \leq n) \text{ and } (B_n = B_0 \cdot F)\}.$$

DEFINITION. Let $b, b' \in \mathbf{B}^+$. We define an equivalence of étale sites

$$D_b^{(bb')} \colon X_{bb'} \longrightarrow X_{b'b}$$

as follows.

- We may think of an element of $X(bb')$ as of the shape (x, x') where $x \in \mathcal{O}(b)$, $x' \in \mathcal{O}(b')$, $\rho(x) = \lambda(x')$, $\rho(x') = \lambda(x) \cdot F$.
- We then set :

$$(x, x') \cdot D_b^{(bb')} := (x', x \cdot F), \tag{6.2}$$

since we have $(x', x \cdot F) \in X_{b'b}$.

The following properties are immediate :

6.3. LEMMA.
(1) *For $b, b', c \in \mathbf{B}^+$, we have*

$$D_b^{(bb'c)} \cdot D_{b'}^{(b'cb)} = D_{bb'}^{(bb'c)} \colon X_{bb'c} \longrightarrow X_{cbb'}.$$

(2) *We have $D_b^{(b)} = F \colon X_b \longrightarrow X_b$.*

Varieties associated with roots of $\boldsymbol{\pi}$.

The variety $X_{\boldsymbol{\pi}}$.

Since $\boldsymbol{\pi} = \mathbf{w}_0^2$, any element $\mathbf{s}$ of $\mathbf{S}$ is a left divisor of $\boldsymbol{\pi}$ in $\mathbf{B}^+_{\mathrm{red}}$: there exists $\boldsymbol{\pi}_{\mathbf{s}} \in \mathbf{B}^+_{\mathrm{red}}$ such that $\boldsymbol{\pi} = \mathbf{s}\boldsymbol{\pi}_{\mathbf{s}}$.

Since $\boldsymbol{\pi}$ is central in the braid group, the definition 6.2 shows that $D_{\mathbf{s}}^{(\boldsymbol{\pi})}$ is an automorphism of étale site of $X_{\boldsymbol{\pi}}$. Then lemma 6.3 shows that the operators $(D_{\mathbf{s}}^{(\boldsymbol{\pi})})_{\mathbf{s} \in \mathbf{S}}$ satisfy the braid relations, hence we get

6.4. PROPOSITION. *The map $\mathbf{s} \mapsto D_{\mathbf{s}}^{(\boldsymbol{\pi})}$ extends to a group morphism*

$$B \mapsto \mathrm{Aut}(\mathrm{R}\Gamma_c(X_{\boldsymbol{\pi}}, \mathbb{Z}_\ell)).$$

Some arguments in favor of the following conjectures may be found in [**BrMi2**], §2.B, and in [**DMR**].

Let

$$\mathrm{R}\Gamma(X_{\boldsymbol{\pi}}, \overline{\mathbb{Q}}_\ell) := \bigoplus_{n=0}^{n=2N} \mathrm{H}^n(X_{\boldsymbol{\pi}}, \overline{\mathbb{Q}}_\ell)$$

be the graded module of the ℓ-adic cohomology of $X_{\boldsymbol{\pi}}$, seen as a graded $\overline{\mathbb{Q}}_\ell \mathbf{G}^F$-module.

Let $\overline{\mathbb{Q}}_\ell\mathcal{H}_q(W)$ denote the ordinary Hecke algebra, *i.e.*, the algebra of $\overline{\mathbb{Q}}_\ell\mathbf{G}^F$-endomorphisms of the module $\overline{\mathbb{Q}}_\ell[\mathbf{G}^F/B_0^F]$, where B_0 denotes an F–stable Borel subgroup of $\mathbf{G}^F$.

6.5. CONJECTURES.

(1) *The operators D_b $(b \in B)$ generate the algebra of $\overline{\mathbb{Q}}_\ell\mathbf{G}^F$-endomorphisms of* $\mathrm{R\Gamma}(X_{\boldsymbol{\pi}}, \overline{\mathbb{Q}}_\ell)$.

(2) *The morphism $b \mapsto D_b$ factorizes to an isomorphism*

$$\overline{\mathbb{Q}}_\ell\mathcal{H}_q(W)^{\mathrm{op}} \xrightarrow{\sim} \mathrm{End}_{\overline{\mathbb{Q}}_\ell\mathbf{G}^F}(\mathrm{R\Gamma}_c(X_{\boldsymbol{\pi}}, \overline{\mathbb{Q}}_\ell))$$

which defines a graded version of the representation of $\overline{\mathbb{Q}}_\ell\mathcal{H}_q(W)$ on the module $\overline{\mathbb{Q}}_\ell[\mathbf{G}^F/B_0^F]$.

(3) *The graded module $\mathrm{R\Gamma}(X_{\boldsymbol{\pi}}, \overline{\mathbb{Q}}_\ell)$ has only even degrees components. Two distinct (even degrees) components are disjoint as $\overline{\mathbb{Q}}_\ell\mathbf{G}^F$–modules, i.e.,*

$$\mathrm{Hom}_{\overline{\mathbb{Q}}_\ell\mathbf{G}^F}(\mathrm{R\Gamma}(X_{\boldsymbol{\pi}}, \overline{\mathbb{Q}}_\ell), \mathrm{R\Gamma}_c(X_{\boldsymbol{\pi}}, \overline{\mathbb{Q}}_\ell)[n]) = 0 \quad \textit{if} \quad n \neq 0\,.$$

(4) *Given $\gamma \in \mathrm{Uch}(\mathbb{G})$, the degree n such that $\psi_q^{\mathbb{G}}(\gamma)$ is a constituent of the $\overline{\mathbb{Q}}_\ell\mathbf{G}^F$-module $\mathrm{H}_c^n((X_{\boldsymbol{\pi}}, \overline{\mathbb{Q}}_\ell)$ does not depend on q (hence is "generic").*

The preceding construction and conjectures may be generalized to varieties associated to roots of $\boldsymbol{\pi}$.

The varieties $X_{\mathbf{w}}$ for $\mathbf{w}$ a root of $\boldsymbol{\pi}$.

Let $d \geq 1$ be an integer, and let $\mathbf{w} \in \mathbf{B}^+_{\mathrm{red}}$ such that $\mathbf{w}^d = \boldsymbol{\pi}$.

The following proposition is essentially due to [**BrMi2**], §5. Its proof is immediate.

6.6. PROPOSITION.

(1) *The map*

$$\begin{cases} X_{\mathbf{w}}^{(F)} \longrightarrow X_{\boldsymbol{\pi}}^{(F^d)} \\ x \mapsto (\lambda(x), \lambda(x).F, \lambda(x).F^2, \ldots, \lambda(x).F^d) \end{cases}$$

is an embedding, which identifies $X_{\mathbf{w}}^{(F)}$ with the closed subvariety $X_{\mathbf{w}}^{(F,d)}$ of $X_{\boldsymbol{\pi}}^{(F^d)}$ defined by

$$X_{\mathbf{w}}^{(F,d)} := \{y \in X_{\boldsymbol{\pi}}^{(F^d)} \mid (y \cdot F = y \cdot D_{\mathbf{w}}^{(\boldsymbol{\pi})})\}\,.$$

(2) *The restriction $D_b^{(\boldsymbol{\pi})}) \mapsto D_b^{(\boldsymbol{\pi})})|_{X_{\mathbf{w}}^{(F,d)}}$ defines an operation of the centralizer $C_{\mathbf{B}^+}(\mathbf{w})$ of $\mathbf{w}$ in $\mathbf{B}^+$.*

Arguments in favor of the following conjectures may be found, for example, in [**Lu2**], [**BrMi2**] (§5), and [**DMR**].

For the first assertion below, see above 2.26.

6.7. CONJECTURES.

(0) *The monoid $C_{\mathbf{B}^+}(\mathbf{w})$ generates the centralizer $C_B(\mathbf{w})$ of $\mathbf{w}$ in the braid group, and the natural morphism $B(w) \mapsto C_B(\mathbf{w})$ is an isomorphism.*

(1) *The operators $D_b^{(\mathbf{w})}$ $(b \in B(\mathbf{w}))$ generate the algebra of $\overline{\mathbb{Q}}_\ell\mathbf{G}^F$-endomorphisms of $\mathrm{R\Gamma}(X_{\boldsymbol{\pi}}, \overline{\mathbb{Q}}_\ell)$.*

(2) *There exists a ζ_d-cyclotomic specialization ϕ of the Hecke algebra of the group $W(w)$ such that the morphism $b \mapsto D_b^{(\mathbf{w})}$ factorizes to an isomorphism*

$$\overline{\mathbb{Q}}_\ell \mathcal{H}_\phi(W(w))^{\mathrm{op}} \xrightarrow{\sim} \mathrm{End}_{\overline{\mathbb{Q}}_\ell \mathbf{G}^F}(\mathrm{R}\Gamma_c(X_{\mathbf{w}}, \overline{\mathbb{Q}}_\ell)) .$$

(3) *Two distinct components are disjoint as $\overline{\mathbb{Q}}_\ell \mathbf{G}^F$-modules, i.e.,*

$$\mathrm{Hom}_{\overline{\mathbb{Q}}_\ell \mathbf{G}^F}(\mathrm{R}\Gamma(X_{\mathbf{w}}, \overline{\mathbb{Q}}_\ell), \mathrm{R}\Gamma_c(X_{\mathbf{w}}, \overline{\mathbb{Q}}_\ell)[n]) = 0 \quad \text{if} \quad n \neq 0 .$$

(4) *Given $\gamma \in \mathrm{Uch}(\mathbb{G}, \mathbb{T}_w)$, the degree n such that $\psi_q^{\mathbb{G}}(\gamma)$ is a constituent of $\mathrm{H}_c^n((X_{\mathbf{w}}, \overline{\mathbb{Q}}_\ell)$ does not depend on q (hence is "generic").*

APPENDIX 1

DIAGRAMS AND TABLES

Here are some definitions, notation, conventions, which will allow the reader to understand the diagrams.

The groups have presentations given by diagrams $\mathcal{D}$ such that

- the nodes correspond to pseudo-reflections in W, the order of which is given inside the circle representing the node,
- two distinct nodes which do not commute are related by "homogeneous" relations with the same "support" (of cardinality 2 or 3), which are represented by links beween two or three nodes, or circles between three nodes, weighted with a number representing the degree of the relation (as in Coxeter diagrams, 3 is omitted, 4 is represented by a double line, 6 is represented by a triple line). These homogeneous relations are called the *braid relations* of $\mathcal{D}$.

More details are provided below.

Meaning of the diagrams.

This paragraph provides a list of examples which illustrate the way in which diagrams provide presentations for the attached groups.

- The diagram (d)—e—(d) with nodes s, t corresponds to the presentation

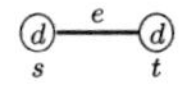

$$s^d = t^d = 1 \text{ and } \underbrace{stst\cdots}_{e \text{ factors}} = \underbrace{tstst\cdots}_{e \text{ factors}}$$

- The diagram (5)═(3) with nodes s, t corresponds to the presentation

$$s^5 = t^3 = 1 \text{ and } stst = tsts\,.$$

- The diagram s(a), (b)t, (c)u around a circle labelled e corresponds to the presentation

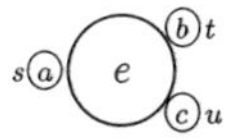

$$s^a = t^b = u^c = 1 \text{ and } \underbrace{stustu\cdots}_{e \text{ factors}} = \underbrace{tustus\cdots}_{e \text{ factors}} = \underbrace{ustust\cdots}_{e \text{ factors}}\,.$$

- The diagram with nodes (2) s, (2) u joined by a circle to (2) t, and the line (2) v — (2) t — (2) w, with s linked to v and u linked to w, corresponds to the presentation

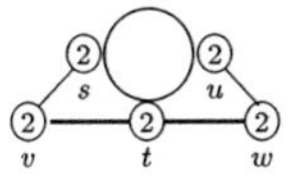

$$\begin{aligned}
&s^2 = t^2 = u^2 = v^2 = w^2 = 1\,,\\
&uv = vu\,,\ sw = ws\,,\ vw = wv\,,\\
&sut = uts = tsu\,,\\
&svs = vsv\,,\ tvt = vtv\,,\ twt = wtw\,, wuw = uwu\,.
\end{aligned}$$

- The diagram 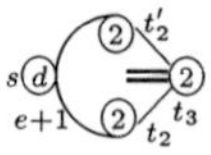 corresponds to the presentation

$$\begin{aligned}
&s^d = {t'_2}^2 = t_2^2 = t_3^2 = 1\,,\ st_3 = t_3 s\,,\\
&st'_2t_2 = t'_2t_2s\,,\\
&t'_2t_3t'_2 = t_3t'_2t_3\,,\ t_2t_3t_2 = t_3t_2t_3\,,\ t_3t'_2t_2t_3t'_2t_2 = t'_2t_2t_3t'_2t_2t_3\,,\\
&\underbrace{t_2st'_2t_2t'_2t_2t'_2\cdots}_{e+1 \text{ factors}} = \underbrace{st'_2t_2t'_2t_2t'_2t_2\cdots}_{e+1 \text{ factors}}\,.
\end{aligned}$$

- The diagram 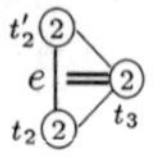 corresponds to the presentation

$$\begin{aligned}
&{t'_2}^2 = t_2^2 = t_3^2 = 1\,,\\
&t'_2t_3t'_2 = t_3t'_2t_3\,,\ t_2t_3t_2 = t_3t_2t_3\,,\ t_3t'_2t_2t_3t'_2t_2 = t'_2t_2t_3t'_2t_2t_3\,,\\
&\underbrace{t_2t'_2t_2t'_2t_2t'_2\cdots}_{e \text{ factors}} = \underbrace{t'_2t_2t'_2t_2t'_2t_2\cdots}_{e \text{ factors}}\,.
\end{aligned}$$

- The diagram 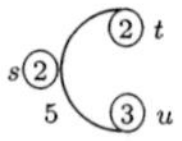 corresponds to the presentation

$$s^2 = t^2 = u^3 = 1\,,\ stu = tus\,,\ ustut = stutu\,.$$

- The diagram 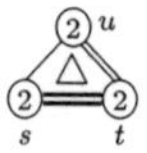 corresponds to the presentation

$$s^2 = t^2 = u^2 = 1\,,\ stst = tsts\,,\ tutu = utut\,,\ utusut = sutusu\,,\ sus = usu\,.$$

- The diagram 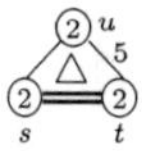 corresponds to the presentation

$$s^2 = t^2 = u^2 = 1\,,\ stst = tsts\,,\ tutut = ututu\,,\ utusut = sutusu\,,\ sus = usu\,.$$

- The diagram 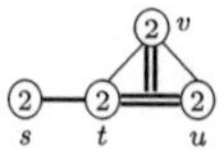 corresponds to the presentation

$$\begin{aligned}
&s^2 = t^2 = u^2 = v^2 = 1\,,\ sv = vs\,,\ su = us\,,\\
&sts = tst\,,\ vtv = tvt\,,\ uvu = vuv\,,\ tutu = utut\,,\ vtuvtu = tuvtuv\,.
\end{aligned}$$

- The diagram 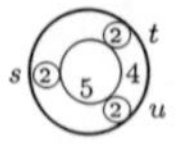 corresponds to the presentation

$$s^2 = t^2 = u^2 = 1\,,\ ustus = stust\,,\ tust = ustu\,.$$

- The diagram 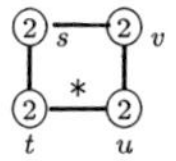 corresponds to the presentation

$$s^2 = t^2 = u^2 = v^2 = 1\,,\; su = us\,,\; tv = vt\,,$$
$$sts = tst\,,\; tut = utu\,,\; uvu = vuv\,,\; vsv = svs\,,\; stuvstuvs = tuvstuvst\,.$$

In the following tables, we denote by $H \rtimes K$ a group which is a non-trivial split extension of K by H. We denote by $H \cdot K$ a group which is a non-split extension of K by H. We denote by p^n an elementary abelian group of order p^n.

A diagram where the orders of the nodes are "forgotten" and where only the braid relations are kept is called a *braid diagram* for the corresponding group.

The groups have been ordered by their diagrams, by collecting groups with the same braid diagram. Thus, for example,

- G_{15} has the same braid diagram as the groups $G(4d,4,2)$ for all $d \geq 2$,
- G_4, G_8, G_{16}, G_{25}, G_{32} all have the same braid diagrams as groups $\mathfrak{S}_3$, $\mathfrak{S}_4$ and $\mathfrak{S}_5$,
- G_5, G_{10}, G_{18} have the same braid diagram as the groups $G(d,1,2)$ for all $d \geq 2$,
- G_7, G_{11}, G_{19} have the same braid diagram as the groups $G(2d,2,2)$ for all $d \geq 2$,
- G_{26} has the same braid diagram as $G(d,1,3)$ for $d \geq 2$.

The element β (generator of $Z(W)$) is given in the last column of our tables. Notice that the knowledge of degrees and codegrees allows then to find the order of $Z(W)$, which is not explicitely provided in the tables.

The tables provide diagrams and data for all irreducible reflection groups.

- Tables 1 and 2 collect groups corresponding to infinite families of braid diagrams,
- Table 3 collects groups corresponding to exceptional braid diagrams (notice that the fact that the diagram for G_{31} provides a braid diagram is only conjectural), but G_{24}, G_{27}, G_{29}, G_{33}, G_{34},
- The last table (table 4) provides diagrams for the remaining cases (G_{24}, G_{27}, G_{29}, G_{33}, G_{34}). It is not known nor conjectural whether these diagrams provide braid diagrams for the corresponding braid groups.

Degrees and codegrees of a braid diagram.

The following property may be noticed on the tables. It generalizes a property already noticed by Orlik and Solomon for the case of Coxeter–Shephard groups (see [**OrSo3**], (3.7)).

7.1. Theorem. *Let $\mathcal{D}$ be a braid diagram of rank r. There exist two families*

$$(\mathbf{d}_1, \mathbf{d}_2, \ldots, \mathbf{d}_r) \quad \textit{and} \quad (\mathbf{d}_1^\vee, \mathbf{d}_2^\vee, \ldots, \mathbf{d}_r^\vee)$$

of r integers, depending only on $\mathcal{D}$, and called respectively the degrees and the codegrees of $\mathcal{D}$, with the following property: whenever W is a complex reflection *group with $\mathcal{D}$ as a braid diagram, its degrees and codegrees are given by the formulae*

$$d_j = |Z(W)|\mathbf{d}_j \quad \textit{and} \quad d_j^\vee = |Z(W)|\mathbf{d}_j^\vee \quad (j = 1, 2, \ldots, r).$$

The zeta function of a braid diagram.

In [**DeLo**], Denef and Loeser compute the *zeta function of local monodromy of the discriminant* of a complex reflection group W, which is the element of $\mathbb{Q}[q]$ defined by the formula

$$Z(q, W) := \prod_j \det(1 - q\mu, H^j(F_0, \mathbb{C}))^{(-1)^{j+1}},$$

where F_0 denotes the Milnor fiber of the discriminant at 0 and μ denotes the monodromy automorphism (see [**DeLo**]).

Putting together the tables of [**DeLo**] and our braid diagrams, one may notice the following fact.

7.2. Theorem. *The zeta function of local monodromy of the discriminant of a complex reflection group W depends only on the braid diagram of W.*

Remark. Two different braid diagrams may be associated to isomorphic braid groups. For example, this is the case for the following rank 2 diagrams (where the sign "$\sim$" means that the corresponding groups are isomorphic) :

For e even,

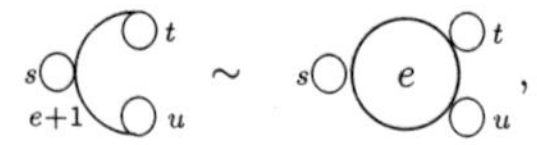

,

for e odd,

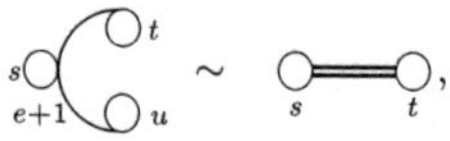

,

and

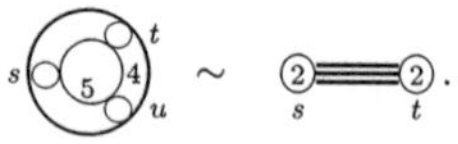

.

It should be noticed, however, that the above pairs of diagrams do not have the same degrees and codegrees, nor do they have the same zeta function. Thus, degrees, codegrees and zeta functions are indeed attached to the braid diagrams, not to the braid groups.

name	diagram	degrees	codegrees	β	field	$G/Z(G)$
$G(de,e,r)$ $e>2,d,r\geq 2$	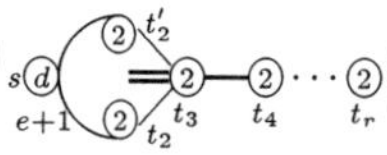 s (d) joined to (2) t_2' and (2) t_2 (edge label $e+1$); (2) t_2' and (2) t_2 joined to (2) t_3 (double edge); (2) t_3 — (2) t_4 $\cdots$ (2) t_r	$(ed,2ed,\ldots,(r-1)ed,rd)$	$(0,ed,\ldots,(r-1)ed)$	$s^{\frac{r}{(e\wedge r)}}(t_2't_2t_3\cdots t_r)^{\frac{e(r-1)}{(e\wedge r)}}$	$\mathbb{Q}(\zeta_{de})$	
G_{15}	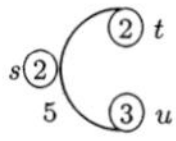 s (2) joined to (2) t and (3) u (edge label 5)	$12,24$	$0,24$	$ustut=s(tu)^2$	$\mathbb{Q}(\zeta_{24})$	$\mathfrak{S}_4$
$\mathfrak{S}_{r+1}$	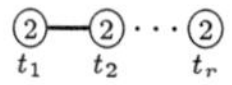 (2) — (2) $\cdots$ (2); t_1, t_2, t_r	$(2,3,\ldots,\ldots,r+1)$	$(0,1,\ldots,\ldots,r-1)$	$(t_1\cdots t_r)^{r+1}$	$\mathbb{Q}$	
G_4	 (3) — (3); s, t	$4,6$	$0,2$	$(st)^3$	$\mathbb{Q}(\zeta_3)$	$\mathfrak{A}_4$
G_8	 (4) — (4); s, t	$8,12$	$0,4$	$(st)^3$	$\mathbb{Q}(i)$	$\mathfrak{S}_4$
G_{16}	 (5) — (5); s, t	$20,30$	$0,10$	$(st)^3$	$\mathbb{Q}(\zeta_5)$	$\mathfrak{A}_5$
G_{25}	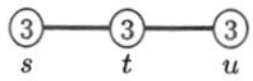 (3) — (3) — (3); s, t, u	$6,9,12$	$0,3,6$	$(stu)^4$	$\mathbb{Q}(\zeta_3)$	$3^2\rtimes SL_2(3)$
G_{32}	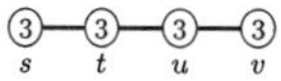 (3) — (3) — (3) — (3); s, t, u, v	$12,18,24,30$	$0,6,12,18$	$(stuv)^5$	$\mathbb{Q}(\zeta_3)$	$PSp_4(3)$
$G(d,1,r)$ $d\geq 2$	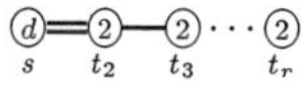 (d) = (2) — (2) $\cdots$ (2); s, t_2, t_3, t_r	$(d,2d,\ldots,\ldots,rd)$	$(0,d,\ldots,\ldots,(r-1)d)$	$(st_2t_3\cdots t_r)^r$	$\mathbb{Q}(\zeta_d)$	
G_5	 (3) = (3); s, t	$6,12$	$0,6$	$(st)^2$	$\mathbb{Q}(\zeta_3)$	$\mathfrak{A}_4$
G_{10}	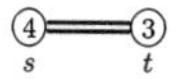 (4) = (3); s, t	$12,24$	$0,12$	$(st)^2$	$\mathbb{Q}(\zeta_{12})$	$\mathfrak{S}_4$
G_{18}	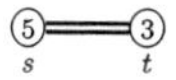 (5) = (3); s, t	$30,60$	$0,30$	$(st)^2$	$\mathbb{Q}(\zeta_{15})$	$\mathfrak{A}_5$
G_{26}	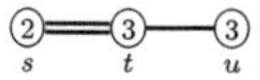 (2) = (3) — (3); s, t, u	$6,12,18$	$0,6,12$	$(stu)^3$	$\mathbb{Q}(\zeta_3)$	$3^2\rtimes SL_2(3)$

TABLE 1

name	diagram	degrees	codegrees	β	field	$G/Z(G)$
$G(2d,2,r)$ $d,r\geq 2$	s ⓓ, t'_2 ②, t_2 ②, t_3 ②, t_4 ② $\cdots$ t_r ②	$(2d,4d,\ldots$ $2(r-1)d,rd)$	$(0,2d,\ldots$ $2(r-1)d)$	$s^{\frac{r}{(2\wedge r)}}(t'_2t_2t_3\cdots t_r)^{\frac{2(r-1)}{(2\wedge r)}}$	$\mathbb{Q}(\zeta_{2d})$	
G_7	s ②, t ③, u ③	$12,12$	$0,12$	stu	$\mathbb{Q}(\zeta_{12})$	$\mathfrak{A}_4$
G_{11}	s ②, t ③, u ④	$24,24$	$0,24$	stu	$\mathbb{Q}(\zeta_{24})$	$\mathfrak{S}_4$
G_{19}	s ②, t ③, u ⑤	$60,60$	$0,60$	stu	$\mathbb{Q}(\zeta_{60})$	$\mathfrak{A}_5$
$G(e,e,r)$ $e\geq 2, r>2$	t'_2 ②, e, t_2 ②, t_3 ②, t_4 ② $\cdots$ t_r ②	$(e,2e,\ldots,$ $(r-1)e,r)$	$(0,e,\ldots,(r-2)e,$ $(r-1)e-r)$	$(t'_2t_2t_3\cdots t_r)^{\frac{e(r-1)}{(e\wedge r)}}$	$\mathbb{Q}(\zeta_e)$	
$G(e,e,2)$ $e\geq 3$	s ② —e— ② t	$2,e$	$0,e-2$	$(st)^{e/(e\wedge 2)}$	$\mathbb{Q}(\zeta_e+\zeta_e^{-1})$	
G_6	s ③ ≡ ② t	$4,12$	$0,8$	$(st)^3$	$\mathbb{Q}(\zeta_{12})$	$\mathfrak{A}_4$
G_9	s ④ ≡ ② t	$8,24$	$0,16$	$(st)^3$	$\mathbb{Q}(\zeta_8)$	$\mathfrak{S}_4$
G_{17}	s ⑤ ≡ ② t	$20,60$	$0,40$	$(st)^3$	$\mathbb{Q}(\zeta_{20})$	$\mathfrak{A}_5$
G_{14}	s ③ —8— ② t	$6,24$	$0,18$	$(st)^4$	$\mathbb{Q}(\zeta_3,\sqrt{-2})$	$\mathfrak{S}_4$
G_{20}	s ③ —5— ③ t	$12,30$	$0,18$	$(st)^5$	$\mathbb{Q}(\zeta_3,\sqrt{5})$	$\mathfrak{A}_5$
G_{21}	s ③ —10— ② t	$12,60$	$0,48$	$(st)^5$	$\mathbb{Q}(\zeta_{12},\sqrt{5})$	$\mathfrak{A}_5$

TABLE 2

name	diagram	degrees	codegrees	β	field	$G/Z(G)$
G_{12}	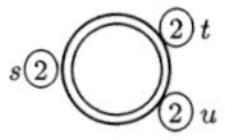	6,8	0,10	$(stu)^4$	$\mathbb{Q}(\sqrt{-2})$	$\mathfrak{S}_4$
G_{13}	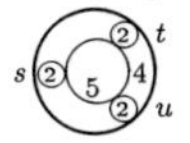	8,12	0,16	$(stu)^3$	$\mathbb{Q}(\zeta_8)$	$\mathfrak{S}_4$
G_{22}	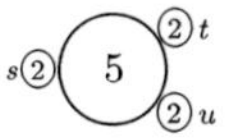	12,20	0,28	$(stu)^5$	$\mathbb{Q}(i,\sqrt{5})$	$\mathfrak{A}_5$
G_{23}	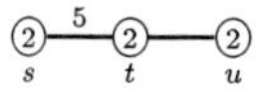	2,6,10	0,4,8	$(stu)^5$	$\mathbb{Q}(\sqrt{5})$	$\mathfrak{A}_5$
G_{28}	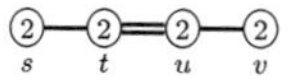	2,6, 8,12	0,4, 6,10	$(stuv)^6$	$\mathbb{Q}$	$2^4\rtimes(\mathfrak{S}_3\times\mathfrak{S}_3)$ †
G_{30}	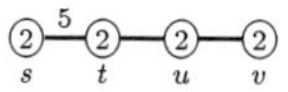	2,12, 20,30	0,10, 18,28	$(stuv)^{15}$	$\mathbb{Q}(\sqrt{5})$	$(\mathfrak{A}_5\times\mathfrak{A}_5)\rtimes 2$ ‡
G_{35}	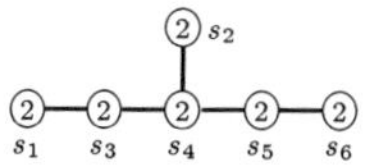	2,5,6,8, 9,12	0,3,4,6, 7,10	$(s_1\cdots s_6)^{12}$	$\mathbb{Q}$	$SO_6^-(2)'$
G_{36}	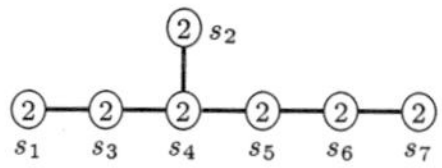	2,6,8, 10,12, 14,18	0,4,6, 8,10, 12,16	$(s_1\cdots s_7)^9$	$\mathbb{Q}$	$SO_7(2)$
G_{37}	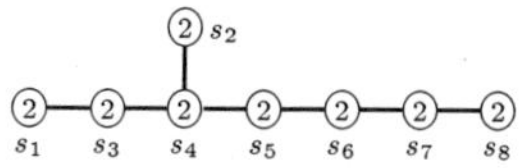	2,8,12, 14,18,20, 24,30	0,6,10, 12,16,18, 22,28	$(s_1\cdots s_8)^{15}$	$\mathbb{Q}$	$SO_8^+(2)$
G_{31}	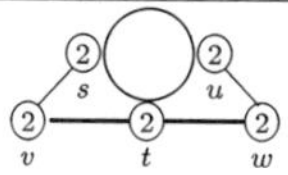	8,12, 20,24	0,12, 16,28	$(stuvw)^6$	$\mathbb{Q}(i)$	$2^4\rtimes\mathfrak{S}_6$ ⋆

TABLE 3

It is still conjectural whether the corresponding braid diagram for G_{31} provides a presentation for the associated braid group.

† The action of $\mathfrak{S}_3\times\mathfrak{S}_3$ on 2^4 is irreducible.

‡ The automorphism of order 2 of $\mathfrak{A}_5\times\mathfrak{A}_5$ permutes the two factors.

⋆ The group $G_{31}/Z(G_{31})$ is not isomorphic to the quotient of the Weyl group D_6 by its center.

name	diagram	degrees	codegrees	β	field	$G/Z(G)$
G_{24}	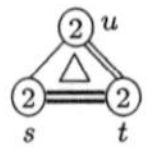	4,6,14	0,8,10	$(stu)^7$	$\mathbb{Q}(\sqrt{-7})$	$GL_3(2)$
G_{27}	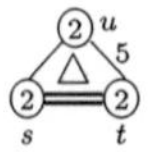	6,12,30	0,18,24	$(stu)^5$	$\mathbb{Q}(\zeta_3,\sqrt{5})$	$\mathfrak{A}_6$
G_{29}	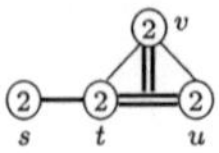	4,8,12,20	0,8,12,16	$(stuv)^5$	$\mathbb{Q}(i)$	$2^4 \rtimes \mathfrak{S}_5$ †
G_{33}	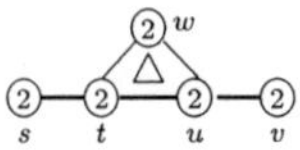	4,6,10, 12,18	0,6,8, 12,14	$(ustvw)^9$	$\mathbb{Q}(\zeta_3)$	$SO_5(3)'$
	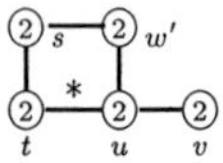					
G_{34}	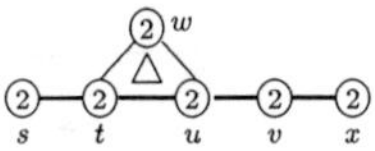	6,12,18,24, 30,42	0,12,18,24, 30,36	$(stuvwx)^7$	$\mathbb{Q}(\zeta_3)$	$PSO_6^-(3)'\cdot 2$
	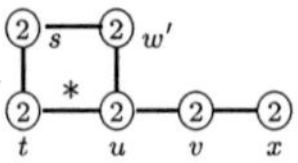					

TABLE 4

These diagrams provide presentations for the corresponding finite groups. It is not known nor conjectural whether they provide presentations for the corresponding braid groups.

† The group $G_{29}/Z(G_{29})$ is not isomorphic to the Weyl group D_5.

name	diagram	degrees	codegrees	β
$B(de,e,r)$ $e\geq 2, r\geq 2, d>1$	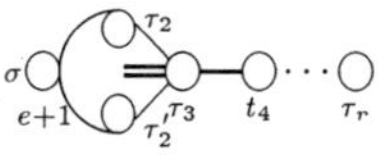 σ, $e+1$, τ_2, τ_2', τ_3, t_4, τ_r	$e, 2e, \ldots, (r-1)e, r$	$0, e, \ldots, (r-1)e$	$\sigma^{\frac{r}{(e\wedge r)}}(\tau_2\tau_2'\tau_3\cdots\tau_r)^{\frac{e(r-1)}{(e\wedge r)}}$
$B(1,1,r)$	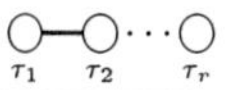 τ_1, τ_2, τ_r	$2, 3, \ldots, r+1$	$0, 1, \ldots, r-1$	$(\tau_1\cdots\tau_r)^{r+1}$
$B(d,1,r)$ $d>1$	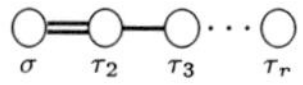 σ, τ_2, τ_3, τ_r	$1, 2, \ldots, r$	$0, , \ldots, (r-1)$	$(\sigma\tau_2\tau_3\cdots\tau_r)^r$
$B(e,e,r)$ $e\geq 2, r\geq 2$	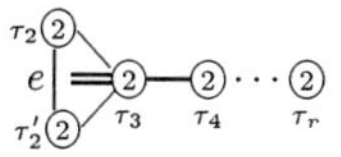 τ_2 2, e, τ_2' 2, τ_3 2, τ_4 2, τ_r 2	$e, 2e, \ldots, (r-1)e, r$	$0, e, \ldots, (r-2)e, (r-1)e-r$	$(\tau_2\tau_2'\tau_3\cdots\tau_r)^{\frac{e(r-1)}{(e\wedge r)}}$

TABLE 5 : BRAID DIAGRAMS

This table provides a complete list of the infinite families of braid diagrams and corresponding data. Note that the braid diagram $B(de,e,r)$ for $e=2, d>1$ can also be described by a diagram as the one used for $G(2d,2,r)$ in Table 2. Similarly, the diagram for $B(e,e,r)$, $e=2$, can also be described by the Coxeter diagram of type D_r. The list of exceptional diagrams (but those associated with G_{24}, G_{27}, G_{29}, G_{33}, G_{34}) is identical with table 3.

APPENDIX 2

SYMMETRIC ALGEBRAS

For the convenience of the reader, we collect and prove here various results about specializations of algebras and about symmetric algebras. Most of these results are known. The presentation given here is inspired by various ideas in [**Bro1**], [**BrKi**], [**Ge**], [**GePf**], [**GeRo**] *and* [**Ro1**].

In what follows the following notation and assumptions will be in force:

- We denote by R an integrally closed commutative noetherian domain, with field of fractions denoted by F.
- We denote by $\mathcal{H}$ an R–algebra which is a free R–module of finite rank.
  [Actually, for most of the results proved below it would be enough to assume that $\mathcal{H}$ is a finitely generated projective R–module.]
- The R–dual of the module $\mathcal{H}$ is $\mathcal{H}^\vee := \mathrm{Hom}(\mathcal{H}, R)$. The map

$$\mathcal{H}^\vee \otimes \mathcal{H} \to \mathrm{End}_R(\mathcal{H}) \ , \ \phi \otimes h \mapsto (h' \mapsto \phi(h')h)$$

is an isomorphism. The trace $\mathrm{tr}_\mathcal{H} \colon \mathrm{End}_R(\mathcal{H}) \to R$ is defined, through the preceding isomorphism, by the natural pairing $\mathcal{H}^\vee \otimes \mathcal{H} \to R$. We still denote by

$$\mathrm{tr}_\mathcal{H} \colon \mathcal{H} \to R$$

the composition of the trace with the natural monomorphism $\mathcal{H} \hookrightarrow \mathrm{End}_R(\mathcal{H})$ defined by the left regular representation of $\mathcal{H}$.

- We set $F\mathcal{H} := F \otimes_R \mathcal{H}$. More generally, if $R \to R'$ is a ring morphism, we set $R'\mathcal{H} := R' \otimes_R \mathcal{H}$.

Values of characters.

8.1. PROPOSITION. *Let χ be the character of some finite dimensional $F\mathcal{H}$–module.*
(1) *For all $h \in \mathcal{H}$, we have $\chi(h) \in R$.*
(2) *Assume moreover that R is a $\mathbb{Z}$–graded ring, and that $\mathcal{H}$ is a graded R–module which has an R–basis consisting of homogeneous elements. Then, for all $h \in \mathcal{H}$ with degree n, $\chi(h)$ is an element with degree n of R.*

PROOF OF 8.1.
(1) For $h \in \mathcal{H}$, the characteristic polynomial $\mathrm{Cha}(h)(x)$ of the left multiplication by h in $\mathcal{H}$ is a unitary element of $R[x]$. Hence its roots are all integral over R.

In order to prove 8.1, we may assume that χ is the character of an irreducible $F\mathcal{H}$–module. In this case, the characteristic polynomial $\mathrm{Cha}_\chi(h)(x)$ of h acting on this module divides $\mathrm{Cha}(h)(x)$ and is unitary, from which it follows that its coefficients are integral over R. Since R is integrally closed, we see that $\mathrm{Cha}_\chi(h)(x) \in R[x]$. In particular, we have $\chi(h) \in R$.

(2) Keeping the same notation as above, we see now that, if

$$\mathrm{Cha}(h)(x) = x^d - a_{d-1}(h)x^{d-1} + \cdots + (-1)^d a_0(h) \,,$$

then for each j the coefficient $a_j(h)$ is a homogeneous element of degree $n(d-j)$ in R (since the matrix of the left multiplication by h on a basis consisting of homogeneous elements must have homogeneous entries all of degree n).

Let us define a graduation on $R[x]$ by assigning the degree $\ell + nm$ to λx^m, for λ homogeneous of degree ℓ. Then we see that $\mathrm{Cha}(h)(x)$ is a homogeneous element of degree dn in $R[x]$.

Any divisor of a homogeneous element is again homogeneous. So $\mathrm{Cha}_\chi(h)(x)$ is homogeneous, and since its degree as a polynomial in x is $\chi(1)$, it is homogeneous of degree $n\chi(1)$. This proves in particular that the coefficient of $x^{\chi(1)-1}$, namely $\chi(h)$, is homogeneous of degree n. □

Let $p\colon R \twoheadrightarrow k$ be a morphism from R onto a field k. For $\varphi\colon \mathcal{H} \to R$ a linear form, we denote by $\varphi_k\colon k\mathcal{H} \to k$ the function such that the following diagram is commutative:

$$\begin{array}{ccc} \mathcal{H} & \xrightarrow{\varphi} & R \\ {\scriptstyle p}\downarrow & & \downarrow{\scriptstyle p} \\ k\mathcal{H} & \xrightarrow{\varphi_k} & k\,. \end{array}$$

In particular, by 8.1 above, any character χ of a $F\mathcal{H}$-module defines a central function $\chi_k\colon k\mathcal{H} \to k$.

Under certain hypotheses, the function χ_k is again a character, as shown in particular by the following classical result, known as "Tits deformation theorem".

8.2. THEOREM. *Assume that $F\mathcal{H}$ and $k\mathcal{H}$ are split semi-simple. Then the map $\chi \mapsto \chi_k$ defines a bijection from the set $\mathrm{Irr}(F\mathcal{H})$ of irreducible characters of $F\mathcal{H}$ onto the set $\mathrm{Irr}(k\mathcal{H})$ of irreducible characters of $k\mathcal{H}$, which preserves the degrees of characters.*

PROOF OF 8.2. We only sketch a proof of this well known result.

Let n be the rank of $\mathcal{H}$ over R. We extend the scalars to the polynomial algebra $R[x_1, x_2, \dots, x_n]$ to get the algebra $R[x_1, x_2, \dots, x_n]\mathcal{H}$. Choose a basis $(h_1, h_2, \dots, h_n)$ of $\mathcal{H}$ over R and consider the generic element $g := \sum_{j=1}^{j=n} x_j h_j$. Let $\mathrm{Cha}(g)(x)$ be the characteristic polynomial of the left multiplication by g. Then we have the following factorization in $F(x_1, x_2, \dots, x_n)[x]$ (hence in $R[x_1, x_2, \dots, x_n][x]$ since $R[x_1, x_2, \dots, x_n][x]$ is integrally closed):

$$(*) \qquad \mathrm{Cha}(g)(x) = \prod_{\chi \in \mathrm{Irr}(F\mathcal{H})} \mathrm{Cha}_\chi(g)(x)^{d_\chi}\,,$$

where $\mathrm{Cha}_\chi(g)(x)$ is the characteristic polynomial of g in a representation with character χ and where d_χ is the degree of χ, *i.e.*, the degree (in x) of $\mathrm{Cha}_\chi(g)(x)$. It is easy to see that each $\mathrm{Cha}_\chi(g)(x)$ is irreducible.

Applying the morphism p, we get a decomposition of $p(\mathrm{Cha}(g))(x)$ in the algebra $k[x_1, x_2, \dots, x_n][x]$:

$$p(\mathrm{Cha}(g))(x) = \prod_{\chi \in \mathrm{Irr}(F\mathcal{H})} p(\mathrm{Cha}_\chi(g))(x)^{d_\chi}\,.$$

On the other hand, since $p(\mathrm{Cha}(g))(x)$ is the characteristic polynomial of the left multiplication by g in $k[x_1, x_2, \dots, x_n]\mathcal{H}$, this polynomial has a decomposition into irreducible factors in $k[x_1, x_2, \dots, x_n][x]$ which is analogous to (*) above:

$$p(\mathrm{Cha}(g))(x) = \prod_{\phi \in \mathrm{Irr}(k\mathcal{H})} \mathrm{Cha}_\phi(g)(x)^{d_\phi}\,,$$

where d_ϕ is the degree of $\mathrm{Cha}_\phi(g)(x)$. It is easy to see that the decompositions must coincide, whence the theorem 8.2 follows. □

Symmetric algebras.

8.3. Definitions.

- *A linear form $t\colon \mathcal{H} \to R$ is called central if $t(hh') = t(h'h)$ for all $h, h' \in \mathcal{H}$.*
- *A symmetrizing form on $\mathcal{H}$ is a central linear form $t\colon \mathcal{H} \to R$ such that the morphism*

$$\widehat{t}\colon \mathcal{H} \to \mathcal{H}^\vee \ , \ h \mapsto (h' \mapsto t(hh'))$$

is an isomorphism.

- *The algebra $\mathcal{H}$ is said to be symmetric if there exists a symmetrizing form.*
- *If the trace form $\mathrm{tr}_\mathcal{H}\colon \mathcal{H} \to R$ is symmetrizing, we say that $\mathcal{H}$ is trace symmetric.*

Let us note some elementary facts about specializations and symmetrizing forms.

8.4. Proposition. *Let $p\colon R \twoheadrightarrow k$ be a morphism from R onto a field k. Assume that there is a central form $t\colon \mathcal{H} \to R$ which defines a symmetrizing form on $k\mathcal{H}$. Then t is a symmetrizing form for $F\mathcal{H}$.*

In particular, if $k\mathcal{H}$ is trace symmetric, then $F\mathcal{H}$ is trace symmetric.

Proof of 8.4. Let $\mathfrak{m}$ be the kernel of p. We set $\overline{t} := p \cdot t\colon \mathcal{H} \to k$. The discriminant of the form $\overline{t}$ on an R–basis of $\mathcal{H}$ does not belong to $\mathfrak{m}$, since its image through p has to be non zero. This proves that t is a symmetrizing form for the algebra $R_\mathfrak{m}\mathcal{H}$ (where $R_\mathfrak{m}$ is the localization of R at $\mathfrak{m}$), hence in particular is a symmetrizing form for $F\mathcal{H}$. □

Symmetrizing forms and separability.

If the algebra $F\mathcal{H}$ is trace symmetric, then it is separable, *i.e.*, whenever L is an extension of F, the algebra $L\mathcal{H}$ is semi–simple. This is a particular case of a more general result which we prove below (see 8.7).

Let us first introduce the notion of "pseudo–character" (see for example [**Ro1**]).

Whenever $\tau\colon \mathcal{H} \to R$ is a central function on $\mathcal{H}$, for all natural integers n we construct a symmetric function $S_n(\tau)\colon \mathcal{H}^n \to R$ by the formula

$$S_n(\tau) := \sum_{\sigma \in \mathfrak{S}_n} \varepsilon(\sigma)(\sigma \cdot \tau)\,,$$

where, for $\sigma = \sigma_1\sigma_2\cdots\sigma_r$ the decomposition of σ into a product of disjoint cycles, and $\sigma_j = (j_1, \ldots, j_m)$, we set

$$(\sigma \cdot \tau)(x_1, x_2, \ldots, x_n) := \tau(\sigma_1(x))\tau(\sigma_2(x))\cdots\tau(\sigma_r(x))$$

where

$$\sigma_j(x) := x_{j_1}\cdots x_{j_m}\,.$$

We say that the function τ is a *pseudo–character* of $\mathcal{H}$ if there exists an integer d such that $S_d(\tau) = 0$.

The following result (see [**Ro1**], 3.1) justifies the terminology.

8.5. Lemma. *Any character of a $F\mathcal{H}$–module is a pseudo–character.*

Indeed, one reduces easily to the case where $\mathcal{H}$ is a matrix algebra over a characteristic zero field. Then the assertion becomes a formal property of traces of matrices.

Pseudo–characters inherit from characters the property of vanishing on nilpotent elements (see [**Ro1**], 2.6).

8.6. Lemma. *Any pseudo–character vanishes on nilpotent elements of $\mathcal{H}$.*

8.7. PROPOSITION. *Assume that the algebra $F\mathcal{H}$ has a symmetrizing form which is a linear combination of pseudo-characters. Then $F\mathcal{H}$ is separable.*

PROOF OF 8.7. Let t be a symmetrizing form on $F\mathcal{H}$ which is a linear combination of pseudo-characters. Let L be an extension of F. Since the elements of the Jacobson radical of $L\mathcal{H}$ are nilpotent, they are contained in the kernel of all pseudo-characters, hence in the kernel of the map $\widehat{t}$, which proves that the Jacobson radical of $L\mathcal{H}$ equals $\{0\}$. □

The Casimir element.

From now on we assume that

$\mathcal{H}$ *is endowed with a symmetrizing form t.*

The isomorphisms $\mathcal{H}^\vee \otimes \mathcal{H} \xrightarrow{\sim} \mathrm{End}_R(\mathcal{H})$ and $\widehat{t}\colon \mathcal{H} \xrightarrow{\sim} \mathcal{H}^\vee$ define an isomorphism $\mathcal{H} \otimes \mathcal{H} \xrightarrow{\sim} \mathrm{End}_R(\mathcal{H})$.

We denote by $C_\mathcal{H}$ and call *the Casimir element* of $(\mathcal{H}, t)$ the element of $\mathcal{H}\otimes\mathcal{H}$ corresponding to the identity on $\mathcal{H}$ through the preceding isomorphism. If J is a finite set and if $(h_j)_{j\in J}$ and $(h'_j)_{j\in J}$ are two families of elements of $\mathcal{H}$ indexed by J such that $C_\mathcal{H} = \sum_j h'_j \otimes h_j$, we get the following characterization of $C_\mathcal{H}$:

$$h = \sum_{j\in J} t(hh'_j)h_j \quad \text{for all } h \in \mathcal{H}. \tag{8.8}$$

REMARK. If $(e_j)_j$ is an R-basis of $\mathcal{H}$, and if $(e'_j)_j$ is the dual basis (defined by the condition $t(e_j e'_{j'}) = \delta_{j,j'}$), then we have $C_\mathcal{H} = \sum_{j\in J} e'_j \otimes e_j$.

The following properties of the element $C_\mathcal{H}$ are straightforward.

8.9. LEMMA.

(1) *For all $h \in \mathcal{H}$, we have $\sum_{j\in J} hh'_j \otimes h_j = \sum_{j\in J} h_j \otimes h'_j h$.*
(2) *We have $\sum_{j\in J} h'_j \otimes h_j = \sum_{j\in J} h_j \otimes h'_j$, and $hC_\mathcal{H} = C_\mathcal{H} h$ for all $h \in \mathcal{H}$.*
(3) *For all $h \in \mathcal{H}$, we have*

$$h = \sum_{j\in J} t(hh'_j)h_j = \sum_{j\in J} t(hh_j)h'_j = \sum_{j\in J} t(h'_j)h_j h = \sum_{j\in J} t(h_j)h'_j h.$$

Let us prove (1). By 8.8, we have $hh'_j = \sum_{i\in J} t(hh'_j h'_i)h_i$, hence

$$\begin{aligned}\sum_{j\in J} hh'_j \otimes h_j &= \sum_{i,j} t(hh'_j h'_i)h_i \otimes h_j \\ &= \sum_i h_i \otimes \sum_j t(h'_i hh'_j)h_j \\ &= \sum_i h_i \otimes h'_i h.\end{aligned}$$

The assertions (2) and (3) follow immediately from (1).

For $\tau\colon \mathcal{H} \to R$ a linear form, we denote by $\tau^\vee$ the element of $\mathcal{H}$ defined by the condition

$$t(\tau^\vee h) = \tau(h) \quad \text{for all } h \in \mathcal{H}.$$

It is easy to check the following set of properties.

8.10. LEMMA.
(1) *τ is central if and only if $\tau^\vee$ is central in $\mathcal{H}$.*
(2) *We have $\tau^\vee = \sum_{j\in J} \tau(h'_j)h_j = \sum_{j\in J} \tau(h_j)h'_j$, and more generally, for all $h \in \mathcal{H}$, we have $\tau^\vee h = \sum_{j\in J} \tau(h'_j h)h_j = \sum_{j\in J} \tau(h_j h)h'_j$.*

Let χ_{reg} denote the character of the regular representation of $\mathcal{H}$, *i.e.*, the linear form on $\mathcal{H}$ defined by

$$\chi_{\mathrm{reg}}(h) := \mathrm{tr}_{\mathcal{H}/R}(\lambda_h)$$

where λ_h is the endormorphism of $\mathcal{H}$ defined by $\lambda_h \colon \mathcal{H} \longrightarrow \mathcal{H}\ ,\ x \mapsto hx$.

8.11. PROPOSITION. *For all $h \in \mathcal{H}$, we have*

$$\chi_{\mathrm{reg}}(h) = t(hc_{\mathcal{H}})\,,\quad \text{or, in other words}\quad \chi_{\mathrm{reg}}^\vee = c_{\mathcal{H}}\,.$$

PROOF OF 8.11. The characterization of the Casimir element $C_{\mathcal{H}}$ shows that, through the isomorphism $\mathcal{H}^\vee \otimes \mathcal{H} \xrightarrow{\sim} \mathrm{Hom}_R(\mathcal{H},\mathcal{H})$, the endomorphism λ_h corresponds to the element $\sum_{j\in J} \widehat{t}(h'_j h) \otimes h_j$. Thus the trace of λ_h is

$$\mathrm{tr}(\lambda_h) = \sum_{j\in J} \widehat{t}(h'_j h)(h_j) = \sum_{j\in J} t(h'_j h h_j) = t(c_{\mathcal{H}} h)\,.$$

□

Casimir element and projectivity.

Let V be an $\mathcal{H}$–module, which is a projective R–module (hence, the natural morphism $\mathrm{Hom}_R(V,R) \otimes_R V \longrightarrow \mathrm{Hom}_R(V,V)$ is an isomorphism). The $\mathcal{H}$–morphism

$$t_V \colon \begin{cases} \mathrm{Hom}_{\mathcal{H}}(V,\mathcal{H}) \longrightarrow \mathrm{Hom}_R(V,R) \\ \phi \mapsto t\cdot\phi \end{cases}$$

is an isomorphism, whose inverse is the morphism $t_V^{-1} \colon \mathrm{Hom}_R(V,R) \longrightarrow \mathrm{Hom}_{\mathcal{H}}(V,\mathcal{H})$ defined by

$$\forall h \in \mathcal{H}\,,\ v \in V\ ,\ \psi(hv) = t(ht_V^{-1}\psi)(v))\,.$$

As a consequence, the natural morphism $\mathrm{Hom}_{\mathcal{H}}(V,\mathcal{H}) \otimes_R V \longrightarrow \mathrm{Hom}_{\mathcal{H}}(V,V)$ can be factorized as follows

$$\mathrm{Hom}_{\mathcal{H}}(V,\mathcal{H}) \otimes_R V \xrightarrow{\sim} \mathrm{Hom}_R(V,R) \otimes_R V \xrightarrow{\sim} \mathrm{Hom}_R(V,V) \xrightarrow{H_V} \mathrm{Hom}_{\mathcal{H}}(V,V)$$

where the map H_V is defined by :

$$H_V(\alpha)(v) := \sum_{j\in J} h'_j \alpha(h_j v)\,.$$

8.12. PROPOSITION. *Let V be an $\mathcal{H}$–module which is a finitely generated projective R–module. The $\mathcal{H}$–module V is projective if and only if the image of the map*

$$H_V \colon \begin{cases} \mathrm{Hom}_R(V,V) \longrightarrow \mathrm{Hom}_{\mathcal{H}}(V,V) \\ \alpha \mapsto \sum_{j\in J} h'_j \alpha(h_j v) \end{cases}$$

contains the identity endomorphism of V.

PROOF OF 8.12. Indeed we know that V is a projective $\mathcal{H}$–module if and only if the natural morphism $\mathrm{Hom}_{\mathcal{H}}(V,\mathcal{H}) \otimes_R V \longrightarrow \mathrm{Hom}_{\mathcal{H}}(V,V)$ reaches the identity endomorphism of V. □

Schur elements.

Let χ be the character of an absolutely irreducible $F\mathcal{H}$–module V_χ. We denote by $\rho_\chi \colon F\mathcal{H} \twoheadrightarrow \mathrm{End}_F(V_\chi)$ the natural (epi)morphism. It restricts to a morphism $\omega_\chi \colon Z\mathcal{H} \to R$ (where we identify R with a subring of the center of $\mathrm{End}_F(V)$).

The *Schur element of* χ is by definition the element of R defined by

(8.13) $$S_\chi := \omega_\chi(\chi^\vee)\,.$$

Notice that by 8.10, (2), we have

$$S_\chi \chi(1) = \sum_{j\in J} \chi(h'_j)\chi(h_j)\,.$$

8.14. PROPOSITION. *Assume that $F\mathcal{H}$ is split semi–simple.*

(1) *For each irreducible character χ of $F\mathcal{H}$, let e_χ be the primitive idempotent of the center $ZF\mathcal{H}$ of the algebra $F\mathcal{H}$ associated with χ. Then we have*

$$\chi^\vee = S_\chi e_\chi\,.$$

(2) *We have*

$$t = \sum_{\chi\in\mathrm{Irr}(F\mathcal{H})} \frac{1}{S_\chi}\chi\,.$$

PROOF OF 8.14.

(1) Since, for all $h \in \mathcal{H}$, we have $\chi(e_\chi h) = \chi(h)$, we see that $t(\chi^\vee e_\chi h) = t(\chi^\vee h)$, which proves that $\chi^\vee = \chi^\vee e_\chi$. The desired equality results from the fact that, for all $z \in ZF\mathcal{H}$, we have $z = \sum_{\chi\in\mathrm{Irr}(F\mathcal{H})} \omega_\chi(z) e_\chi$.

(2) Through the isomorphism between $\mathcal{H}$ and its dual, the equality

$$t = \sum_{\chi\in\mathrm{Irr}(F\mathcal{H})} \frac{1}{S_\chi}\chi$$

is equivalent to

$$1 = \sum_{\chi\in\mathrm{Irr}(F\mathcal{H})} \frac{1}{S_\chi}\chi^\vee\,,$$

which is obvious by (1) above. □

8.15. THEOREM. *Assume that $\mathcal{H}$ is endowed with a symmetrizing form t, and that $F\mathcal{H}$ is split semi-simple. For each $\chi \in \mathrm{Irr}(F\mathcal{H})$, let us denote by S_χ the corresponding Schur element.*

Let $p \colon R \twoheadrightarrow k$ be a morphism from R onto a field k. Assume that $k\mathcal{H}$ is split. Then the following assertions are equivalent:

(i) *For all $\chi \in \mathrm{Irr}(F\mathcal{H})$, we have $p(S_\chi) \neq 0$.*

(ii) *$k\mathcal{H}$ is (split) semi–simple.*

PROOF OF 8.15. Notice that the form t defines a symmetrizing form t_k on $k\mathcal{H}$.

(i) ⇒ (ii): We have

$$t_k = \sum_{\chi\in\mathrm{Irr}(F\mathcal{H})} \frac{1}{p(S_\chi)}\chi_k\,.$$

In particular, we see that t_k is a linear combination of pseudo–characters of $k\mathcal{H}$, hence $k\mathcal{H}$ is separable by 8.7 above.

(ii) ⇒ (i): By 8.2, we know that $\mathrm{Irr}(k\mathcal{H}) = \{\chi_k \mid \chi \in \mathrm{Irr}(F\mathcal{H})\}$. It is then clear that $S_{\chi_k} = p(S_\chi)$, which proves that $p(S_\chi) \neq 0$. □

REFERENCES

[AlBr] J.L. Alperin and M. Broué, *Local Methods in Block Theory*, Ann. of Math. **110** (1979), 143–157.

[AlLu] D. Alvis and G. Lusztig, *The representations and generic degrees of the Hecke algebra of type H_4*, J. reine angew. Math. **336** (1982), 201–212.

[Ar] S. Ariki, *Representation theory of a Hecke algebra of $G(r,p,n)$*, J. Algebra **177** (1995), 164–185.

[ArKo] S. Ariki and K. Koike, *A Hecke algebra of $(\mathbb{Z}/r\mathbb{Z})\wr S_n$ and construction of its irreducible representations*, Advances in Math. **106** (1994), 216–243.

[Ar] E. Artin, *Theory of braids*, Ann. of Math. **48** (1947), 101–126.

[Ba] E. Bannai, *Fundamental groups of the spaces of regular orbits of the finite unitary reflection groups of dimension 2*, J. Math. Soc. Japan **28** (1976), 447–454.

[Ben] M. Benard, *Schur indices and splitting fields of the unitary reflection groups*, J. Algebra **38** (1976), 318–342.

[Bes1] D. Bessis, *Sur le corps de définition d'un groupe de réflexions complexe*, Comm. in Algebra **25** (8) (1997), 2703–2716.

[Bes2] ———, *Groupes de tresses et éléments réguliers*, J. reine angew. Mathematik (Crelle) **518** (2000), 1–40.

[Bes3] ———, *Zariski theorems and diagrams for braid groups*, Preprint (2000), available at http://www.math.yale.edu/users/db278.

[Bi] J. Birman, *Braids, links and mapping class groups*, Princeton University Press, Princeton, 1974.

[Bou1] N. Bourbaki, *Groupes et algèbres de Lie, chap. 4, 5 et 6*, Hermann, Paris, 1968.

[Bou2] ———, *Algèbre Commutative, chap. V*, Hermann, Paris, 1968.

[BreMa] K. Bremke and G. Malle, *Reduced words and a length function for $G(e,1,n)$*, Indag. Mathem. **8** (1997), 453–469.

[Br1] E. Brieskorn, *Die Fundamentalgruppe des Raumes der regulären Orbits einer endlichen komplexen Spiegelungsgruppe*, Invent. Math. **12** (1971), 37–61.

[Br2] ———, *Sur les groupes de tresses [d'après V.I. Arnold]*, Séminaire Bourbaki, 24ème année, 1971/72, Lecture Notes in Math., vol. 317, Springer-Verlag, Berlin, 1973.

[BrSa] E. Brieskorn and K. Saito, *Artin-Gruppen und Coxeter-Gruppen*, Invent. Math. **17** (1972), 245–271.

[Bro1] M. Broué, *Isométries parfaites, types de blocs, catégories dérivées*, Astérisque **181–182** (1990), 61–92.

[Bro2] ———, *On representations of symmetric algebras: an introduction*, Notes by Markus Stricker, Forschungsinstitut für Mathematik ETH Zürich (1991).

[Bro3] ———, *Rickard equivalences and block theory*, Groups '93 Galway-St Andrews, London Math. Soc., Lecture Notes Series 211, Cambridge University Press, Cambridge, U.K., 1995, pp. 58–79.

[BrKi] M. Broué and S. Kim, *Sur les blocs de Rouquier des algèbres de Hecke cyclotomiques*, Preprint (2000), Publ. de l'Institut de Math. de Jussieu.

[BrMa1] M. Broué and G. Malle, *Théorèmes de Sylow génériques pour les groupes réductifs sur les corps finis*, Math. Ann. **292** (1992), 241–262.

[BrMa2] ———, *Zyklotomische Heckealgebren*, Astérisque **212** (1993), 119–189.

[BrMa3] ———, *Generalized Harish-Chandra Theory*, Representation Theory of Alg. Groups and Related Finite Groups (R. Carter & M. Geck, eds.), Cambridge University Press, 1997.

[BMM1] M. Broué, G. Malle, J. Michel, *Generic blocks of finite reductive groups*, Astérisque **212** (1993), 7–92.

[BMM2] ———, *Towards Spetses I*, Trans. Groups **4 2–3** (1999), 157–218.

[BMM3] ———, *Towards Spetses II*, En préparation (2000).

[BMR] M. Broué, G. Malle et R. Rouquier, *Complex reflection groups, braid groups, Hecke algebras*, J. reine angew. Math. **500** (1998), 127–190.

[BrMi1] M. Broué et J. Michel, *Blocs et séries de Lusztig dans un groupe réductif fini*, J. reine angew. Math. **395** (1989), 56–67.

[BrMi2] ———, *Sur certains éléments réguliers des groupes de Weyl et les variétés de Deligne-Lusztig associées*, Finite Reductive Groups: Related Structures and Representations (M. Cabanes, ed.), Progress in Math., vol. 141, Birkhäuser, 1997, pp. 73–140.

[Ca] R. W. Carter, *Finite groups of Lie type: Conjugacy classes and complex characters*, John Wiley and Sons, Chichester, 1985.

[Ch1] I. Cherednik, *On a calculation of monodromy of certain W-invariant local systems of type B, C and D*, Funct. Anal. Appl. **23** (1989), 91–92.

[Ch2] ———, *Generalized braid groups and local r-matrix systems*, Soviet Math. Dokl. **40** (1990), 43–48.

[Ch3] ———, *Monodromy representations for generalized Knizhnik-Zamolodchikov equations and Hecke algebras*, Publ. RIMS Kyoto Univ. **27** (1991), 711–726.

[Ch] C. Chevalley, *Invariants of finite groups generated by reflections*, Amer. J. Math **77** (1955), 778–782.

[Cho] W.-L. Chow, *On the algebraic braid group*, Ann. of Math. **49** (1948), 654–658.

[Co] A. M. Cohen, *Finite complex reflection groups*, Ann. scient. Éc. Norm. Sup. **9** (1976), 379–436.

[Cor] R. Corran, *On monoids related to braid groups*, Thesis, The University of Sydney, Sydney, Australia, 2000.

[Cx] H. S. M. Coxeter, *Finite groups generated by unitary reflections*, Abh. math. Sem. Univ. Hamburg **31** (1967), 125–135.

[De1] P. Deligne, *Équations différentielles à points singuliers réguliers*, Lect. Notes in Math., vol. 163, Springer-Verlag, Berlin, 1970.

[De2] ———, *Les immeubles des groupes de tresses généralisés*, Invent. Math. **17** (1972), 273–302.

[De3] ———, *Action du groupe des tresses sur une catégorie*, Invent. Math. **128** (1997), 159–175.

[DeLu] P. Deligne and G. Lusztig, *Representations of reductive groups over finite fields*, Annals of Math. **103** (1976), 103–161.

[DeMo] P. Deligne and G.D. Mostow, *Commensurabilities among lattices in* $\mathrm{PU}(1,n)$, Annals of mathematics studies, vol. 132, Princeton University Press, Princeton, 1993.

[DeLo] J. Denef and F. Loeser, *Regular elements and monodromy of discriminants of finite reflection groups*, Indag. Mathem. **6 (2)** (1995), 129–143.

[DiMi] F. Digne and J. Michel, *Fonctions $\mathcal{L}$ des variétés de Deligne-Lusztig et descente de Shintani*, Mémoires de la S.M.F., vol. 20, 1985.

[DMR] F. Digne, J. Michel and R. Rouquier, *Cohomologie de certaines variétés de Deligne-Lusztig attachées à des éléments réguliers*, Preprint (2000).

[Du] C.F. Dunkl, *Differential-difference operators and monodromy representations of Hecke algebras*, Pacific J. of Math. **159** (1993), 271–298.

[FoSr1] P. Fong and B. Srinivasan, *The blocks of finite general linear groups and unitary groups*, Invent. Math. **69** (1982), 109–153.

[FoSr2] ———, *The blocks of finite classical groups*, J. reine angew. Math. **396** (1989), 122–191.

[FoSr3] ———, *Generalized Harish-Chandra theory for unipotent characters of finite classical groups*, J. Algebra **104** (1986), 301–309.

[FoNe] R.H. Fox and L. Neuwirth, *The braid groups*, Math. Scand. **10** (1962), 119–126.

[Ge] M. Geck, *Beiträge zur Darstellungstheorie von Iwahori-Hecke-Algebren*, RWTH Aachen, Habilitationsschrift (1993).

[GIM] M. Geck, L. Iancu et G. Malle, *Weights of Markov traces and generic degrees*, Indag. Mathem. (2000) (to appear).

[GePf] M. Geck et G. Pfeiffer, *Characters of finite Coxeter groups and Iwahori-Hecke algebras*, London Mathematical Society Monographs, New series, vol. 21, Oxford University Press, Oxford, 2000.

[GeRo] M. Geck et R. Rouquier, *Centers and simple modules for Iwahori-Hecke algebras*, Finite Reductive Groups : Related Structures and Representations (M. Cabanes, eds.), Progress in Mathematics, vol. 141, Birkhäuser, 1997, pp. 251–272.

[Gu] E.A. Gutkin, *Matrices connected with groups generated by mappings*, Func. Anal. and Appl. (Funkt. Anal. i Prilozhen) **7** (1973), 153–154 (81–82).

[Ha] A. Haefliger, *Local theory of meromorphic connections in dimension one (Fuchs theory)*, A. Borel et al., Algebraic D-Modules, Perspectives in Math., vol. 2, Academic Press, Boston, 1987, pp. 129–149.

[Hu] M. C. Hughes, *Complex reflection groups*, Comm. Algebra **18** (1990), 3999–4029.

[Ko1] T. Kohno, *Monodromy representations of braid groups and Yang-Baxter equations*, Ann. Inst. Fourier **37** (1987), 139–160.

[Ko2] ———, *Integrable connections related to Manin and Schechtman's higher braid groups*, Illinois J. of Math. **34** (1990), 476–484.

[Le] G. Lehrer, *Poincaré polynomials for unitary reflection groups*, Invent. Math. **120** (1995), 411–425.

[LeSp1] G. Lehrer and T.A. Springer, *Intersection multiplicities and reflection subquotients of unitary reflection groups I*, Preprint (1998).

[LeSp2] ———, *Reflection subquotients of unitary reflection groups*, Canad. J. Math. **51** (1999), 1175–1193.

[Lu0] G. Lusztig, *Representations of finite Chevalley groups*, C.B.M.S. Regional Conference Series in Mathematics, vol. 39, A.M.S., Providence, 1977.

[Lu1] ———, *On the finiteness of the number of unipotent classes*, Inventiones math. **34** (1976), 201–213.

[Lu2] ———, *Coxeter Orbits and Eigenspaces of Frobenius*, Inventiones math. **38** (1976), 101–159.

[Lu3] ———, *Characters of reductive groups over a finite field*, Annals of Mathematical Studies, no 107, Princeton University Press, Princeton, New Jersey, 1984.

[Lu4] ———, *Intersection cohomology complexes on a reductive group*, Invent. Math. **75** (1984), 205–272.

[Lu5] ———, *Leading coefficients of character values of Hecke algebras*, Proc. Symp. Pure Math., vol. 47(2), Amer.Math.Soc., 1987, pp. 235-262.

[Lu6] ———, *Green functions and character sheaves*, Ann. of Math. **131** (1990), 355–408.

[Lu7] ———, *Coxeter groups and unipotent representations*, Astérisque **212** (1993), 191–203.

[Lu8] ———, *Exotic Fourier transform*, Duke J. Math. **73** (1994), 243–248.

[Ma1] G. Malle, *Unipotente Grade imprimitiver komplexer Spiegelungsgruppen*, J. Algebra **177** (1995), 768–826.

[Ma2] ———, *Degrés relatifs des algèbres cyclotomiques associées aux groupes de réflexions complexes de dimension deux*, Finite Reductive Groups: Related Structures and Representations (M. Cabanes, ed.), Progress in Math., vol. 141, Birkhäuser, 1997, pp. 311–332.

[Ma3] ———, *Spetses*, ICM II, Doc. Math. J. DMV, 1998, pp. 87–96.

[Ma4] ———, *On the rationality and fake degrees of characters of cyclotomic algebras*, J. Math. Sci. Univ. Tokyo **6** (1999), 647–677.

[Ma5] ———, *On the generic degrees of cyclotomic algebras*, Representation Theory **4** (2000).

[MM] G. Malle and A. Mathas, *Symmetric cyclotomic Hecke algebras*, J. Algebra **205** (1998), 275–293.

[Mat] A. Mathas, *Iwahori-Hecke algebras and Schur algebras of the symmetric group*, University lecture series, vol. 15, AMS, Providence, 1999.

[Na] T. Nakamura, *A note on the $K(\pi,1)$ property of the orbit space of the unitary reflection group $G(m,\ell,n)$*, Sci. Papers College of Arts and Sciences Univ. Tokyo **33** (1983), 1–6.

[Op1] E.M. Opdam, *Dunkl operators, Bessel functions and the discriminant of a finite Coxeter group*, Comp. Math. **85** (1993), 333–373.

[Op2] ———, *A remark on the irreducible characters and fake degrees of finite real reflection groups*, Invent. Math. **120** (1995), 447–454.

[Op3] ———, *Complex Reflection Groups and Fake Degrees*, Preprint (1998).

[OrSo1] P. Orlik and L. Solomon, *Combinatorics and topology of complements of hyperplanes*, Invent. Math. **56** (1980), 167–189.

[OrSo2] ———, *Unitary reflection groups and cohomology*, Invent. Math. **59** (1980), 77–94.

[OrSo3] ———, *Braids and discriminants*, Braids (J. S. Birman and A. Lebgober, eds.), Contemp. Math., vol. 78, AMS, Providence, 1988, pp. 605–613.

[OrTe] P. Orlik and H. Terao, *Arrangements of hyperplanes*, Springer, Berlin - Heidelberg, 1992.

[Ri] J. Rickard, *Finite group actions and étale cohomology*, Publ. de l'I.H.E.S. **80** (1995), 81–94.

[Ro1] R. Rouquier, *Caractérisation des caractères et pseudo-caractères*, J. Algebra **180** (1996), 571–586.

[Ro2] ———, *Familles et blocs d'algèbres de Hecke*, Comptes Rendus Acad. Sc. **329** (1999), 1037–1042.

[Sch] K. -D. Schewe, *Blöcke exzeptioneller Chevalley-Gruppen*, Dissertation, vol. 165, Bonner Mathem. Schriften, Bonn, 1985.

[ShTo] G. C. Shephard and J. A. Todd, *Finite unitary reflection groups*, Canad. J. Math. **6** (1954), 274–304.

[Sp] T. A. Springer, *Regular elements of finite reflection groups*, Invent. Math. **25** (1974), 159–198.

[St1] R. Steinberg, *Differential equations invariant under finite reflection groups*, Trans. Amer. Math. Soc. **112** (1964), 392–400.

[St2] ———, *Endomorphisms of linear algebraic groups,*, Memoirs of the AMS, number 80, American Mathematical Society, Providence, 1968.

Institut Henri–Poincaré, 11 rue Pierre et Marie Curie, F–75231 Paris Cédex 05, France, et Institut de Mathématiques de Jussieu–Chevaleret (CNRS UMR 7586), Université Paris 7, Case 7012, 2 Place Jussieu F–75251 Paris Cédex 05, France

E-mail address: broue@ihp.jussieu.fr

URL: http://www.math.jussieu.fr/~broue

Current Developments in Mathematics, 2000
Pages 109 - 147

Stochastic Hydrodynamics

Weinan E

ABSTRACT. We review the progress made recently on the mathematical understanding of prototypical stochastic partial differential equations in fluid mechanics, including the stochastic Burgers equation, stochastic passive scalar equations and stochastic Navier-Stokes equation, with particular emphasis on issues related to the problem of turbulence. Questions addressed include the existence and uniqueness of invariant measures, construction of the invariant measures and typical behavior of stationary solutions. Also discussed is the interplay between regularity and dissipation in the inviscid limit. The concepts and methods reviewed here include the one force – one solution principle, variational methods, master equations, generalized flows, reduction to Gibbsian dynamics, hypoellipticity, etc.

CONTENTS

Based on talks given at the meeting: *Current Development in Mathematics*, Harvard University, December 2000.

1. Introduction

The fundamental problem in stochastic hydrodynamics is the problem of turbulence, which is widely regarded as the last unsolved problem in classical physics. The physical issues are very well summarized in the book of Monin and Yaglom "Statistical Fluid Mechanics" [**86**]. The mathematical issues, together with a report on recent progress on the understanding of these issues, are the subject of this review.

We start with a physical description of the problem. We are interested in the dynamics of a viscous fluid governed by the Navier-Stokes equation

$$\begin{cases} \mathbf{u}_t + (\mathbf{u} \cdot \nabla)\mathbf{u} + \nabla p = \nu \Delta \mathbf{u} + \mathbf{f} \\ \nabla \cdot \mathbf{u} = 0 \end{cases} \tag{1}$$

Here $\mathbf{u} = (u_1, u_2, u_3)$ is the velocity field, p is the pressure, $\mathbf{f}$ represents the forcing on the fluid, ν is the viscosity of the fluid. It is helpful to non-dimensionalize the system (1). Let U, L be the typical velocity and length scale respectively. Define $T = \frac{L}{U}$, and perform a change of variable $\mathbf{x} = \mathbf{x}'L$, $t = t'T$, $\mathbf{u} = \mathbf{u}'U$. Omitting the primes, we get

$$\begin{cases} \mathbf{u}_t + (\mathbf{u} \cdot \nabla)\mathbf{u} + \nabla p = \frac{1}{Re}\Delta \mathbf{u} + \mathbf{f} \\ \nabla \cdot \mathbf{u} = 0 \end{cases} \tag{2}$$

where $Re = \frac{UL}{\nu}$ is the dimensionless Reynolds number, which is the key parameter in our problem. We are interested in the situation when Re is very large. Take a typical car on the highway, we have U= 65 miles/hour $\doteq$ 3000 cm/s, L=length of the car $\doteq$ 800 cm, ν=viscosity of air = 0.132 cm^2/s, which gives $Re \doteq 1.8 \times 10^7$.

In a high Reynolds number flow, the most ubiquitous phenomenon is the presence of a wide range of length scales and a hierarchy of eddies (swirls, vortices). The size of the range of scales was estimated by Kolmogorov [**59, 60**] to be $O(Re^{3/4})$, on the assumption that there is a disparity between the length scales at which energy is supplied to the system, and the scale ℓ_d at which energy is dissipated through viscosity – the viscous cut-off scale. In between these two scales, energy is transported but not dissipated, resulting in a constant energy flux $\bar{\varepsilon}$. This range is called the inertial range.

Kolmogorov further postulated that the inertial range properties of a fully developed turbulent flow is universal. Using a scaling argument, he predicted that

$$\left\langle |\mathbf{u}(\mathbf{x}+\mathbf{r},t) - \mathbf{u}(\mathbf{x},t)|^n \right\rangle \sim C_n |\mathbf{r}|^{\frac{n}{3}} \bar{\varepsilon}^{\frac{n}{3}} \tag{3}$$

for $\ell_d \ll |\mathbf{r}| \ll L$, where C_n should be universal constants, $\langle \quad \rangle$ means either time average or ensemble average over many realizations of the experiment. (3) means that typical turbulent velocity fields are Hölder continuous with exponent 1/3.

Most notable special cases of (3) are found when $n = 2$ and 3. When $n = 2$, one can represent (3) in Fourier space. Let $\hat{\mathbf{u}}$ be the Fourier transform of $\mathbf{u}$, and $E(k) = \langle \sum_{k<|\mathbf{k}|\leq k+1} |\hat{\mathbf{u}}(\mathbf{k})|^2 \rangle$ be the energy spectrum. Then (3) translates to

$$E(k) \doteq C_k \bar{\varepsilon}^{\frac{2}{3}} k^{-\frac{5}{3}}, L^{-1} \ll k \ll k_d = \ell_d^{-1} \tag{4}$$

This is the well-known $-\frac{5}{3}$ spectrum. When $n = 3$, Kolmogorov derived, from the Navier-Stokes equation, the relation

$$\left\langle (u_{\|}(\mathbf{x}+\mathbf{r},t) - u_{\|}(\mathbf{x},t))^3 \right\rangle \sim -\frac{4}{5}\bar{\varepsilon}|\mathbf{r}| \tag{5}$$

for $\ell_d \ll |\mathbf{r}| \ll L$, assuming that the inertial range properties are homogenous and isotropic. Here $u_{\|}$ is the component of the velocity field parallel to $\mathbf{r}$. It is worth emphasizing that (5) is among the very few results in turbulence theory that are actually derived from the Navier-Stokes equation.

Kolmogorov's theory was immediately challenged by Landau [**48**], who argued that dissipation processes in turbulence are tied with the non-universal large scales, hence the small scale inertial range properties cannot be truly universal. In reality, dissipation is intermittent, dominated by strong but rare events. This means that there are nontrivial corrections, to a mean-field theory such as Kolmogorov's due to fluctuation. Specifically, (3) should be changed to

$$\left\langle |\mathbf{u}(\mathbf{x}+\mathbf{r},t) - \mathbf{u}(\mathbf{x},t)|^n \right\rangle \sim C_n L^{\frac{n}{3}-\alpha_n} |\mathbf{r}|^{\alpha_n} \bar{\varepsilon}^{\frac{n}{3}} \tag{6}$$

$\mathbf{u}$ is said to have multi-fractal scaling if α_n is not linear in n. Work on such intermittency corrections has been a central theme in modern turbulence theory [**3, 4, 20, 48, 61**].

We now turn to the mathematical formulation of the problem. To begin with, we have to understand the sense of averaging, the bracket $\langle \quad \rangle$. Since a turbulent flow is intrinsically unstable and stochastic – a small perturbation in the initial data leads to large differences in the solution, it is natural to use a statistical description. Therefore we will be interested in statistical steady states, the invariant measures of the Navier-Stokes equation (2). We can now formulate the mathematical questions as follows.

1. The existence and uniqueness of an invariant measure for the dynamics of (2), for both finite and infinite Reynolds numbers. The existence of an invariant measure at infinite Reynolds number ensures that there is sufficient dissipation in the system even in the limit of zero viscosity. Uniqueness guarantees that the dynamics is ergodic in the phase space, as is commonly assumed in turbulence theory.
2. Let μ_0 be the invariant measure of (2) at infinite Reynolds number. We can then study the regularities of the statistically stationary solutions. In particular, we can study the decay of their energy spectrum.

At the present time, neither question is answered. Worse than that, we still face the well-known problem that uniqueness of solutions to the three-dimensional Navier-Stokes equation is still yet to be proved, leaving us with no starting point to address these questions.

However, it is important to realize that, from a physical viewpoint, the problem of hydrodynamic turbulence is just one of many problems exhibiting turbulent behavior, e.g. a wide range of active scales, slow algebraic decay of the energy spectrum, intermittent dissipation processes, etc. The mathematical issues discussed above are equally important in these problems. A list of such "generalized turbulence problems" include:

1. Two-dimensional hydrodynamic turbulence.
2. Wave turbulence.
3. Shell models of turbulence.
4. Burgers turbulence.
5. Turbulent transport of passive scalars.

These problems are closely related to the original turbulence problem and are expected to be simpler to deal with. There are many more such problems in non-equilibrium statistical physics outside the traditional areas of hydrodynamics. However, here we will restrict ourselves to problems in hydrodynamics.

Below we will discuss progress made recently on the problems 1, 4 and 5. Our basic strategy is to consider the appropriate PDEs under a large scale stochastic forcing and study the properties of the statistical steady states. For problems 4 and 5, we obtain rather precise descriptions of the small scales. For problem 1, we prove ergodicity of the dynamics.

Before proceeding further, let us remark on the stochastic setting of our approach. Traditionally problems in PDEs have been studied using deterministic methods. However, deterministic methods usually predict the worst-case scenario and it is awkward to avoid atypical behavior such as starting from an unstable fixed point. Therefore if our main interest is the *typical* behavior of solutions of a complex system in phase space, it is much more advantageous to use a stochastic setting which provides a natural framework to describe typical behavior of the solutions. On the other hand, it is important, in a stochastic setting, to distinguish results that are tied with the randomness in the formulation and those that are intrinsic properties of the underlying PDEs.

2. Stochastic Burgers Equation

In this section, we consider the Burgers equation with stochastic forcing

$$\frac{\partial u}{\partial t} + u\frac{\partial u}{\partial x} = \nu\frac{\partial^2 u}{\partial x^2} + f \tag{7}$$

Besides being a prototype for nonlinear waves, (7) is also a canonical example in non-equilibrium statistical physics. As such, it describes the statistical mechanics of strings in a random potential. The string is assumed to be directed, i.e. there exists a time axis such that the position of the string is a single-valued function over this time axis. Vortex lines in high temperature superconductors [**10**], charge density waves [**43**], directed polymers and stochastic interfaces in $1+1$ dimensional SOS models [**58**] are all examples of such strings. To see this connection, define the partition function for the configurations of the strings over the time interval $[0,t]$ assuming that they are pinned at time t at location x:

$$Z(x,t) = \langle \exp\Big(-\beta \int_0^t \left(\frac{1}{2}|\dot{\xi}(\tau)|^2 + V(\xi(\tau),\tau)\right) d\tau\Big) | \xi(t) = x \rangle, \tag{8}$$

where $\beta = 1/kT$, k is the Boltzmann constant, T is the temperature. The averaging $\langle\ \rangle$ is over all path ξ such that $\xi(t) = x$. The first term in the exponent is the elastic energy and the second term is the potential energy with V being the potential. The associated free energy $\varphi = kT \log Z$ satisfies

$$\frac{\partial \varphi}{\partial t} + \frac{1}{2}|\nabla \varphi|^2 = kT\Delta\varphi + V. \tag{9}$$

In one dimension, let $u = \partial\varphi/\partial x$, we obtain (8) with $\nu = kT$. When V is the space-time white noise, (9) is the well-known KPZ equation [**58**].

2.1. Invariant Measures. The existence of an invariant measure for (2) is only understood so far on finite domains. Consider, for example, (7) on $[0, 2\pi]$ with periodic boundary condition. We will sometimes identify the domain as S^1, the unit circle. The forcing function can be expressed as:

$$f(x,t)dt = \sum_k f_k(x)dW_k(t), \tag{10}$$

where the $\{W_k(\cdot)\}$'s are independent Wiener processes. We will assume that

$$f_k(\cdot) \in C^3(S^1), \qquad |\frac{\partial f_k}{\partial x}(x)| \le \frac{C}{k^2}, \tag{11}$$

for all k. We will use $(\Omega, \mathcal{F}, P)$ to denote the probability space for the forcing functions f, and $\omega \in \Omega$ to denote a typical realization of the force. $\mathcal{F}_t$ denotes the σ-algebra generated by the forces up to time t.

When $\nu = 0$, (7) is understood in the weak sense with solutions satisfying the entropy condition. The precise definition for the random case is given in [**30**]. In this case, we write (7) as

$$\frac{\partial u}{\partial t} + \frac{1}{2}\frac{\partial}{\partial x}u^2 = f. \tag{12}$$

In accordance with the conservative nature of the Burgers equation, we will assume that the f_k's satisfy: $\int_{S^1} f_k(x)dx = 0$. Then (12) admits a conservation law: $\int_{S^1} u(x,t)dx = \int_{S^1} u(x,0)dx$. Without loss of generality, we will assume $\int_{S^1} u(x,t)dx = 0$ and restrict ourselves to the space of functions satisfying this condition.

A natural phase space for (12) is the Skorohod space D on S^1 which consists of functions admitting only discontinuities of the first kind [**9**]. Let $\mathcal{D}$ be the Borel σ-algebra on D. (12) can then be viewed as a Markov process on D with transition probability

$$P_t(u, A) = \int_\Omega \chi_A(u,\omega)P(d\omega) \tag{13}$$

where $u \in D$, $A \in \mathcal{D}$, and

$$\chi_A(u,\omega) = \begin{cases} 1 & \text{if } S_\omega(t)u \in A \\ 0 & \text{otherwise} \end{cases} \tag{14}$$

Here $S_\omega(t)$ denotes the solution operator of (12) at time t with forcing ω.

DEFINITION 1. *An invariant measure $\mu_0(du)$ of the Markov process (12) is a measure on $(D, \mathcal{D})$ satisfying*

$$\mu_0(A) = \int_D P_t(u, A)\mu_0(du) \tag{15}$$

for any $A \in \mathcal{D}$ and any $t > 0$.

THEOREM 2. [**30**]. *(12) admits a unique invariant measure on the space D.*

The proof of Theorem 2 was based on the following variational characterization of weak solutions of (12) satisfying the entropy condition. Let $\xi : [t_1, t_2] \to R^1$ be a Lipschitz continuous curve. Define the action functional

$$\begin{aligned} \mathcal{A}_{t_1,t_2}(\xi) &= \int_{t_1}^{t_2} \Big(\frac{1}{2}(\dot\xi(s))^2 - \sum_k f_k(\xi(s))\dot\xi(s)(W_k(s) - W_k(t_1))\Big) ds \\ &+ \sum_k f_k(\xi(t_2))(W_k(t_2) - W_k(t_1)) \end{aligned} \tag{16}$$

Then we have for $\tau < t$,

$$u(x,t) = \frac{\partial}{\partial x} \inf_{\xi(t)=x} \Big\{ \mathcal{A}_{\tau,t}(\xi) + \int_0^{\xi(\tau)} u(z,\tau)dz \Big\} \tag{17}$$

Minimizers of the functional in (16) satisfy the following Euler-Lagrange equation:

$$d\gamma(s) = v(s)ds, \qquad dv(s) = \sum_{k=1}^{\infty} f_k(\gamma(s))dW_k(s) \tag{18}$$

Using (17), we can construct solutions of (12) as follows. Fix t. For values of x such that the minimizer to the functional (16) is unique, say $\xi(\cdot)$, then $u(\cdot,t)$ is continuous at x, and $u(x,t) = \dot\xi(t)$. On the other hand, for values of x such that the minimizers to the variational problem are not unique, then $u(\cdot,t)$ is discontinuous at x with $u(x+,t) = \inf_\alpha\{\dot\xi_\alpha(t)\}$ and $u(x-,t) = \sup_\alpha\{\dot\xi_\alpha(t)\}$ where $\{\xi_\alpha\}$ denotes the family of minimizers of (16) such that $\xi_\alpha(t) = x$.

This construction is just a reformulation of the method of characteristics for weak solutions. It is a variational formulation of the backward characteristics defined in [**23**]. In particular, the Euler-Lagrange equation (18) is nothing but the equation for the characteristics of (12).

Of particular interest to the construction of the invariant measure is a special class of minimizers called the one-sided minimizers (OSM).

DEFINITION 3. *A piecewise C^1-curve $\xi : (-\infty, 0] \to S^1$ is a one-sided minimizer if for any Lipschitz continuous $\tilde\xi : (-\infty, 0] \to S^1$ such that $\tilde\xi(0) = \xi(0)$ and $\tilde\xi = \xi$ on $(-\infty, \tau]$ for some $\tau < 0$, we have*

$$\mathcal{A}_{s,0}(\xi) \le \mathcal{A}_{s,0}(\tilde\xi), \tag{19}$$

for all $s \le \tau$. Similarly, we define one-sided minimizers on $(-\infty, t]$, for $t \in R^1$.

Next we ask the following question: Given (x,t), how many OSMs ξ exist such that $\xi(t) = x$? This question is answered by studying the intersection properties of OSMs. As a general fact in the calculus of variations, two different OSMs cannot intersect more than once. In other words, if ξ_1, ξ_2 are two OSMs that there exist $t_1, t_2 \in R^1, t_1 \ne t_2$, and $\xi_1(t_1) = \xi_2(t_1), \xi_1(t_2) = \xi_2(t_2)$, then $\xi_1 \equiv \xi_2$ on their common domain of definition [**80**]. However, more is true in the random case. If ξ_1 and ξ_2 are two different OSMs, such that $\xi_1(s) = \xi_2(s)$ for some s, then neither ξ_1 nor ξ_2 can be extended as an OSM beyond the interval $(-\infty, s]$. This is because that in the random case, two OSMs always have an effective intersection at $t = -\infty$. The precise formulation of this property is given in [**30**].

These intersection properties have far-reaching consequences. Let us fix $t = 0$. By considering the image of all OSMs at $t = -1$, it is easy to see that the set

$$I_\omega = \{x \in S^1 : \text{there exist more than one OSM } \xi \text{ such that } \xi(0) = x\},$$

can at most be a countable set for almost all $\omega \in \Omega$. Therefore we can define:

$$u_\omega(x,0) = \dot{\xi}(0), \tag{20}$$

where ξ is the OSM such that $\xi(0) = x$. $u_\omega(\cdot,0)$ is a well-defined function in $L^\infty(S^1)$ for almost all ω.

Similar construction can be carried out for any $t \in R^1$ which defines $u_\omega(\cdot,t)$. Furthermore it is easy to conclude from the variational principle (17) that

$$u_\omega(\cdot,t) = S_\omega(t)u_\omega(\cdot,0), \tag{21}$$

for $t > 0$. In other words, u_ω is a solution of (12). This is a special solution of (12) for the given realization of the force ω. This construction is an example of the so-called "one force, one solution" principle, namely for almost every realization of the force, we can associate one and only one special solution to that force. In other words, the random attractor consists of one and only one trajectory. It is straightforward to check that the distribution of the mapping:

$$\Phi_0 : \omega \to u_\omega(\cdot,0), \tag{22}$$

is an invariant measure for (12). Moreover Φ satisfies the invariance principle:

$$\Phi_0(\theta_t\omega) = S_\omega(t)\Phi_0(\omega) \tag{23}$$

Therefore we have

THEOREM 4. [**30**]. *There exists an invariant mapping $\{\Phi_0\} : \Omega \to D$ satisfying (23). Furthermore the invariant measure μ_0 is the distribution of Φ_0.*

Theorem 4 states that the statistically stationary solutions are functionals of the forcing. Uniqueness of the invariant measure follows from the fact that the OSMs are largely unique.

It is shown in [**93, 94**] that for $\nu > 0$, (12) also admits a unique invariant mapping Φ_0^ν whose distribution is the unique invariant measure for (12), denoted by μ_ν. We have

THEOREM 5. [**30**].

$$\Phi_0^\nu(\omega) \to \Phi_0(\omega), \tag{24}$$

in $L^1(S^1)$ for almost every $\omega \in \Omega$. Consequently

$$\mu_\nu \to \mu_0, \tag{25}$$

weakly.

We remark that (24), (25) are a different kind of convergence statement from standard theorems on zero-viscosity limits of (12) studied in the PDE literature [**95, 97**]. There we are given a sequence of initial data that converge in the inviscid limit, and we ask whether convergence still holds at later times. Here we are not given any initial data, and we proceed to establish convergence with the only information that the solutions are defined for all $t \in R^1$ in a special way using the OSMs. Consequently the techniques used to prove Theorem 5 are very different from the ones used in the PDE literature to study inviscid limits. See [**30**].

Our next task is to characterize the statistically stationary solutions. This requires a non-degeneracy condition to the effect that the process (18) starting at any $x \in S^1$ is transitive on S^1. This condition is generically satisfied. However it is violated if the sum in (10) contains only one term. We refer to [**30**] for a detailed formulation and examination of this condition. Under this non-degeneracy condition, we have

THEOREM 6. **[30]**. *For almost all ω, u_ω satisfies the following:*

1. *There exists a unique two-sided minimizer (TSM, defined below) $y_\omega : R^1 \to S^1$ which is a characteristic of (12) associated with the solution u_ω.*
2. *There exists a unique so-called main shock $\gamma_\omega : R^1 \to S^1$, which is a continuous shock curve defined for all $t \in R^1$.*
3. *The TSM is a hyperbolic trajectory for the dynamical systems (18).*
4. *For any $t \in R^1$, there exist global stable and unstable manifolds associated with y_ω at time t, denoted by $W^s_\omega(t)$ and $W^u_\omega(t)$ respectively, on the phase space $S^1 \times R^1$.*
5. *The graph of $u_\omega(\cdot, t)$ is a subset of $W^u_\omega(t)$.*

As a corollary, we have that for each fixed t, u_ω is almost surely a piecewise smooth function.

DEFINITION 7. *A piecewise C^1-curve $\xi : (-\infty, +\infty) \to S^1$ is a two-sided minimizer if for any Lipschitz continuous $\tilde{\xi} : (-\infty, +\infty) \to S^1$ such that $\tilde{\xi} = \xi$ away from a compact set, we have*

$$\mathcal{A}_{-s,s}(\xi) \leq \mathcal{A}_{-s,s}(\tilde{\xi}), \tag{26}$$

for large enough s.

The possibility of establishing hyperbolicity of TSMs in the random case comes from the following:

BASIC COLLISION LEMMA 8. *Assuming that the non-degeneracy condition holds. Then there exists a constant p_0, depending only on the $\{f_k\}$'s, with the following property: Given an arbitrary pair of points $(x_1, x_2) \in [0,1]^2$ at $t = 0$ whose positions are $\mathcal{F}_0$-measurable,*

$$P\{x_1, x_2 \text{ merge before } t = 1\} > p_0. \tag{27}$$

Heuristically two points merge before $t = 1$ if the forward characteristics emanating from them intersect before $t = 1$. This of course depends on the forces as well as the solution at $t = 0$. The lemma states that independent of the solution at $t = 0$, one can always find a set of forces with positive measure under which the two points merge.

The proof of the Basic Collision Lemma relies on PDE techniques and is given in the Appendix D of **[30]**.

The Basic Collision Lemma provides the mechanism for the origin of the hyperbolicity. In particular, the uniqueness of the TSM and the main shock is a simple consequence of the Basic Collision Lemma.

The regularity and structural properties described here are used in **[29, 34, 35]** to study the statistical behavior of the Burgers equation. A summary of these results is given below.

2.2. Statistical Theory. We now address the questions frequently asked in the physics literature regarding (12), building upon the regularity results described in §2.1. Since we have established the existence of μ_0 which is the statistical steady state at the inviscid limit, we can restrict our attention to this case. We warn the reader that some of the results described in this section are not fully rigorously proved. This section is a summary of [**34, 35**]

We will denote by B the spatial correlation function of the forcing:

$$\langle f(x,t)f(y,s)\rangle = 2B(x-y)\delta(t-s)$$

2.2.1. *Structure Functions.* The fact that u_ω is piecewise smooth implies easily that

$$S_p(r) = \left\langle |u(x+r)-u(x)|^p \right\rangle = \begin{cases} r^p\langle|\xi|^p\rangle + o(r^p) & \text{if } 0 \le p < 1, \\ r\rho\langle|s|^p\rangle + o(r) & \text{if } 1 < p, \end{cases} \tag{28}$$

where ρ is the number density of the shocks, which is finite from the results described earlier, s is the jump across the shocks, ξ is the regular part of the velocity gradient:

$$\frac{\partial}{\partial x}u(x,t) = \xi(x,t) + \sum_j s(y_j,t)\delta(x-y_j), \tag{29}$$

with $\xi(\cdot,t) \in L^1(S^1)$.

If we write $S_p(r) = C_p r^{\alpha_p}$, for $r \ll 1$, then $\alpha_p = p$ for $0 \le p < 1$, and $\alpha_p = 1$, for $p \ge 1$. This is a reflection of the fact that as far as regularity is concerned, almost everywhere the solution is either Lipschitz continuous or discontinuous. The linear part in the graph of α_p probes the Lipschitz continuous part of the solution. The flat part probes the discontinuous part of the solution. Such a situation is sometimes referred to as "bifractal".

2.2.2. *Velocity Gradient PDF.* More difficult are the questions of probability density functions (PDFs) for quantities such as u, $\partial u/\partial x$ and $\delta u(x,t) = u(x+y,t) - u(y,t)$. In particular, the PDF of $\xi = \partial u/\partial x$ (suitably defined), $Q(\xi)$, has attracted a lot of attention in recent years. In the inviscid limit, it is agreed that $Q(\xi)$ has the behavior

$$Q(\xi) \sim \begin{cases} C_-|\xi|^{-\alpha} & \text{as } \xi \to -\infty, \\ C_+\xi^\beta e^{-\xi^3/(3B_1)} & \text{as } \xi \to +\infty. \end{cases} \tag{30}$$

where C_-, C_+ are constants, $B_1 = -B_{xx}(0)$. But many different values of α and β have been proposed (see [**35**] for a review).

A priori, there is even an issue how to define Q. In the inviscid limit one can define the PDF for the velocity divided difference $(u(y+\delta,t) - u(y,t))/\delta$, $Q^\delta(\xi,t)$ and then take the limit as $\delta \to 0$. An alternative is to first define the PDF of $\partial u/\partial x$ for the viscous problem, call it $Q^\nu(\xi,t)$, and then take the limit as $\nu \to 0$:

$$\underline{Q}(\xi,t) = \lim_{\nu\to 0} Q^\nu(\xi,t), \qquad Q(\xi,t) = \lim_{\delta\to 0} Q^\delta(\xi,t). \tag{31}$$

While a fully rigorous proof was not constructed, [**35**] presented strong evidence that for the case studied here,

$$Q = \underline{Q}. \tag{32}$$

Below we will concentrate on Q.

Using the calculus for functions of bounded variation [**101**], we can derive an equation for Q

$$\frac{\partial Q}{\partial t} = \xi Q + \frac{\partial}{\partial \xi}\left(\xi^2 Q\right) + B_1\frac{\partial^2 Q}{\partial \xi^2} + F(\xi,t), \tag{33}$$

where

$$F(\xi,t)=\rho\int_{-\infty}^{0}sV(s,\xi,t)ds\leq 0. \tag{34}$$

Here $V(s,\xi,t)=V_{+}(s,\xi,t)+V_{-}(s,\xi,t)$, $V_{\pm}(s,\xi_{\pm},t)$ are the PDFs of $(s(y,t),\xi_{\pm}(y,t)=\partial u(y\pm,t)/\partial y)$ conditional on y being a shock position. This equation should be compared with the equation for Q^{ν}:

$$\frac{\partial Q^{\nu}}{\partial t}=\xi Q^{\nu}+\frac{\partial}{\partial\xi}(\xi^{2}Q^{\nu})+B_{1}\frac{\partial^{2}Q^{\nu}}{\partial\xi^{2}}-\nu\frac{\partial}{\partial\xi}\Big(\Big\langle\frac{\partial^{2}\xi}{\partial x^{2}}\Big|\xi\Big\rangle Q^{\nu}\Big), \tag{35}$$

where $\langle\partial^{2}\xi/\partial x^{2}|\xi\rangle$ is the ensemble-average of $\partial^{2}\xi/\partial x^{2}$ conditional on ξ. We see that

$$F(\xi,t)=-\lim_{\nu\to 0}\nu\frac{\partial}{\partial\xi}\Big(\Big\langle\frac{\partial^{2}\xi}{\partial x^{2}}\Big|\xi\Big\rangle Q^{\nu}\Big),$$

Even though we are primarily interested in statistical steady states, we have written down the master equations for the more general case including transients.

Integrating (33), we get

$$\frac{d}{dt}\int_{R^{1}}Q(\xi,t)d\xi=\langle\xi\rangle+\rho\langle s\rangle=0. \tag{36}$$

Consequently

$$\lim_{\nu\to 0}\int_{R^{1}}\xi Q^{\nu}(\xi,t)d\xi\neq\int_{R^{1}}\xi Q(\xi,t)d\xi, \tag{37}$$

since the left hand side vanishes due to homogeneity.

Even though (33) is not a closed equation since the form of F is unknown, we can already obtain from it non-trivial information. As an example, we have

$$\lim_{|\xi|\to+\infty}|\xi|^{3}Q(\xi,t)=0, \tag{38}$$

i.e. Q decays faster than $|\xi|^{-3}$ as $\xi\to-\infty$.

This result rules out all the proposed value of α except that of **[29]** which gives $\alpha=7/2$.

(38) is obtained by combining (36) together with the following:

1. Take the first moment of (33) gives

$$\frac{d}{dt}\langle\xi\rangle=[\xi^{3}Q]_{-\infty}^{+\infty}+\frac{\rho}{2}\Big(\langle s\xi_{-}\rangle+\langle s\xi_{+}\rangle\Big).$$

2. Along the shock, we have

$$\frac{d}{dt}(\rho\langle s\rangle)=-\frac{\rho}{2}(\langle s\xi_{-}\rangle+\langle s\xi_{+}\rangle),$$

which is a consequence of the equations for the dynamics of the shocks.

3. By the definition of F,

$$\int_{R^{1}}\xi F(\xi,t)d\xi=\frac{\rho}{2}\Big(\langle s\xi_{-}\rangle+\langle s\xi_{+}\rangle\Big),$$

It can also be shown that $\xi F(\xi,t)$ is absolutely integrable on R^{1} for all t.

2.2.3. *Asymptotics for the Statistical Stationary State.* In statistical steady state, (33) becomes

$$\xi Q + \frac{d}{d\xi}(\xi^2 Q) + B_1 \frac{d^2 Q}{d\xi^2} + F(\xi) = 0. \tag{39}$$

This is a second order ODE with an inhomogeneous term F. We can write the general solutions of (39) as

$$Q = Q_s + C_1 Q_1 + C_2 Q_2, \tag{40}$$

where Q_1 and Q_2 are two linearly independent solutions of the homogeneous equation, and Q_s is a particular solution. For example, we can take:

$$Q_1(\xi) = \xi \mathrm{e}^{-\Lambda}, \tag{41}$$

$$Q_2(\xi) = 1 - \frac{\xi \mathrm{e}^{-\Lambda}}{B_1} \int_{-\infty}^{\xi} \xi' \mathrm{e}^{\Lambda'} d\xi', \tag{42}$$

$$Q_s(\xi) = \frac{1}{B_1} \int_{-\infty}^{\xi} \xi' F(\xi') d\xi' - \frac{\xi \mathrm{e}^{-\Lambda}}{B_1} \int_{-\infty}^{\xi} \mathrm{e}^{\Lambda'} G(\xi') d\xi', \tag{43}$$

where

$$\Lambda = \frac{\xi^3}{3B_1}, \qquad G(\xi) = F(\xi) + \frac{\xi}{B_1} \int_{-\infty}^{\xi} d\xi' \, \xi' F(\xi'). \tag{44}$$

2.2.4. *Bounds from Realizability Constraints.* Using the realizability constraint $Q \in L^1(R^1)$, $Q \geq 0$, we can show that

$$Q = Q_s, \tag{45}$$

and

$$\lim_{\xi \to +\infty} \xi^{-2} \mathrm{e}^{\Lambda} F(\xi) = 0. \tag{46}$$

Indeed, starting from $Q = Q_s + C_1 Q_1 + C_2 Q_2$, we have:

1. $Q_s \in L^1(R^1)$, $Q_2 \in L^1(R^1)$, but $Q_1 \notin L^1(R^1)$. Therefore C_1 must be zero.
2. As $|\xi| \to +\infty$, $|Q_2| > |Q_s|$, but $Q_2 > 0$ as $\xi \to -\infty$, and $Q_2 < 0$ as $\xi \to +\infty$. Therefore C_2 must be zero.
3. As $\xi \to +\infty$, $Q_s \geq 0$, *iff* (46) holds.

These statements can be established by evaluating Q_2, Q_s using Laplace's method. We arrive at:

$$Q(\xi) \sim \begin{cases} |\xi|^{-3} \int_{-\infty}^{\xi} \xi' F(\xi') d\xi' & \text{as } \xi \to -\infty, \\ C_+ \xi \mathrm{e}^{-\Lambda} & \text{as } \xi \to +\infty. \end{cases} \tag{47}$$

Once again, we obtain (38). Furthermore we get $\beta = 1$ in (30).

2.2.5. *The Exponent 7/2.* Let $W(s, \xi_+, \xi_-, x, t)$ be the PDF of

$$(u(y+x,t) - u(y-x,t), \xi(y+x,t), \xi(y-x,t)),$$

condition on y being a shock location. Then

$$V_\pm(s, \xi_\pm, t) = \int_{R^1} W(s, \xi_+, \xi_-, 0, t) d\xi_\mp. \tag{48}$$

W satisfies an equation of the form [**35**]

$$\frac{\partial W}{\partial t} = \mathcal{O}W + S, \tag{49}$$

where $\mathcal{O}$ is a differential operator in x, s, $\xi_\pm$, S is a source term accounting for shock creation and collision. Information on the source term S can be obtained using local analysis around shock creation and collision points. Upon using this information in (48), we get:

$$F(\xi) \sim C|\xi|^{-5/2} \qquad \text{as } \xi \to -\infty, \tag{50}$$

Therefore

$$Q(\xi) \sim |\xi|^{-3} \int_{-\infty}^{\xi} \xi' F(\xi') d\xi' = C_-|\xi|^{-7/2} \qquad \text{as } \xi \to -\infty. \tag{51}$$

which confirms the prediction of [**29**].

The analysis in [**35**] that accomplishes this last step is quite involved. The result of that is a confirmation of the geometric picture proposed in [**29**], namely the leading order contribution to the left tail of Q comes from the boundary of the set of the shocks, here the points of shock creation. This geometric picture may have interesting consequences in higher dimension. The analysis in [**35**] is also a success in working with the master equation without making closure assumptions. In a sense closure in [**35**] is achieved through dimension reduction. The PDF of ξ is first expressed in terms of the lower dimensional dissipative structures, here the shocks. The PDF for the shocks and the shock environment is further expressed in terms of the singular structures on the shocks, namely the points of shock creation and collision, which are then amenable to local analysis. Clearly this approach should be applicable to high dimensions and should yield more interesting results.

As we mentioned earlier, the value of α has been a point of controversy for sometime. Using operator product expansion, Polyakov [**90**] suggested the value $\alpha = \frac{5}{2}$. Boldyrev [**11**] extended Polyakov's analysis and suggested that α should be in the interval $[2, 3]$. The value of $\alpha = 3$ was obtained by Gotoh and Kraichnan [**52**] who argued that the dissipation term can be neglected in the tails of the PDF. Other values of α are proposed in [**12**]. E, Khanin, Mazel and Sinai [**29**] suggested the value $\alpha = \frac{7}{2}$ by arguing that the dominant contribution comes from the region near the points of shock creation. Very strong analytical support to this result was given in [**34, 35**]. Numerical verification was hindered by the fact that most numerical schemes contain too much numerical viscosity which pollutes the tails of the PDF. Using an innovative Lagrangian approach, Bec was finally able to obtain clean numerical results which give overwhelming support to the value $\alpha = \frac{7}{2}$ [**7, 6**]. However, we should remark that even though the prediction for the value of α in [**90**] is likely to be incorrect, the master equation approach initiated there may have strong impact in the field of turbulence theory.

2.3. Stochastic Hamilton-Jacobi Equation in High Dimensions. Here we briefly summarize the work of Iturriaga and Khanin [**55**] which extends some of the results in [**30**] to high dimensions. The most natural setting is in the context of stochastically forced Hamilton-Jacobi equations on a d-dimensional torus $\mathbb{T}^d = \mathbb{R}^d/\mathbb{Z}^d$:

$$\varphi_t + \frac{1}{2}|\nabla\varphi|^2 + F^\omega(\mathbf{x}, t) = 0 \tag{52}$$

where $F^\omega(\mathbf{x}, t) = \sum_{k=1}^N F_k(\mathbf{x})\dot{W}_k(t)$, and $\{\dot{W}_k(t)\}$ are independent white-noises. If we let $\mathbf{u} = \nabla\varphi$, then formally we obtain the d-dimensional Burgers system for $\mathbf{u}$:

$$\mathbf{u}_t + (\mathbf{u}\cdot\nabla)\mathbf{u} + \nabla F^\omega(\mathbf{x}, t) = 0 \tag{53}$$

This connection can be made rigorous if we add a viscous term. However this only gives irrotational solutions of (53). In principle (53) can have solutions which are not in the form of $\mathbf{u} = \nabla\varphi$ and these solutions may have entirely different properties.

The proper notion of weak solutions is that of the viscosity solution which can also be constructed via the variational principle [**75**]:

$$\varphi(\mathbf{x}, t) = \inf_{\gamma(t)=\mathbf{x}} \left(\varphi(\gamma(0), 0) + \mathcal{A}^\omega_{0,t}(\gamma)\right)$$

where

$$\mathcal{A}^\omega_{s,t} = \int_s^t \left(\frac{1}{2}\dot{\gamma}(\tau)^2 - F^\omega(\gamma(\tau), \tau)\right) d\tau$$

and the infimum is taken overall absolutely continuous curves γ on $[0, t]$ such that $\gamma(t) = \mathbf{x}$. Here as in §2.1 we have restricted our attention to solutions that satisfy $\int_{\mathbb{T}^d} \mathbf{u}(\mathbf{x}, t)dx = 0$, but extension to the general case of $\int_{\mathbb{T}^d} \mathbf{u}(\mathbf{x}, t)dx = \mathbf{b} \neq 0$ is straightforward (see [**55**]).

As in the one dimensional case, one can define the notion of OSMs and TSMs (also called global minimizers). The existence of a unique invariant measure can be proved by studying OSMs (see [**55**]).

The next natural question is the uniqueness and hyperbolicity of the TSM. In this context, we have

THEOREM 9. [**55**]. *Assume that the mapping*

$$(F_1(\mathbf{x}), \cdots F_N(\mathbf{x})) : \mathbb{T}^d \to \mathbb{R}^N$$

is an embedding. Then for almost all ω there exists a unique two-sided minimizer for the action $\mathcal{A}^\omega$.

The issue of hyperbolicity still remains open.

This theory is intimately related to the theory of homogenization for Hamilton-Jacobi equation [**76, 28, 40**], especially the connection between the stationary solutions to the Hamilton-Jacobi equation and minimizers of the Lagrangian systems (the Aubry-Mather theory). In the deterministic context, such a connection has been exploited in [**27, 42, 56**].

3. Stochastic Passive Scalar Equation

Consider the transport equation for the scalar field $\theta^\kappa(\mathbf{x}, t)$ in $\mathbb{R}^d$:

$$\frac{\partial\theta^\kappa}{\partial t} + (\mathbf{u}(\mathbf{x}, t)\cdot\nabla)\theta^\kappa = \kappa\Delta\theta^\kappa \tag{54}$$

where $\mathbf{u}$ is a given (turbulent) velocity field. (54) is the basic equation describing the turbulent transport of a scalar field. Even though this problem is linear, it shares many of the complexities of fully developed turbulence. In the past several years, remarkable progress has been made on the understanding of the structures of such scalar fields [**2, 19, 50, 51, 92**]. Much of this is based on sophisticated field-theoretic techniques. Here we will discuss the mathematical work on this problem, drawing mainly from the work in [**36**].

We will be interested in θ^κ in the limit as $\kappa \to 0$. It is known from classical results that if $\mathbf{u}$ is Lipschitz continuous in $\mathbf{x}$, then as $\kappa \to 0$, θ^κ converges to θ, the solution of

$$\frac{\partial \theta}{\partial t} + (\mathbf{u}(\mathbf{x},t)\cdot\nabla)\theta = 0. \tag{55}$$

Furthermore, we can solve this PDE by the method of characteristics. If we define $\{\varphi_{s,t}(\mathbf{x})\}$ as the flow generated by the velocity field $\mathbf{u}$, satisfying the ordinary differential equations (ODEs)

$$\frac{d\varphi_{s,t}(\mathbf{x})}{dt} = \mathbf{u}(\varphi_{s,t}(\mathbf{x}),t), \qquad \varphi_{s,s}(\mathbf{x}) = \mathbf{x}, \tag{56}$$

for $s < t$, then the solution of the transport equation in (55) for the initial condition $\theta^\kappa(\mathbf{x},0) = \theta_0(\mathbf{x})$ is given by

$$\theta(\mathbf{x},t) = \theta_0(\varphi_{0,t}^{-1}(\mathbf{x})) = \theta_0(\varphi_{t,0}(\mathbf{x})). \tag{57}$$

This classical scenario breaks down when $\mathbf{u}$ fails to be Lipschitz continuous in $\mathbf{x}$, which is precisely the case for fully developed turbulent velocity fields. In this case Kolmogorov's theory of turbulent flows suggests that $\mathbf{u}$ is only Hölder continuous with an exponent roughly equal to $\frac{1}{3}$ for $d = 3$. In such a situation the solution of the ODEs in (56) may fail to be unique, and we then have to consider probability distributions on the set of solutions in order to solve the transport equation in (55). In terms of the ODE (56), this means that we have to consider the notion of generalized flows [**15, 16, 36, 38**].

When the regularity condition on $\mathbf{u}$ fails, it is natural to consider the limit of physical regularizations. There are at least two natural ways to regularize (56) or (55). The first is to add diffusion:

$$d\varphi_{s,t}^\kappa(\mathbf{x}) = \mathbf{u}(\varphi_{s,t}^\kappa(\mathbf{x}),t)dt + \sqrt{2\kappa}d\beta(t), \tag{58}$$

and consider the limit as $\kappa \to 0$. We will call this the κ-limit. For $\kappa > 0$, the diffusion is non-degenerate, and has a unique solution according to Stroock and Varadhan [**96**]. The second is to smooth out the velocity field. Let ψ_ε be defined as $\psi_\varepsilon(\mathbf{x}) = \varepsilon^{-d}\psi(\mathbf{x}/\varepsilon)$, where ψ is a standard mollifier: $\psi \geq 0$, $\int_{\mathbb{R}^d}\psi dx = 1$, ψ decays fast at infinity. Let $\mathbf{u}^\varepsilon = \mathbf{u} \star \psi_\varepsilon$ and consider

$$\frac{d\varphi_{s,t}^\varepsilon(\mathbf{x})}{dt} = \mathbf{u}^\varepsilon(\varphi_{s,t}^\varepsilon(\mathbf{x}),t), \tag{59}$$

in the limit as $\varepsilon \to 0$. We will call this the ε-limit. Physically κ plays the role of molecular diffusivity, ε can be thought of as a crude model of the viscous cut-off scale. In terms of the Prandtl number $Pr = \frac{\varepsilon}{\kappa}$ the κ-limit corresponds to the situation when $Pr \to 0$, whereas the ε-limit corresponds to the situation when $Pr \to \infty$.

There are two ways to think about probability distributions on the solutions of the ODEs in (56). We can either think of it as probability measures on the path-space (functions of t) supported by paths which are solutions of (56), or we can think of it as transition probability at time t if the starting position at time s is $\mathbf{x}$. In the classical situation when $\mathbf{u}$ is Lipschitz continuous, this transition probability degenerates to a point mass centered at the unique solution of (56). When Lipschitz condition fails, this transition probability may be non-degenerate and the system in (56) is intrinsically stochastic. We will adopt this latter approach and use this transition probability to define the "generalized flows".

Definition. [**15, 16, 38, 72**] Let $\{g(\mathbf{x}, s|d\mathbf{y}, t)\}$ be a family of probability measures, which are continuous in $(\mathbf{x}, s, t)$, and satisfy

1.

$$\int_s^t \int_{\mathbb{R}^d} \left\{ \frac{\partial \phi}{\partial t}(\mathbf{y}, \tau) + \mathbf{u}(\mathbf{y}, \tau) \cdot \nabla_y \phi(\mathbf{y}, \tau) \right\} g(\mathbf{x}, s|d\mathbf{y}, \tau) d\tau - \int \phi(\mathbf{y}, t) g(\mathbf{x}, s|d\mathbf{y}, t) + \phi(\mathbf{x}, s) = 0 \tag{60}$$

for all test functions $\phi \in C^\infty([0, \infty) \times \mathbb{R}^d$.

2. The Chapman-Kolmogorov equation

$$g(\mathbf{x}, s|d\mathbf{y}, t) = \int_{\mathbb{R}^d} g(\mathbf{x}, s|d\mathbf{z}, \tau) g(\mathbf{z}, \tau|d\mathbf{y}, t)$$

for all $\tau \in [s, t]$.

Then $\{g(\mathbf{x}, s|d\mathbf{y}, t)\}$ is called a generalized flow.

This definition only takes into account the so-called separable generalized flows. Non-separable generalized flows are considered in [**38**].

Before proceeding further, we relate the regularized flows in (58), (59) to the solutions of the transport equations. Consider the κ- regularization first. It is convenient to introduce the backward transition probability

$$g^\kappa(\mathbf{x}, t|d\mathbf{y}, s) = \mathbf{E}_{\boldsymbol{\beta}} \delta(\mathbf{y} - \varphi^\kappa_{t,s}(\mathbf{x})) d\mathbf{y}, \quad s < t, \tag{61}$$

where the expectation is taken with respect to $\boldsymbol{\beta}(t)$, and $\varphi^\kappa_{t,s}(\mathbf{x})$ is the flow inverse to $\varphi^\kappa_{s,t}(\mathbf{x})$ defined in (58) (i.e. $\varphi^\kappa_{s,t}(\mathbf{x})$ is the forward flow and $\varphi^\kappa_{t,s}(\mathbf{x})$ is the backward flow). The action of g^κ generates a semi-group of transformation

$$S^\kappa_{t,s}\psi(\mathbf{x}) = \int_{\mathbb{R}^d} \psi(\mathbf{y}) g^\kappa(\mathbf{x}, t|d\mathbf{y}, s), \tag{62}$$

for all test functions ψ. $\theta^\kappa(\mathbf{x}, t) = S^\kappa_{t,s}\psi(\mathbf{x})$ solve the transport equation in (54) for the initial condition $\theta^\kappa(\mathbf{x}, s) = \psi(\mathbf{x})$. Similarly, for the flow in (59), define

$$S^\varepsilon_{t,s}\psi(\mathbf{x}) = \psi(\varphi^\varepsilon_{t,s}(\mathbf{x})), \qquad s < t. \tag{63}$$

$\theta^\varepsilon(\mathbf{x}, t) = S^\varepsilon_{t,s}\psi(\mathbf{x})$ solves the transport equation

$$\frac{\partial \theta^\varepsilon}{\partial t} + (\mathbf{u}^\varepsilon(\mathbf{x}, t) \cdot \nabla)\theta^\varepsilon = 0, \tag{64}$$

with initial condition $\theta(\mathbf{x}, s) = \psi(\mathbf{x})$. Similar definitions can be given for forward flows but we will restrict attention to the backward ones since we are primarily interested in scalar transport. The results given below generalize trivially to forward flows.

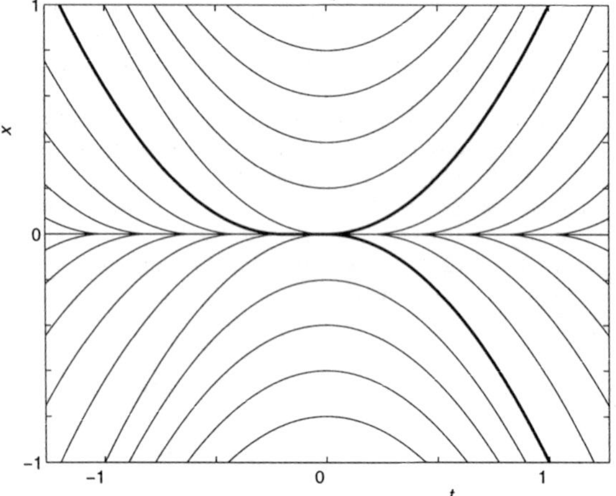

FIGURE 1. In the κ-limit, the particle coming from the left along the highlighted trajectory has equal probablity to take the two highlighted trajectories on the right.

3.1. A Simple Example of Generalized Flows. Let us consider the following example in the κ-limit:

$$dx = \operatorname{sgn}(xt)x^{\frac{1}{2}}dt + \sqrt{2\kappa}d\beta(t), \quad x(0) = y$$

To avoid the discontinuity at $t = 0$, we replace the drift term by $2t$ for $x > \frac{1}{4}t^2$ and by $-2t$ for $x < -\frac{1}{4}t^2$. We will denote the solutions as $x_t^\kappa(y)$. It is easy to see that as $\kappa \to 0$, $x_t^\kappa(y)$ converges to the unique solution of $\frac{dx}{dt} = x^{\frac{1}{2}}$ if $y \neq 0$, and when $y = 0$, $x_t^\kappa(y)$ converges to the random process that takes the extreme trajectories $x_1(t) = \frac{1}{4}t^2$ and $x_2(t) = -\frac{1}{4}t^2$, with equal probability (see Figure 1).

A slightly more interesting example is described in Figure 2. Here the velocity field is time-dependent and fails to be Lipschitz continuous at the nodes. In Figure 2, instead of plotting u, we plotted the possible trajectories that are solutions of the ODE $\frac{dx}{dt} = u$. It is easy to check that in the κ-limit, we obtain a symmetric random walk on the separatrices, which are highlighted in Figure 2 by heavy lines.

3.2. Delta-correlated Gaussian Velocity Fields. In [**63**] Kraichnan introduced a simple model of passive scalar transport by considering the advection by a Gaussian, spatially non-smooth, Kolmogorov-like and white-in-time velocity field. This class of velocity fiels is now referred to as the Kraichnan model. Definitive progress has been made in the last several years on the understanding of the structure functions and intermittency associated with the Kraichnan model [**8, 19, 50, 51, 78, 92**]. Much of these are done using fairly sophisticated perturbative field theory. We will first present the mathematical framework, followed by a summary of the statistical theory. We consider a generalization of the Kraichnan model introduced in [**51**] that takes into account compressibility effects (see also [**72**]). The velocity field $\mathbf{u}$ is assumed to be a statistically homogeneous, isotropic

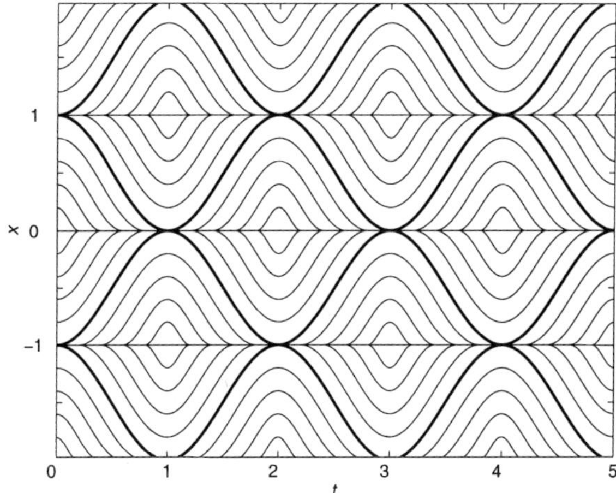

FIGURE 2. In the κ-limit, a tracer particle performs a symmetric random walk on the network formed by the highlighted trajectories.

and stationary Gaussian field with mean zero and covariance

$$\mathbf{E}\, u_\alpha(\mathbf{x},t)u_\beta(\mathbf{y},s) = (C_0\delta_{\alpha\beta} - c_{\alpha\beta}(\mathbf{x}-\mathbf{y}))\delta(t-s). \tag{65}$$

We assume that $\mathbf{u}$ has a correlation length ℓ_0, i.e. the covariance in (65) decays fast for $|\mathbf{x}-\mathbf{y}| > \ell_0$. Consequently $c_{\alpha\beta}(\mathbf{x}) \to C_0\delta_{\alpha\beta}$ as $|\mathbf{x}|/\ell_0 \to \infty$. On the other hand, we will be mainly interested in small scale phenomena for which $|\mathbf{x}| \ll \ell_0$. In this range, we take $c_{\alpha\beta}(\mathbf{x}) = d_{\alpha\beta}(\mathbf{x}) + O(|\mathbf{x}|^2/\ell_0^2)$ with

$$d_{\alpha\beta}(\mathbf{x}) = A d^P_{\alpha\beta}(\mathbf{x}) + B d^S_{\alpha\beta}(\mathbf{x}), \tag{66}$$

and

$$\begin{aligned} d^P_{\alpha\beta}(\mathbf{x}) &= D\left(\delta_{\alpha\beta} + \xi\frac{x_\alpha x_\beta}{|\mathbf{x}|^2}\right)|\mathbf{x}|^\xi, \\ d^S_{\alpha\beta}(\mathbf{x}) &= D\left((d+\xi-1)\delta_{\alpha\beta} - \xi\frac{x_\alpha x_\beta}{|\mathbf{x}|^2}\right)|\mathbf{x}|^\xi. \end{aligned} \tag{67}$$

Here D is just a parameter. The parameters A and B measure the divergence and rotation of the field $\mathbf{u}$. $A = 0$ corresponds to incompressible fields with $\nabla\cdot\mathbf{u} = 0$. $B = 0$ corresponds to irrotational fields with $\nabla\times\mathbf{u} = 0$. The parameter ξ measures the spatial regularity of $\mathbf{u}$. For $\xi \in (0,2)$, the local characteristic of $\mathbf{u}$ fails to be twice differentiable. This means roughly that $\mathbf{u}$ is spatially only Hölder continuous with exponent $\frac{\xi}{2}$. We call such velocity fields Kolmogorov-like. This spatial behavior has important consequences on both the transport equation (55) and the systems of ODEs (56).

Let $\mathcal{S}^2 = A + (d-1)B$, $\mathcal{C}^2 = A$, $\mathcal{P} = \mathcal{C}^2/\mathcal{S}^2$. $\mathcal{P} \in [0,1]$ is a measure of the degree of compressibility of $\mathbf{u}$.

The following result is a classification of the different regimes we obtain when we consider the κ- and ε-limits.

THEOREM 10. *There exist three different regimes:*

1. $\mathcal{P} \geq d/\xi^2$, *the strongly compressible regime. In this case, there exists a family of random maps* $\{\varphi^{\omega}_{t,s}(\mathbf{x})\}$, *such that for all smooth test functions* ψ *and for all* $(s,t,\mathbf{x})$, $s<t$,

$$\mathbf{E}\left(S^{\omega,\kappa}_{t,s}\psi(\mathbf{x}) - \psi(\varphi^{\omega}_{t,s}(\mathbf{x}))\right)^2 \to 0, \tag{68}$$

as $\kappa \to 0$, *and*

$$\mathbf{E}\left(\psi(\varphi^{\omega,\varepsilon}_{t,s}(\mathbf{x})) - \psi(\varphi^{\omega}_{t,s}(\mathbf{x}))\right)^2 \to 0, \tag{69}$$

as $\varepsilon \to 0$. *Moreover, the limiting flow* $\{\varphi^{\omega}_{t,s}(\mathbf{x})\}$ *coalesces in the sense that for almost all* $(t,\mathbf{x},\mathbf{y})$, $\mathbf{x}\neq\mathbf{y}$, *we can define a time* τ *such that* $-\infty<\tau<t$ *a.s. and*

$$\varphi^{\omega}_{t,s}(\mathbf{x}) = \varphi^{\omega}_{t,s}(\mathbf{y}) \quad \text{for} \quad s\leq\tau. \tag{70}$$

2. $\mathcal{P} \leq (d+\xi-2)/2\xi$, *the weakly compressible regime. In this case there exists a random family of generalized flows* $g_\omega(\mathbf{x},t|d\mathbf{y},s)$, *such that for all test function* ψ,

$$S^{\omega}_{t,s}\psi(\mathbf{x}) = \int_{\mathbb{R}^d}\psi(\mathbf{y})g_\omega(\mathbf{x},t|d\mathbf{y},s), \tag{71}$$

satisfies

$$\mathbf{E}\left(S^{\omega,\kappa}_{t,s}\psi(\mathbf{x}) - S^{\omega}_{t,s}\psi(\mathbf{x})\right)^2 \to 0, \tag{72}$$

as $\kappa\to 0$ *for all* $(s,t,\mathbf{x})$, $s<t$, *and*

$$\mathbf{E}\Big(\int_{\mathbb{R}^d}\eta(\mathbf{x})\left(\psi(\varphi^{\omega,\varepsilon}_{t,s}(\mathbf{x})) - S^{\omega}_{t,s}\psi(\mathbf{x})\right)dx\Big)^2 \to 0, \tag{73}$$

as $\varepsilon\to 0$ *for all* (s,t), $s<t$, *and for all test functions* η. *Moreover,* $g_\omega(\mathbf{x},t|d\mathbf{y},s)$ *is non-degenerate in the sense that there exists a* ψ, *such that there exists a* ψ, *such that*

$$S^{\omega}_{t,s}\psi^2(\mathbf{x}) - \left(S^{\omega}_{t,s}\psi(\mathbf{x})\right)^2 > 0 \quad a.s. \tag{74}$$

3. $(d+\xi-2)/2\xi < P < d/\xi^2$, *the intermediate regime. In this case there exists a random family of generalized flows* $g_\omega(\mathbf{x},t|d\mathbf{y},s)$, *such that for all test function* ψ *and for all* $(s,t,\mathbf{x})$, $s<t$,

$$\mathbf{E}\left(S^{\omega,\kappa}_{t,s}\psi(\mathbf{x}) - S^{\omega}_{s,t}\psi(\mathbf{x})\right)^2 \to 0 \tag{75}$$

as $\kappa\to 0$. *In the* ε*-limit, the flows* $\varphi^{\omega,\varepsilon}_{t,s}(\mathbf{x})$ *converges in the sense of distributions, i.e. there exists a family of probability densities* $\{G_n(\mathbf{x}_1,\dots,\mathbf{x}_n,t|\mathbf{y}_1,\dots,\mathbf{y}_n,s)d\mathbf{y}_1\cdots d$ $n=1,2,\dots$, *such that*

$$\mathbf{E}\psi(\varphi^{\omega,\varepsilon}_{t,s}(\mathbf{x}_1),\cdots,\varphi^{\omega,\varepsilon}_{t,s}(\mathbf{x}_n)) \to \int\limits_{\mathbb{R}^d\times\cdots\times\mathbb{R}^d}\psi(\mathbf{y}_1,\cdots,\mathbf{y}_n) \times G_n(\mathbf{x}_1,\cdots,\mathbf{x}_n,t|\mathbf{y}_1,\dots,\mathbf{y}_n,s)d\mathbf{y}_1\cdots d\mathbf{y}_n, \tag{76}$$

as $\varepsilon\to 0$ *for any continuous function* ψ *with compact support. Furthermore, there is coalesces in the* $\varepsilon-$*limit in the sense that*

$$\begin{aligned} G_2(\mathbf{x}_1,\mathbf{x}_2,t|\mathbf{y}_1,\mathbf{y}_2,s) &= \tilde{G}_2(\mathbf{x}_1,\mathbf{x}_2,t|\mathbf{y}_1,\mathbf{y}_2,s) \\ &\quad + A(y_1,\mathbf{x}_1,\mathbf{x}_2,t,s)\delta(\mathbf{y}_1-\mathbf{y}_2), \end{aligned} \tag{77}$$

with $A > 0$ when $t > s$. Here $\tilde{G}_2$ is the absolutely continuous part of G_2 with respect to the Lebesgue measure. Similar statements hold for the other G_n's. In particular, the $\{G_n\}$'s differ from the moments of the κ-limit g_ω defined in (75).

The family of probability densities $\{G_n\}$ are consistent in the sense of Kolmogorov. Therefore there exists a stochastic flow defined on some probability space with $\{G_n\}$ as its distribution [**69**].

Rephrasing the content of this result, we have strong convergence to a family of flow maps in the strongly compressible regime for both the κ-limit and the ε-limit. In the weakly compressible regime, we have strong convergence to a family of generalized flows for the κ-limit, but weak convergence to the same limit for the ε-regularization. In fact, using the terminology of Young measures [**98**], the limiting generalized flow $\{g_\omega(\mathbf{x}, t|d\mathbf{y}, s)\}$ is nothing but the Young measure for the sequence of oscillating flow maps $\{\varphi_{s,t}^{\omega,\varepsilon}(\mathbf{x})\}$. Finally, in contrast to what is observed in the other two regimes, the ε-limit and κ-limit are not the same in the intermediate regime.

Theorem 10 in its present form is found in [**36**]. Earlier Gawedzki, Vergassola [**51**] and independently Le Jan and Raimond [**72**] considered the κ-limit. In that case there is only the difference of convergence to a family of flow maps in the strongly compressible and intermediate regimes, and convergence to generalized stochastic flows in the weakly compressible regime. Joint limits $\kappa, \varepsilon \to 0$ were studied in [**37**] and it was concluded that the limiting behavior depends on a new "turbulent Prandtl number."

¿From Theorem 10, it is natural to define the solution of the transport equation in (55) for the initial condition $\theta_\omega(\mathbf{x}, s) = \theta_0(\mathbf{x})$ as

$$\theta_\omega(\mathbf{x}, t) = S_{t,s}^\omega \theta_0(\mathbf{x}) = \int_{\mathbb{R}^d} \theta_0(\mathbf{y}) g_\omega(\mathbf{x}, t|d\mathbf{y}, s), \tag{78}$$

for the weakly compressible and the intermediate regimes in the κ-limit (non-degenerate cases), and as

$$\theta_\omega(\mathbf{x}, t) = \theta_0(\varphi_{t,s}^\omega(\mathbf{x})), \tag{79}$$

for the strongly compressible regime. In the intermediate regime in the ε-limit, it makes sense to look at the limiting moments of $\theta_\omega^\varepsilon(\mathbf{x}, t)$ since we have as $\varepsilon \to 0$

$$\begin{aligned} \mathbf{E}(\theta_\omega^\varepsilon(\mathbf{x}_1, t) \cdots \theta_\omega^\varepsilon(\mathbf{x}_n, t)) \to \int_{\mathbb{R}^d \times \cdots \times \mathbb{R}^d} & \theta_0(\mathbf{y}_1) \cdots \theta_0(\mathbf{y}_n) \\ & \times G_n(\mathbf{x}_1, \cdots, \mathbf{x}_n, t|\mathbf{y}_1, \ldots, \mathbf{y}_n, s) d\mathbf{y}_1 \cdots d\mathbf{y}_n. \end{aligned} \tag{80}$$

An outline of the proof of Thoerem 10 is as follows. Define $P(\rho|r, s)$ through ε-regularization as

$$\int_0^\infty \eta(r) P(\rho|r, s-t) dr = \lim_{\varepsilon \to 0} \mathbf{E}\, \eta(|\varphi_{t,s}^{\omega,\varepsilon}(\mathbf{y}) - \varphi_{t,s}^{\omega,\varepsilon}(\mathbf{z})|), \tag{81}$$

where η is a test function, and similarly through κ-regularization. Here $\rho = |\mathbf{y} - \mathbf{z}|$ and $s < t$. $P(\rho|r, s)$ can be thought of as the probability density that two particles have distance r at time $s < t$ if their final distance is ρ at time t. For the Kraichnan model, P satisfies the backward equation

$$-\frac{\partial P}{\partial s} = -\frac{\partial}{\partial r}(b(r)P) + \frac{\partial^2}{\partial r^2}(a(r)P), \tag{82}$$

for the final condition $\lim_{s\to 0-} P(\rho|r,s) = \delta(r-\rho)$, and $a(r)$, $b(r)$ have the behavior

$$
\begin{aligned}
a(r) &= D(\mathcal{S}^2 + \xi\mathcal{C}^2)r^\xi + O(r^2/\ell_0^2),\\
b(r) &= D((d-1+\xi)\mathcal{S}^2 - \xi\mathcal{C}^2)r^{\xi-1} + O(r/\ell_0^2).
\end{aligned}
\tag{83}
$$

For $r \gg \ell_0$, $a(r)$ tends to C_0, and $b(r)$ tends to $C_0(d-1)/r$. The equation (82) then reduces to a diffusion equation with constant coefficients. At $r = 0$ the equation (82) is singular. The proof of Theorem 10 is essentially reduced to the study of this singular diffusion equation.

The behavior of the structure functions in these different limits is discussed in [**36, 37, 51**].

3.3. Invariant Measures. Here we study the existence of an invariant measure for the transport equation with appropriate forcing. We restrict attention to the non-degenerate cases which include the weakly compressible regime and the intermediate regime in the κ-limit. In these regimes the limiting generalized flow is non-degenerate and hence behaves as a non-trivial diffusion process. The consequences of this is that there is a dissipation mechanism present (as we will discuss further in Section 5), and the system losses memory. We show that the anomalous dissipation is strong enough for the forced transport equation to have a unique invariant measure for both the weakly compressible regime and the intermediate regime in the κ-limit.

Consider (compare with (55))

$$
\frac{\partial\theta}{\partial t} + (\mathbf{u}(\mathbf{x},t)\cdot\nabla)\theta = b(\mathbf{x},t). \tag{84}
$$

where b is a white-noise forcing such that

$$
\mathbf{E}\, b(\mathbf{x},t)b(\mathbf{y},s) = B(|\mathbf{x}-\mathbf{y}|)\delta(t-s). \tag{85}
$$

$B(r)$ is assumed to be smooth and rapidly decaying to zero for $r \gg L$; L will be referred to as the forcing scale. The solution of (84) for the initial condition $\theta_\omega(\mathbf{x},s) = \theta_0(\mathbf{x})$ is understood as

$$
\theta_\omega(\mathbf{x},t) = S^\omega_{t,s}\theta_0(\mathbf{x}) + \int_s^t S^\omega_{t,\tau} b(\mathbf{x},\tau)d\tau. \tag{86}
$$

Define the product probability space $(\Omega_u \times \Omega_b, \mathcal{F}_u \times \mathcal{F}_b, \mathcal{P}_u \times \mathcal{P}_b)$, and the shift operator $T_\tau\omega(t) = \omega(t+\tau)$, with $\omega = (\omega_u, \omega_b)$. We have

THEOREM 11. *For $d > 2$, in the weakly compressible regime and in the intermediate regime in the κ-limit, for almost all ω, there exists a unique solution of (84) defined on $\mathbb{R}^d \times (-\infty,\infty)$. This solution can be expressed as*

$$
\theta^\star_\omega(\mathbf{x},t) = \int_{-\infty}^t S^\omega_{t,s} b(\mathbf{x},s)ds. \tag{87}
$$

Furthermore the map $\omega \to \theta^\star_\omega$ satisfies the invariance property

$$
\theta^\star_{T_\tau\omega}(\mathbf{x},t) = \theta^\star_\omega(\mathbf{x},t+\tau). \tag{88}
$$

Theorem 11 is the "one force, one solution" principle articulated in [**30**], and discussed earlier. Because of the invariance property (88), the map in (87) leads to a natural invariant measure. As a consequence we have

COROLLARY 1. *For $d > 2$, in the weakly compressible regime and in the intermediate regime in the κ-limit, there exists a unique invariant measure on $L^2_{loc}(\mathbb{R}^d \times \Omega)$ for the dynamics defined by (84).*

To prove Theorem 11, we verify that the dissipation in the system is strong enough in the sense that

$$\mathbf{E}\Big(\int_{T_1}^{T_2}\int_{\mathbb{R}^d} S^{\omega}_{t,t+s} b(\mathbf{x},s)ds\Big)^2 \to 0, \tag{89}$$

as T_1, $T_2 \to -\infty$ for fixed x and t. The average in (89) is given explicitly by

$$\int_{T_1}^{T_2}\int_0^{\infty} B(\rho)P(0|\rho,s)d\rho ds, \tag{90}$$

where P satisfies (82). The convergence of the integral in (89) depends on the rate of decay in $|s|$ of $P(0|\rho,s)$. The latter can be estimated by studying the equation in (82) [**36**], which yields $P(0|\rho,s) \sim C\rho^{\alpha}|s|^{-d/2}$ with $\alpha = (d-1-\xi(\xi+1)\mathcal{P})/(1+\xi\mathcal{P})$ for $|s|$ large and $\rho \ll \ell_0$. Hence, the integral in s in (90) tends to zero as T_1, $T_2 \to -\infty$ if $d > 2$. It follows that the invariant measure in (87) exists provided that $d > 2$.

Since no anomalous dissipation is present in the coalescence cases, i.e the strongly compressible regime and the intermediate regime in the ε-limit, no invariant measure for the temperature field exists in these regimes. It makes sense, however, to ask about the existence of an invariant measure for the temperature difference, i.e. to consider

$$\delta\theta_{\omega}(\mathbf{x},\mathbf{y},t) = \int_T^t S^{\omega}_{t,s}(b(\mathbf{x},s) - b(\mathbf{y},s))ds, \tag{91}$$

in the limit as $T \to -\infty$. When $\theta^{\star}_{\omega}$ exists, one has

$$\delta\theta^{\star}_{\omega}(\mathbf{x},\mathbf{y},t) = \lim_{T\to-\infty} \delta\theta_{\omega}(\mathbf{x},\mathbf{y},t) = \theta^{\star}_{\omega}(\mathbf{x},t) - \theta^{\star}_{\omega}(\mathbf{y},t), \tag{92}$$

but it is conceivable that $\delta\theta^{\star}_{\omega}$ exists in the coalescence cases even though $\theta^{\star}_{\omega}$ is not defined. The reason is that coalescence of the generalized flow implies that the temperature field flattens with time, which is a dissipation mechanism as far as the temperature difference is concerned. Of course, this effect has to overcome the fluctuations produced by the forcing, and the existence of the limit in (91) will depend on how fast particles coalesce under the flow.

For finite ℓ_0, if we consider two particles separated by a distance much longer than the correlation length ℓ_0, the dynamics of their distance under the flow is governed by the equation in (82) for $r \gg \ell_0$, i.e. by a diffusion equation with constant coefficient on the scale of interest. It follows that no tendency of coalescence is observed before the distance becomes smaller than ℓ_0, which does not happen fast enough in order to overcome the the fluctuations produced by the forcing. In other words,

LEMMA 2. *In the coalescence cases, for finite ℓ_0, there is no invariant measure with finite energy for the temperature difference.*

Therefore it is natural to consider the limit as $l_0 \to \infty$. However we have to be careful because the velocity field with the covariance in (65) diverges as $\ell_0 \to \infty$.

The right way to proceed is to consider an alternative velocity $\mathbf{v}$, taken to be Gaussian, white-in-time, but *non-homogeneous*, with covariance

$$\begin{aligned}\mathbf{E}\, v_\alpha(\mathbf{x},t)v_\beta(\mathbf{y},s)\\ = (c_{\alpha\beta}(\mathbf{x}-\mathbf{a}) + c_{\alpha\beta}(\mathbf{a}-\mathbf{y}) - c_{\alpha\beta}(\mathbf{x}-\mathbf{y}))\delta(t-s).\end{aligned} \tag{93}$$

For finite ℓ_0, one has $v(\mathbf{x},t) = u(\mathbf{x},t) - u(\mathbf{a},t)$, where $\mathbf{a}$ is arbitrary but fixed. However, $\mathbf{v}$ makes sense in the limit as $\ell_0 \to \infty$. Denote by $\vartheta_\omega(\mathbf{x},t)$ the temperature field advected by $\mathbf{v}$, i.e. the solution of the transport equation (84) with $\mathbf{u}$ replaced by $\mathbf{v}$:

$$\frac{\partial \vartheta}{\partial t} + (\mathbf{v}(\mathbf{x},t)\cdot\nabla)\vartheta = b(\mathbf{x},t). \tag{94}$$

Restricting to zero initial condition, it follows from the homogeneity of the forcing that the single-time moments of θ_ω and ϑ_ω coincide for finite ℓ_0, but in contrast to θ_ω, ϑ_ω makes sense as $\ell_0 \to \infty$. Furthermore, Theorem 10 can be extended to (94) in the limit as $\ell_0 \to \infty$.

Let $\delta\vartheta_\omega(\mathbf{x},\mathbf{y},t) = \vartheta_\omega(\mathbf{x},t) - \vartheta_\omega(\mathbf{y},t)$ where ϑ_ω solves the equation in (94). The temperature difference $\delta\vartheta_\omega$ satisfies the transport equation

$$\frac{\partial \delta\vartheta}{\partial t} + (\mathbf{v}(\mathbf{x},t)\cdot\nabla_{\mathbf{x}} + \mathbf{v}(\mathbf{y},t)\cdot\nabla_{\mathbf{y}})\delta\vartheta = b(\mathbf{x},t) - b(\mathbf{y},t). \tag{95}$$

We have

THEOREM 12. *In the limit as $\ell_0 \to \infty$, for almost all ω, in the strongly and the weakly compressible regimes, as well as in the intermediate regime if the flow is non-degenerate, there exists a unique solution of (95) defined on $\mathbb{R}^d \times (-\infty,\infty)$. This solution can be expressed as*

$$\delta\vartheta^\star_\omega(\mathbf{x},\mathbf{y},t) = \int_{-\infty}^t S^\omega_{t,s}(b(\mathbf{x},s) - b(\mathbf{y},s))ds, \tag{96}$$

where $S^\omega_{s,t}$ is the semi-group for the generalized flow associated with the velocity defined in (93) in the limit as $\ell_0 \to \infty$. Furthermore the map $\omega \to \delta\vartheta^\star_\omega$ satisfies the invariance property

$$\delta\vartheta^\star_{T_\tau\omega}(\mathbf{x},\mathbf{y},t) = \delta\vartheta^\star_\omega(\mathbf{x},\mathbf{y},t+\tau). \tag{97}$$

An immediate consequence of this theorem is

COROLLARY 3. *In the limit as $\ell_0 \to \infty$, in the strongly and the weakly compressible regimes, as well as in the intermediate regime if the flow is non-degenerate, there exists a unique invariant measure on $L^2_{loc}(\mathbb{R}^d \times \Omega)$ for the dynamics defined by (95).*

The proof of Theorem 12 proceeds similarly as the proof of Theorem 11. In the non-degenerate cases, one studies the convergence of (compare with (89))

$$\mathbf{E}\Big(\int_{T_1}^{T_2}\int_{\mathbb{R}^d} S^\omega_{t,t+s}(b(\mathbf{x},s) - b(\mathbf{y},s))ds\Big)^2 \to 0, \tag{98}$$

as T_1, $T_2 \to -\infty$ for fixed x and t. The average in (98) can be expressed in terms of P, and it can be shown that the expression in (98) converges as T_1, $T_2 \to -\infty$

in the non-degenerate cases. In the strongly compressible regimes, because of the existence of a flow of maps, (98) is replaced by

$$\mathbf{E}\Big(\int_{T_1}^{T_2}(b(\varphi_{t,s}^{\omega}(\mathbf{x}),s)-b(\varphi_{t,s}^{\omega}(\mathbf{y}),s))ds\Big)^2. \tag{99}$$

This average can again be expressed in terms of P, and it can be shown that the convergence of the time integral in (99) depends on the rate at P looses mass at $r=0+$ (i.e. the rate at which particles coalesce). The analysis of the equation in (82) shows that the process is fast enough in order that the integral over s in (99) tends to zero as $T_1, T_2 \to -\infty$ in the strongly compressible regime. In contrast, the equivalent of (99) in the intermediate regime in the ε-limit can be shown to diverge as $T_1, T_2 \to -\infty$.

3.4. Correlation Functions and Zero Modes. In this section, we briefly summarize the perturbation theory developed in [**50, 19**] on the behavior of the n-point correlation functions. These results rely on the fact that a closed set of equations for the single time correlation functions can be derived for the Kraichnan model. We will restrict ourselves to the incompressible case when $A=0$.

Consider

$$F_n(\mathbf{x}_1,\cdots,\mathbf{x}_n,t)=\langle\theta(\mathbf{x}_1,t)\cdots(\theta(\mathbf{x}_n,t))\rangle \tag{100}$$

where θ satisfies (84) $F_{2n-1}=0$ by summetry. F_{2n} satisfies

$$\begin{aligned}\frac{\partial}{\partial t}F_{2n}(\mathbf{x}_1,\cdots,\mathbf{x}_{2n},t) &= \sum_{j,k=1}^{2n} C_{\alpha\beta}(\mathbf{x}_j-\mathbf{x}_k)\nabla_\alpha^j\nabla_\beta^k F_{2n}(\mathbf{x}_1,\cdots,\mathbf{x}_{2n},t)\\ &\quad + 2\sum_{\substack{j,k=1\\ j<k}}^{2n} B(|\mathbf{x}_j-\mathbf{x}_k|)F_{2n-2}(\mathbf{x}_1,\underset{\hat{j}\ \hat{k}}{\cdots},\mathbf{x}_{2n},t).\end{aligned} \tag{101}$$

At statistical steady state, these equations reduce to

$$\begin{aligned}&\sum_{j,k=1}^{2n} C_{\alpha\beta}(\mathbf{x}_j-\mathbf{x}_k)\nabla_\alpha^j\nabla_\beta^k F_{2n}(\mathbf{x}_1,\cdots,\mathbf{x}_{2n})\\ &\qquad = -2\sum_{\substack{j,k=1\\ j<k}}^{2n} B(|\mathbf{x}_j-\mathbf{x}_k|)F_{2n-2}(\mathbf{x}_1,\underset{\hat{j}\ \hat{k}}{\cdots},\mathbf{x}_{2n}).\end{aligned} \tag{102}$$

Evaluation of the structures functions $S_{2p}(r)=\langle(\theta(\mathbf{x}+\mathbf{r},t)-\theta(\mathbf{x},t))^{2p}\rangle$ can in principle be carried out once the solution of (102) is known. However, solving these equations is a very difficult task and so far only perturbative methods have been successful in some regimes.

A simple dimensional argument suggests that a Kolmogorov-like theory would predict normal scaling exponents $\alpha_p=\frac{2-\xi}{2}p$. We will see below that this is not true for $p>2$.

Our task is to analyze the behavior of F_{2n} for small $|\mathbf{x}_j-\mathbf{x}_k|$. We will restrict ourselves to translation invariant solutions.

¿From (102), F_{2n} has contributions from the inhomogeneous part F_{2n-2}, and the homogeneous part, which at small distances are solutions of

$$\mathcal{M}_{2n}^{SC}\tilde{F}_{2n} = \sum_{j,k=1}^{2n} d_{\alpha\beta}(\mathbf{x}_j - \mathbf{x}_k)\nabla_\alpha^j \nabla_\beta^k \tilde{F}_{2n} = 0 \tag{103}$$

where

$$d_{\alpha\beta}(\mathbf{x}) = D\left((d+\xi-1)\delta_{\alpha\beta} - \xi\frac{x_\alpha x_\beta}{r^2}\right)r^\xi \tag{104}$$

(103) is obtained as the scaling limit of the homogeneous equation associated with (102). The solutions of (103) are called zero modes [**8, 19, 50**]. This is an important concept that characterizes the origin of the anomalous scaling, and the leading order singularities of the correlation functions.

As an example, let us study the behavior of the 2-point function at $d = 3$. F_2 satisfies

$$-\frac{2}{r^2}\frac{\partial}{\partial r}\left(Dr^{2+\xi}\frac{\partial F_2}{\partial r}\right) = B(r). \tag{105}$$

It is easy to obtain from this equation that

$$F_2(r) = C_0 - C_1 r^{2-\xi} + \cdots \tag{106}$$

where C_0, C_1 are constants depending on B and the neglected terms are of higher order. Hence we have for S_2

$$S_2(r) = C_2 r^{2-\xi} + \cdots \tag{107}$$

This implies that S_2 obeys normal scaling. Note that the constant C_2 is in general non-universal.

Let us now look at the 4-point function $F_4(\mathbf{x}_1, \mathbf{x}_2, \mathbf{x}_3, \mathbf{x}_4)$ and let $\mathbf{x}_{jk} = \mathbf{x}_j - \mathbf{x}_k, j, k = 1, \cdots 4$. F_4 has contributions from the "Gaussian" part:

$$F_2(\mathbf{x}_{12})F_2(\mathbf{x}_{34}) + F_2(\mathbf{x}_{13})F_2(\mathbf{x}_{24}) + F_2(\mathbf{x}_{14})F_2(\mathbf{x}_{23}), \tag{108}$$

which gives rise to normal scaling, as well as the contribution from the zero modes which may be the dominant contribution to S_4. To find the precise form of the zero modes is a very difficult task. Both [**50**] and [**19**] resorted to perturbation techniques in either ξ or $1/d$. Here we follow [**50**] and write

$$\tilde{F} = E_0 + \xi G_0 + O(\xi^2), \tag{109}$$

where E_0 is the zero mode for the case when $\xi = 0$. Using the notations $\nabla_{12} = \nabla_{\mathbf{x}_{12}}$, etc, we can write down an equation for E_0:

$$-(\Delta_{12} + \Delta_{23} + \Delta_{34} - \nabla_{12}\cdot\nabla_{23} - \nabla_{23}\cdot\nabla_{34})E_0 = 0. \tag{110}$$

Substituting (110) into (103), we obtain an equation for G_0:

$$-\Delta_4 G_0 + \mathcal{L}E_0 = 0$$

where

$$\Delta_4 = \Delta_{\mathbf{x}_1} + \cdots + \Delta_{\mathbf{x}_4}, \mathcal{L} = \sum_{j\neq k}\left(\delta_{\alpha\beta} - \frac{1}{2}\frac{x_{jk}^\alpha x_{jk}^\beta}{|\mathbf{x}_{jk}|^2}\right)\nabla_j^\alpha\nabla_k^\beta - \frac{1}{2}\Delta_4$$

The precise form of G_0 is quite complicated. We refer to [**8, 50**] for the details of this calculation from which one obtains that $\tilde{F}$ must be homogeneous of degree

$$\alpha_4 = 4 - \frac{14}{5}\xi + O(\xi^2) = 2\alpha_2 - \rho_2, \tag{111}$$

where $\rho_2 = \frac{4}{5}\xi + O(\xi^2)$ is the anomalous exponent. Numerical results of Frisch et.al [**49**] has found very good agreement with (111).

4. Stochastic Navier-Stokes Equation

We are interested in two mathematical issues:

1. The existence and uniqueness of an invariant measure under physical conditions.
2. The limiting behavior of the invariant measure as viscosity approaches zero.

Presently progress has only been made on the first issue. Here the main question is the condition on the stochastic forcing under which uniqueness of the invariant measure and consequently ergodicity of the dynamics can be proved.

4.1. Ergodicity When All Determining Modes are Forced. Consider

$$\begin{cases} \mathbf{u}_t + (\mathbf{u}\cdot\nabla)\mathbf{u} + \nabla p = \nu\Delta\mathbf{u} + \frac{\partial \mathbf{W}}{\partial t} \\ \nabla\cdot\mathbf{u} = 0 \end{cases} \tag{112}$$

We will restrict ourselves to two dimension, and we will take $\mathbf{W}$ to be of the form

$$\mathbf{W}(\mathbf{x}, t) = \sum_k \sigma_k \beta_k(t, \omega)\mathbf{e}_k(\mathbf{x}) \tag{113}$$

where the β_k's are standard i.i.d complex-valued Brownian motion, and $\sigma_k \in \mathbb{C}$, are the amplitudes of the forcing, $\{\mathbf{e}_k(\mathbf{x}) = \binom{-ik_2}{ik_1}\frac{e^{ik\cdot x}}{|k|}, k \in \mathbb{Z}\}$ are the complex eigen basis of L^2 divergence-free vector fields on $\mathbb{T}^2$, the two dimensional torus.

Define $B(\mathbf{u},\mathbf{w}) = -P_{div}(\mathbf{u}\cdot\nabla)\mathbf{w}$, $\Lambda^2\mathbf{u} = -P_{div}\Delta\mathbf{u}$ where P_{div} is the L^2 projection operator onto the space of divergence-free vector fields. Let $\mathcal{E}_0 = \sum_k |\sigma_k|^2$, $\mathcal{E}_1 = \sum_k |k|^2 \sigma_k|^2$. Writing $\mathbf{u}(\mathbf{x}) = \sum_k u_k \mathbf{e}_k(\mathbf{x})$, we will define $\mathbb{H}^\alpha = \{\mathbf{u} = (\mathbf{u}_k)_{k\in\mathbb{Z}^2}, \sum_k |k|^{2\alpha}|\mathbf{u}_k|^2 < \infty\}$ and $\mathbb{L}^2 = \mathbb{H}^0$.

We will use the standard probabilistic setup, namely we will work on a probability space $(\Omega, \mathcal{F}, \mathcal{F}_t, \mathbb{P}, \theta_t)$, where $(\Omega, \mathcal{F}, \mathbb{P})$ is the canonical probability space generated by all β_k, $\mathcal{F}_t$ is the σ-algebra generated by $\beta_k(s)$ for $s \le t$, and θ_t is the shift operator on Ω defined by $(\theta_t d\beta_k)(s) = d\beta_k(s+t)$. Expectations with respect to $\mathbb{P}$ will be denoted by $\mathbb{E}$. We formally write (113) as an Ito differential equation in $\mathbb{L}^2$:

$$d\mathbf{u}(\mathbf{x},t) + \nu\Lambda^2\mathbf{u}(\mathbf{x},t)dt = B(\mathbf{u},\mathbf{u})dt + d\mathbf{W}(\mathbf{x},t) \tag{114}$$

Even though technically unnecessary sometimes, we will assume that σ_k decays exponentially fast in k. This is reasonable from a physical viewpoint. Under this condition, it can be easily shown that for almost all realizations of the stochastic forcing, the dynamics (112) is well-defined. Furthermore, for $t \ge s$, let $\mathbf{u}(t,\omega; s, \mathbf{u}_0)$ be the solution of (112) at time t satisfying the initial condition $\mathbf{u}(s) = \mathbf{u}_0$, then (112) generates a continuous stochastic semi-flow on $\mathbb{L}^2$

$$\varphi^\omega_{s,t}(\mathbf{u}_0) = \mathbf{u}(t, \omega; s, \mathbf{u}_0)$$

We will take the state space to be $\mathbb{L}^2$ equipped with the Borel σ-algebra.

Definition. A measure $\mu(du)$ on $\mathbb{L}^2$ is invariant for the stochastic flow (114) if for all bounded continuous functionals F on $\mathbb{L}^2$ and $t > 0$

$$\int_{\mathbb{L}^2} F(\mathbf{u})\mu(du) = \int_{\mathbb{L}^2} \mathbb{E}F(\varphi_{0,t}^{\omega}\mathbf{u})\mu(du) \tag{115}$$

Existence of an invariant measure is proved by Vishik and Fursikov [**100**], Flandoli and Maslowski [**46**]. The basic idea is to use the classical Krylov-Bogoliubov theorem asserting the existence of invariant measures for general dynamical systems on a compact state space. Here our state space is not compact, but since (112) is of parabolic type, the solution operator is compact and this turns out to be sufficient.

The first uniqueness result was also proved by Flandoli and Maslowski [**46**]. Their result requires that the forcing amplitude σ_k satisfy

$$C_1|k|^{-1/2} \le |\sigma_k| \le C_2|k|^{-3/8-\varepsilon} \tag{116}$$

In this case the dynamics of the SNS (Stochastic Navier-Stokes) is dominated by noise. The nonlinear term can be viewed as a perturbation of the linear equation with noise. Such a condition is very rarely met in physical situations.

In his thesis, Mattingly proved that if the viscosity is sufficiently large, the one force, one solution principle holds and hence the invariant measure must be unique [**81, 82**].

In the more general case, E, Mattingly and Sinai proved

THEOREM 13. [**33**] *Assume that $\sigma_k \neq 0$ for $|k| \le N$ where $N^2 \ge \frac{\bar{C}^2}{2}\frac{\mathcal{E}_0}{\nu^3}$ and $\bar{C}$ is some absolute constant. Then (112) has a unique invariant measure on $\mathbb{L}^2$.*

Results of a similar nature were also independently obtained by Kuksin and Shirikyan [**65, 66**], Bricmont, Kupiainen and Lefevere [**17**], see also [**86**].

Theorem 13 says that if all the "determining modes" are forced, then the invariant measure is unique and hence the dynamics is ergodic in the phase space.

The main idea of the proof is to show that for solutions that are defined over the time interval $(-\infty, \infty)$, their high modes can be uniquely represented by the past history of the low modes. This means that the dynamics of the Navier-Stokes equation on its infinite-dimensional phase space can be represented by a finite-dimensional system, but with memory. Ergodicity is then established by proving that time averages of a continuous functional along a trajectory is independent of the initial data of the trajectory. This last step is done using the Girsanov formula.

Below we explain these two steps in more detail.

Step 1. Reduction to Gibbsian dynamics.

We show that the Markovian dynamics of the infinite-dimensional Navier-Stokes equations can be reduced to the Gibbsian dynamics of a finite-dimensional system. To this end, we define two subspaces.

$$\mathbb{L}_\ell^2 = \operatorname{span}\{\mathbf{e}_k, |k| \le N\}, \quad \mathbb{L}_h^2 = \operatorname{span}\{\mathbf{e}_k, |k| > N\} \tag{117}$$

We will call $\mathbb{L}_\ell^2$ the set of low modes and $\mathbb{L}_h^2$ the set of high modes. Obviously $\mathbb{L}^2 = \mathbb{L}_\ell^2 \oplus \mathbb{L}_h^2$. Denote by P_ℓ and P_h the projections onto the low and high mode spaces.

We write $\mathbf{u}(t) = \big(\mathbf{l}(t), \mathbf{h}(t)\big), P_\ell d\mathbf{W}(t) = d\mathbf{W}_\ell(t), P_h d\mathbf{W}(t) = d\mathbf{W}_h(t)$

$$\begin{aligned} d\mathbf{l}(t) &= \big[-\nu\Lambda^2\mathbf{l} + P_\ell B(\mathbf{l},\mathbf{l})\big]\,dt \\ &+ \;[P_\ell B(\mathbf{l},\mathbf{h}) + P_\ell B(\mathbf{h},\mathbf{l}) + P_\ell B(\mathbf{h},\mathbf{h})]\,dt + d\mathbf{W}_\ell(t) \end{aligned} \tag{118}$$

$$d\mathbf{h}(t) = \left[-\nu\Lambda^2\mathbf{h} + P_hB(\mathbf{h},\mathbf{h})\right]dt$$
$$+ \quad [P_hB(\mathbf{l},\mathbf{h}) + P_hB(\mathbf{h},\mathbf{l}) + P_hB(\mathbf{l},\mathbf{l})]\,dt + d\mathbf{W}_h(t) \tag{119}$$

Define the set of "nice pasts" $U \subset C\left((-\infty,0],\mathbb{L}^2\right)$ to consist of all $\mathbf{v} : (-\infty,0] \to \mathbb{L}^2$ such that:

1. $\mathbf{v}(t)$ is in $\mathbb{H}^2$ for all $t \le 0$
2. The energy averages correctly. More precisely,
$$\lim_{t\to-\infty} \frac{1}{|t|}\int_t^0 |\Lambda\mathbf{v}(s)|^2_{\mathbb{L}^2}ds = \frac{\mathcal{E}_0}{2\nu}$$
3. The energy fluctuations are typical. More precisely, there exists a $T = T(\mathbf{v})$ such that
$$|\mathbf{v}(t)|^2_{\mathbb{L}^2} \le \mathcal{E}_0 + \max(|t|,T)^{\frac{2}{3}}$$

for $t \le 0$. It can be shown [**33**] that U contains almost all of the trajectories defined on the whole time interval.

Given a function $\mathbf{l}$, defined up to time t, let us denote by $\Phi_{s,t}(\mathbf{l},\mathbf{h}_0)$ the solution to (119) at time t with initial condition $\mathbf{h}_0$ at time s and the "forcing" $\mathbf{l}$.

LEMMA 4. [**33**] *There exists an absolute positive constant $\mathcal{C}$ such that if $N^2 > \mathcal{C}\frac{\mathcal{E}_0}{\nu^3}$ then the following holds:*

If there exists two solutions $\mathbf{u}_1(t) = \left(\mathbf{l}(t),\mathbf{h}_1(t)\right), \mathbf{u}_2(t) = \left(\mathbf{l}(t),\mathbf{h}_2(t)\right)$ corresponding to some (possibly different) realizations of the forcing and such that $\mathbf{u}_1,\mathbf{u}_2 \in U$, then $\mathbf{u}_1 = \mathbf{u}_2$, i.e. $\mathbf{h}_1 = \mathbf{h}_2$.

Furthermore given a solution $\mathbf{u}(t) = \left(\mathbf{l}(t),\mathbf{h}(t)\right) \in U$, any $\mathbf{h}_0 \in \mathbb{L}^2_h$, and $t \le 0$ the following limit exists

$$\lim_{t_0\to-\infty} \Phi_{t_0,t}(\mathbf{l},\mathbf{h}_0) = \mathbf{h}^*$$

and $\mathbf{h}^* = \mathbf{h}(t)$.

PROOF. Let $\boldsymbol{\rho}(t) = \mathbf{h}_1(t) - \mathbf{h}_2(t)$. From (119) we have

$$\begin{aligned}\frac{d\boldsymbol{\rho}}{dt} &= -\nu\Lambda^2\boldsymbol{\rho} + P_hB(\mathbf{h}_1,\mathbf{h}_1) - P_hB(\mathbf{h}_2,\mathbf{h}_2) + P_hB(\mathbf{l},\boldsymbol{\rho}) + P_hB(\boldsymbol{\rho},\mathbf{l})\\ &= -\nu\Lambda^2\boldsymbol{\rho} + P_hB(\mathbf{l}+\mathbf{h}_1,\boldsymbol{\rho}) + P_hB(\boldsymbol{\rho},\mathbf{l}+\mathbf{h}_2)\\ &= -\nu\Lambda^2\boldsymbol{\rho} + P_hB(\mathbf{u}_1,\boldsymbol{\rho}) + P_hB(\boldsymbol{\rho},\mathbf{u}_2)\ .\end{aligned} \tag{120}$$

Taking the inner product with $\boldsymbol{\rho}$, using the fact that $\langle P_hB(\mathbf{u}_1,\boldsymbol{\rho}),\boldsymbol{\rho}\rangle_{\mathbb{L}^2} = 0$, gives

$$\frac{1}{2}\frac{d}{dt}|\boldsymbol{\rho}|^2_{L^2} = -\nu|\Lambda\boldsymbol{\rho}|^2_{L^2} + \langle P_hB(\boldsymbol{\rho},\mathbf{u}_2),\boldsymbol{\rho}\rangle_{\mathbb{L}^2}\ .$$

Using the fact that [**21**]

$$\begin{aligned}|\langle P_hB(\boldsymbol{\rho},\mathbf{u}_2),\boldsymbol{\rho}\rangle_{\mathbb{L}^2}| &\le \hat{C}\,|\Lambda\boldsymbol{\rho}|_{\mathbb{L}^2}\,|\boldsymbol{\rho}|_{\mathbb{L}^2}\,|\Lambda\mathbf{u}_2|_{\mathbb{L}^2}\\ &\le \frac{\nu}{2}\,|\Lambda\boldsymbol{\rho}|^2_{\mathbb{L}^2} + \frac{\hat{C}^2}{2\nu}\,|\boldsymbol{\rho}|^2_{\mathbb{L}^2}\,|\Lambda\mathbf{u}_2|^2_{\mathbb{L}^2}\ ,\end{aligned}$$

we get

$$\frac{1}{2}\frac{d}{dt}\,|\boldsymbol{\rho}|^2_{\mathbb{L}^2} \le -\frac{\nu}{2}\,|\Lambda\boldsymbol{\rho}|^2_{\mathbb{L}^2} + \frac{\hat{C}^2}{2\nu}\,|\Lambda\mathbf{u}_2|^2_{\mathbb{L}^2}\,|\boldsymbol{\rho}|^2_{\mathbb{L}^2}\ .$$

Since $\boldsymbol{\rho}$ only contains modes with $|k| > N$, Poincaré inequality implies

$$\frac{d}{dt}|\boldsymbol{\rho}|^2_{\mathbb{L}^2} \le \left(-\nu N^2 + \frac{\hat{C}^2}{\nu}|\Lambda \mathbf{u}_2|^2_{\mathbb{L}^2}\right)|\boldsymbol{\rho}|^2_{\mathbb{L}^2} \ .$$

Therefore we have, for $t_0 < t < 0$,

$$|\boldsymbol{\rho}(t)|^2_{\mathbb{L}^2} \le |\boldsymbol{\rho}(t_0)|^2_{\mathbb{L}^2} \exp\left\{-\nu N^2(t-t_0) + \frac{\hat{C}^2}{\nu}\int_{t_0}^t |\Lambda \mathbf{u}_2(s)|^2_{\mathbb{L}^2}\, ds\right\} \ . \tag{121}$$

¿From the second assumption on functions in U, we know that

$$\lim \frac{1}{t}\int_{-t}^0 |\Lambda \mathbf{u}_2(s)|^2_{\mathbb{L}^2}\, ds = \frac{\mathcal{E}_0}{2\nu}.$$

Hence for $t_0 < T_1$, where T_1 depends on t and $\mathbf{u}_2$, we have

$$-\nu N^2(t-t_0) + \frac{\hat{C}^2}{\nu}\int_{t_0}^t |\Lambda \mathbf{u}_2(s)|^2_{\mathbb{L}^2}\, ds \le -\frac{\gamma}{2}(t-t_0)$$

where $\gamma = \nu N^2 - \frac{\hat{C}^2 \mathcal{E}_0}{2\nu^2}$. If we set $\mathcal{C} = \frac{\hat{C}^2}{2}$, then our assumption on N implies $\gamma > 0$.

Now using the last property of paths in U we have for any $t_0 \le T_2$,

$$\begin{aligned}|\boldsymbol{\rho}(t)|^2_{\mathbb{L}^2} \le& |\boldsymbol{\rho}(t_0)|^2_{\mathbb{L}^2} \exp\{-\frac{\gamma}{2}(t-t_0)\}\\ \le& 2\left[\mathcal{E}_0 + |t_0|^{\frac{2}{3}}\right]\exp\{-\frac{\gamma}{2}(t-t_0)\} \to 0\end{aligned}$$

as $t_0 \to -\infty$ where T_2 is some finite constant depending on $\mathbf{u}_1$ and $\mathbf{u}_2$. This completes the proof of the first part of the Lemma.

To prove the second part, observe that (121) only required control of $\int_{t_0}^t |\Lambda \mathbf{u}(s)|^2_{\mathbb{L}^2}\, ds$ for one of the two solutions. If we proceed as before letting the given solution $\mathbf{u}(t)$ play the role of $\mathbf{u}_2$ and the solution to (119) starting from $\mathbf{h}_0$ play the role of $\mathbf{u}_1$, then we obtain the estimate

$$|\boldsymbol{\rho}(t)|^2_{\mathbb{L}^2} \le |\mathbf{h}(t_0) - \mathbf{h}_0|^2_{\mathbb{L}^2} \exp\left\{-\nu N^2(t-t_0) + \frac{\hat{C}^2}{\nu}\int_{t_0}^t |\Lambda \mathbf{u}(s)|^2_{\mathbb{L}^2}\, ds\right\} \ . \tag{122}$$

Since $\mathbf{u}(t) = (\mathbf{l}(t), \mathbf{h}(t)) \in U$, the same reasoning as before shows that $\boldsymbol{\rho}(t)$ goes to zero as $t_0 \to -\infty$. Hence the limit exists and equals $\mathbf{h}(t)$. □

Denote by $\mathcal{P}$ the projection of U to low modes. Given an "initial low mode past" L^0 in $\mathcal{P}$, we can then uniquely construct the high mode at $t = 0$, denoted by $\mathbf{h}(0) = \Phi_0(L^0)$, and the solution to the full system (118), (119), denoted by $\mathbf{u}(t) = (\mathbf{l}(t), \mathbf{h}(t))$. We will denote the trajectory of the low mode on the time interval $(-\infty, t]$ by L^t and $\mathbf{h}(t) = \Phi_t(L^t)$. Using these notations we can rewrite (118) and (119) as a history-dependent finite dimensional system

$$d\mathbf{l}(t) = \left[-\nu\Lambda^2 \mathbf{l}(t) + P_\ell B\big(\mathbf{l}(t), \mathbf{l}(t)\big) + G\big(\mathbf{l}(t), \Phi_t(L^t)\big)\right] dt + d\mathbf{W}(t) \tag{123}$$

where

$$G\left(\mathbf{l}, \mathbf{h}\right) = P_\ell B(\mathbf{l}, \mathbf{h}) + P_\ell B(\mathbf{h}, \mathbf{l}) + P_\ell B(\mathbf{h}, \mathbf{h}) \ . \tag{124}$$

Thus we have a closed formulation of the dynamics on the low modes given an initial past in $L^0 \in \mathcal{P}$. We write $L^t = \mathrm{S}_t^\omega L^0$. We reiterate that L^t is the entire trajectory from time t back to $-\infty$ whereas $\mathbf{l}(t)$ is simply the value of the low modes at time t.

(123) is not Markovian, but rather Gibbsian (see the discussions in §4 of [**33**]). Notice that the stochastic forcing in (123) is non-degenerate which implies ergodicity if it were Markovian.

Step 2. Ergodicity of the finite dimensional Gibbsian dynamics.

Our task is reduced to showing that the memory effect in (123) is not strong enough to destroy ergodicity.

In order to show this, we prove that the measures induced by different initial pasts for (123) are equivalent to each other.

The key idea is to realize that the Girsanov transformation can still be used for memory-dependent systems. The main technical point is to cut-off the nonlinear growth term so that the Novikov condition for the application of the Girsanov formula is satisfied. For details we refer to [**33**].

It is proved in [**83**] that the empirical measures, starting from arbitrary initial condition, converge to the unique invariant measure exponentially fast.

4.2. General Stochastic Dissipative PDEs. The philosophy presented in the last section is applicable to general stochastic dissipative PDEs such as the Ginzburg-Landau equation, the Kuramoto-Sivashinsky equation and the Cahn-Hilliard equation. The general result, proved in [**31**], states that if all determining modes are forced, the invariant measure is unique and the dynamics is ergodic. Related results are also found in [**64, 86, 83**].

4.3. Ergodicity with Minimum Stochastic Forcing. The condition in Theorem 13 is still unsatisfactory since the number of modes that need to be forced goes to infinity as $\nu \to 0$. The physically more interesting situation is when only the large scale modes are forced. It is then natural to ask the question: what is the smallest set of modes that have to be forced in order to have ergodicity?

To answer this question, let us consider the two-dimensional case and use the vorticity formulation of the Navier-Stokes equation ($\eta = \frac{\partial u_1}{\partial x_2} - \frac{\partial u_2}{\partial x_1}$).

$$d\eta = \{\nu\Delta\eta - (\mathbf{u}\cdot\nabla)\eta\}dt + df$$

We will consider the case when the stochastic forcing has the following form.

$$df = \sum_{\mathbf{k}\in K} \sigma_{\mathbf{k}} \cos(\mathbf{k}\cdot\mathbf{x})dB_{\mathbf{k}}(t) + \sum_{\mathbf{k}\in K} \gamma_{\mathbf{k}} \sin(\mathbf{k}\cdot\mathbf{x})d\beta_{\mathbf{k}}(t) \tag{125}$$

where the $B_{\mathbf{k}}$'s and $\beta_{\mathbf{k}}$'s are independent standard Wiener processes defined on a probability space $\{\Omega, F, \mathbb{P}\}$. Here $\mu_{\mathbf{k}}$ and $\gamma_{\mathbf{k}}$ are positive constants representing the amplitude of the forcing. K is the set of modes that are forced. We will be primarily interested in the case when K contains very few low modes.

Writing $\eta(\mathbf{x},t) = \sum_{\mathbf{k}} \eta_{\mathbf{k}}(t)\cos(\mathbf{k}\cdot\mathbf{x}) + \sum_{\mathbf{k}} \sigma_{\mathbf{k}}(t)\sin(\mathbf{k}\cdot\mathbf{x})$ where the index $\mathbf{k}$ is in the first quadrant. We can rewrite (125) as

$$\begin{aligned} d\eta_{\mathbf{l}} \;=\; & \Bigg\{ -\nu|\mathbf{l}|^2\eta_{\mathbf{l}} + \sum_{\mathbf{j}+\mathbf{k}=\mathbf{l}} \frac{\mathbf{j}^\perp\cdot\mathbf{k}}{|\mathbf{j}|^2}(-\eta_{\mathbf{j}}\eta_{\mathbf{k}} + \mu_{\mathbf{j}}\mu_{\mathbf{k}}) \\ & + \sum_{\mathbf{j}-\mathbf{k}=\mathbf{l}} \frac{\mathbf{j}^\perp\cdot\mathbf{k}}{|\mathbf{j}|^2}(\eta_{\mathbf{j}}\eta_{\mathbf{k}} + \mu_{\mathbf{j}}\mu_{\mathbf{k}}) \Bigg\} dt + \sigma_{\mathbf{l}}dB_{\mathbf{l}} \end{aligned} \tag{126}$$

$$d\mu_{\mathbf{l}} = \left\{ -\nu|\mathbf{l}|^2\mu_{\mathbf{l}} - \sum_{\mathbf{j}+\mathbf{k}=\mathbf{l}} \frac{\mathbf{j}^\perp \cdot \mathbf{k}}{|\mathbf{j}|^2}(\eta_{\mathbf{j}}\mu_{\mathbf{k}} + \mu_{\mathbf{j}}\eta_{\mathbf{k}}) + \sum_{\mathbf{j}-\mathbf{k}=\mathbf{l}} \frac{\mathbf{j}^\perp \cdot \mathbf{k}}{|\mathbf{j}|^2}(-\eta_{\mathbf{j}}\mu_{\mathbf{k}} + \mu_{\mathbf{j}}\eta_{\mathbf{k}}) \right\} dt + \gamma_{\mathbf{l}} d\eta_{\mathbf{l}} \tag{127}$$

where $\mathbf{j}^\perp = (-j_2, j_1)$. Here and in the rest of this section, we adopt the convention that the summation is done only over terms in which the indices $\mathbf{j}, \mathbf{k}, \mathbf{l}$ are in the first quadrant.

E and Mattingly [**32**] considered finite dimensional approximations of (126) and (127):

$$d\eta_{\mathbf{l}} = \left\{ -\nu|\mathbf{l}|^2\eta_{\mathbf{l}} + \sum_{\mathbf{j}+\mathbf{k}=\mathbf{l}}^{N} \frac{\mathbf{j}^\perp \cdot \mathbf{k}}{|\mathbf{j}|^2}(-\eta_{\mathbf{j}}\eta_{\mathbf{k}} + \mu_{\mathbf{j}}\mu_{\mathbf{k}}) + \sum_{\mathbf{j}-\mathbf{k}=\mathbf{l}}^{N} \frac{\mathbf{j}^\perp \cdot \mathbf{k}}{|\mathbf{j}|^2}(\eta_{\mathbf{j}}\eta_{\mathbf{k}} + \mu_{\mathbf{j}}\mu_{\mathbf{k}}) \right\} dt + \sigma_{\mathbf{l}} dB_{\mathbf{l}} \tag{128}$$

$$d\mu_{\mathbf{l}} = \left\{ -\nu|\mathbf{l}|^2\mu_{\mathbf{l}} - \sum_{\mathbf{j}+\mathbf{k}=\mathbf{l}}^{N} \frac{\mathbf{j}^\perp \cdot \mathbf{k}}{|\mathbf{j}|^2}(\eta_{\mathbf{j}}\mu_{\mathbf{k}} + \mu_{\mathbf{j}}\eta_{\mathbf{k}}) + \sum_{\mathbf{j}-\mathbf{k}=\mathbf{l}}^{N} \frac{\mathbf{j}^\perp \cdot \mathbf{k}}{|\mathbf{j}|^2}(-\eta_{\mathbf{j}}\mu_{\mathbf{k}} + \mu_{\mathbf{j}}\eta_{\mathbf{k}}) \right\} dt + \gamma_{\mathbf{l}} d\eta_{\mathbf{l}} \tag{129}$$

where $\sum^{N}$ means that the sum is over indices $\mathbf{j}, \mathbf{k}, \mathbf{l}$ such that $|\mathbf{j}|_\infty, |\mathbf{k}|_\infty, |\mathbf{l}|_\infty \le N$.

The main result of [**32**] is the following.

THEOREM 14. *Let $K_1 = \{(0,1),(1,1)\}, K_2 = \{(1,0),(1,1)\}$. If $K \supset K_1$, or $K \supset K_2$ then (128) and (129) has a unique invariant measure.*

At the present time, there are still difficulties in extending Theorem 14 to the full Navier-Stokes equation. The difficulty seems to be technical, in adapting Malliavin calculus to the PDE. Some progress has been made in this direction in [**39**].

It is not clear whether the condition in Theorem 14 is sharp. There are two cases that are not covered by Theorem 14. One is when $K = \{(1,0)\}$. The other is when $K = \{(1,0),(0,1)\}$. It would be interesting to understand what happens in these two cases.

Again the existence of an invariant measure follows from standard compactness results. The proof of uniqueness uses the following result of Harris [**53**].

Let $\{x_n, n = 0, 1, \dots\}$ be a Markov process on a topological space $\mathbb{X}$ with Borel σ-algebra $\mathfrak{B}$. The Markov process $\{x_n, n = 0, 1, \dots\}$ is said to satisfy Harris' condition if there exists a σ-finite measure m on $\mathbb{X}$ such that if $m(E) > 0$, $E \in \mathfrak{B}$, then

$$\mathbb{P}_{x_0}\{x_n \in E \text{ infinitely often }\} = 1$$

for all starting points x_0 in $\mathbb{X}$. Under this condition Harris proved that there is a measure Q, unique up to a constant multiplier, which solves the equation

$$Q(E) = \int_{\mathbb{X}} P(x, E) Q(dx), \text{ for } E \in \mathfrak{B}$$

where $P(x, \cdot)$ is the transition probability distribution of the Markov process.

To establish Harris' condition, we prove the following:

1. Starting from any initial position, the dynamics enters any neighborhood of the origin infinitely often.
2. The transition probability distribution has a smooth density.

The main idea for proving (1) is to observe that in the absence of forcing, solutions of (128)-(129) decay exponentially fast due to the viscous term. This decay still holds when the forcing is small. The main idea for proving (2) is to prove that the Fokker-Planck operator associated with (128)-(129) is hypoelliptic.

LEMMA 5. *Assume that $K_1 \subset K$ or $K_2 \subset K$. Then the Markov process* (128)-(129) *has a transition density $p_t(x, y)$ which is C^∞ in (x, y, t) for $t > 0$.*

LEMMA 6. *Fix a $\Delta t > 0$ and an open neighborhood U_0 of the origin. Then given any initial condition $\eta_0 \in \mathbb{R}^{(N+1)^2-1}$,*

$$\mathbb{P}_{\eta_0}\{\eta(n\Delta t) \in U_0 \textit{ for infinitely many } n \} = 1 .$$

Using these lemmas uniqueness of the invariant measure follows quickly (see [**32**]).

We will not give the detailed proofs of these results here. Instead we will reproduce the calculation that leads to the hypoellipticity of the Fokker-Planck operator.

(128)-(129) is a degenerate diffusion process with analytic coefficients. The regularity of the transition density of such processes can be studied using Malliavin calculus. If we write the Fokker-Planck operator (the generator) of a diffusion process in the form

$$L = X_0 + \frac{1}{2}\sum_{j=1}^{l} X_j^2 \tag{130}$$

then the transition density is smooth if at each point of the state space if

$$\begin{aligned} & X_j, j = 1, 2 \cdots l, \\ & [X_j, X_k], j, k = 0, 1, \cdots l \\ & [X_j, [X_{k_1}, X_{k_2}]], j, k_1, k_2 = 0, 1, \cdots l \\ & \cdots\cdots \end{aligned} \tag{H}$$

span the tangent space [**70, 54, 67, 87**]. Notice that we allow X_0 to enter the higher-order brackets. This condition can be restated as: the ideal generated by $X_1, X_2 \cdots X_l$ in the Lie algebra generated $X_0, X_1, X_2, \cdots X_l$ must have full rank at every point. We denote by SPAN(η) the ideal generated by $X_1, X_2, \cdots X_l$ at the point η.

The Fokker-Planck operator for (128)-(129) can be written as

$$L = X_0 + \frac{1}{2}\sum_{\mathbf{k}\in K}^{N} \sigma_{\mathbf{k}}^2 \left(\frac{\partial}{\partial \alpha_{\mathbf{k}}}\right)^2 + \frac{1}{2}\sum_{\mathbf{k}\in K}^{N} \gamma_{\mathbf{k}}^2 \left(\frac{\partial}{\partial \beta_{\mathbf{k}}}\right)^2 \tag{131}$$

where

$$\begin{aligned} X_0 = &\sum_{\mathbf{l}}^{N}\left\{\sum_{\mathbf{j}+\mathbf{k}=\mathbf{l}}^{N} \frac{\mathbf{j}^\perp\cdot\mathbf{k}}{|\mathbf{j}|^2}(-\alpha_{\mathbf{j}}\alpha_{\mathbf{k}} + \beta_{\mathbf{j}}\alpha_{\mathbf{k}}) + \sum_{\mathbf{j}-\mathbf{k}=\mathbf{l}}^{N} \frac{\mathbf{j}^\perp\cdot\mathbf{k}}{|\mathbf{j}|^2}(\alpha_{\mathbf{j}}\alpha_{\mathbf{k}} + \beta_{\mathbf{j}}\beta_{\mathbf{k}}) - \nu|\mathbf{l}|^2\alpha_{\mathbf{l}}\right\}\frac{\partial}{\partial\alpha_{\mathbf{l}}} \\ &+\sum_{\mathbf{l}}^{N}\left\{\sum_{\mathbf{j}+\mathbf{k}=\mathbf{l}}^{N} \frac{\mathbf{j}^\perp\cdot\mathbf{k}}{|\mathbf{j}|^2}(-\alpha_{\mathbf{j}}\beta_{\mathbf{k}} - \beta_{\mathbf{j}}\alpha_{\mathbf{k}}) + \sum_{\mathbf{j}-\mathbf{k}=\mathbf{l}}^{N} \frac{\mathbf{j}^\perp\cdot\mathbf{k}}{|\mathbf{j}|^2}(-\alpha_{\mathbf{j}}\beta_{\mathbf{k}} + \beta_{\mathbf{j}}\alpha_{\mathbf{k}}) - \nu|\mathbf{l}|^2\beta_{\mathbf{l}}\right\}\frac{\partial}{\partial\beta_{\mathbf{l}}} \end{aligned} \tag{132}$$

We can now calculate

$$\begin{aligned} X_{\mathbf{m}} &= \left[X_0, \frac{\partial}{\partial\alpha_{\mathbf{m}}}\right] \\ &= \sum_{\mathbf{l}}^{N}\left\{\alpha_{\mathbf{l}-\mathbf{m}}\left(\frac{1}{|\mathbf{m}|^2} - \frac{1}{|\mathbf{l}-\mathbf{m}|^2}\right) + \alpha_{\mathbf{m}-\mathbf{l}}\frac{1}{|\mathbf{m}|^2} + \alpha_{\mathbf{l}+\mathbf{m}}\frac{1}{|\mathbf{l}+\mathbf{m}|^2}\right\}(\mathbf{l}^\perp\cdot\mathbf{m})\frac{\partial}{\partial\alpha_{\mathbf{l}}} \\ &+ \sum_{\mathbf{l}}^{N}\left\{\beta_{\mathbf{l}-\mathbf{m}}\left(\frac{1}{|\mathbf{m}|^2} - \frac{1}{|\mathbf{l}-\mathbf{m}|^2}\right) - \beta_{\mathbf{m}-\mathbf{l}}\frac{1}{|\mathbf{m}|^2} + \beta_{\mathbf{l}+\mathbf{m}}\frac{1}{|\mathbf{l}+\mathbf{m}|^2}\right\}(\mathbf{l}^\perp\cdot\mathbf{m})\frac{\partial}{\partial\beta_{\mathbf{l}}} \\ &- \nu|\mathbf{m}|^2\frac{\partial}{\partial\alpha_{\mathbf{m}}} \\ Y_{\mathbf{m}} &= \left[X_0, \frac{\partial}{\partial\beta_{\mathbf{m}}}\right] \\ &= \sum_{\mathbf{l}}^{N}\left\{\beta_{\mathbf{l}-\mathbf{m}}\left(-\frac{1}{|\mathbf{m}|^2} + \frac{1}{|\mathbf{l}-\mathbf{m}|^2}\right) + \beta_{\mathbf{m}-\mathbf{l}}\frac{1}{|\mathbf{m}|^2} + \beta_{\mathbf{m}+\mathbf{l}}\frac{1}{|\mathbf{l}+\mathbf{m}|^2}\right\}(\mathbf{l}^\perp\cdot\mathbf{m})\frac{\partial}{\partial\alpha_{\mathbf{l}}} \\ &+ \sum_{\mathbf{l}}^{N}\left\{\alpha_{\mathbf{l}-\mathbf{m}}\left(\frac{1}{|\mathbf{m}|^2} - \frac{1}{|\mathbf{l}-\mathbf{m}|^2}\right) + \alpha_{\mathbf{m}-\mathbf{l}}\frac{1}{|\mathbf{m}|^2} - \alpha_{\mathbf{m}+\mathbf{l}}\frac{1}{|\mathbf{l}+\mathbf{m}|^2}\right\}(\mathbf{l}^\perp\cdot\mathbf{m})\frac{\partial}{\partial\beta_{\mathbf{l}}} \\ &- \nu|\mathbf{m}|^2\frac{\partial}{\partial\beta_{\mathbf{m}}} \end{aligned}$$

We emphasize our convention that the terms are present in the sums only if the indices are in the first quadrant. If we denote by $Q_N = \{\mathbf{j} = (j_1, j_2), 0 \le j_1, j_2 \le N, (j_1, j_2) \ne (0,0)\}$, then the first sum in $X_{\mathbf{m}}$ should be written as

$$\begin{aligned} &\sum_{\mathbf{l},\mathbf{l}-\mathbf{m}\in Q_N} \alpha_{\mathbf{l}-\mathbf{m}}\left(\frac{1}{|\mathbf{m}|^2} - \frac{1}{|\mathbf{l}-\mathbf{m}|^2}\right)(\mathbf{l}^\perp\cdot\mathbf{m})\frac{\partial}{\partial\alpha_{\mathbf{l}}} + \\ + &\sum_{\mathbf{l},\mathbf{m}-\mathbf{l}\in Q_N} \alpha_{\mathbf{m}-\mathbf{l}}\frac{1}{|\mathbf{m}|^2}(\mathbf{l}^\perp\cdot\mathbf{m})\frac{\partial}{\partial\alpha_{\mathbf{l}}} + \\ + &\sum_{\mathbf{l},\mathbf{l}+\mathbf{m}\in Q_N} \alpha_{\mathbf{l}+\mathbf{m}}\frac{1}{|\mathbf{l}+\mathbf{m}|^2}(\mathbf{l}^\perp\cdot\mathbf{m})\frac{\partial}{\partial\alpha_{\mathbf{l}}}, \end{aligned}$$

and similarly for the other terms.

We then have

$$\left[X_{\mathbf{m}}, \frac{\partial}{\partial \beta_{\mathbf{k}}}\right] = (\mathbf{k}^{\perp}\cdot\mathbf{m})\left\{\left(\frac{1}{|\mathbf{m}|^2}-\frac{1}{|\mathbf{k}|^2}\right)\frac{\partial}{\partial \beta_{\mathbf{m}+\mathbf{k}}} + \frac{1}{|\mathbf{m}|^2}\frac{\partial}{\partial \beta_{\mathbf{m}-\mathbf{k}}} + \frac{1}{|\mathbf{k}|^2}\frac{\partial}{\partial \beta_{\mathbf{k}-\mathbf{m}}}\right\}$$

$$\left[X_{\mathbf{m}}, \frac{\partial}{\partial \alpha_{\mathbf{k}}}\right] = (\mathbf{k}^{\perp}\cdot\mathbf{m})\left\{\left(\frac{1}{|\mathbf{m}|^2}-\frac{1}{|\mathbf{k}|^2}\right)\frac{\partial}{\partial \alpha_{\mathbf{m}+\mathbf{k}}} - \frac{1}{|\mathbf{m}|^2}\frac{\partial}{\partial \alpha_{\mathbf{m}-\mathbf{k}}} + \frac{1}{|\mathbf{k}|^2}\frac{\partial}{\partial \alpha_{\mathbf{k}-\mathbf{m}}}\right\}$$

$$\left[Y_{\mathbf{m}}, \frac{\partial}{\partial \beta_{\mathbf{k}}}\right] = (\mathbf{k}^{\perp}\cdot\mathbf{m})\left\{-\left(\frac{1}{|\mathbf{m}|^2}-\frac{1}{|\mathbf{k}|^2}\right)\frac{\partial}{\partial \alpha_{\mathbf{m}+\mathbf{k}}} - \frac{1}{|\mathbf{m}|^2}\frac{\partial}{\partial \alpha_{\mathbf{m}-\mathbf{k}}} + \frac{1}{|\mathbf{k}|^2}\frac{\partial}{\partial \alpha_{\mathbf{k}-\mathbf{m}}}\right\}$$

$$\left[Y_{\mathbf{m}}, \frac{\partial}{\partial \alpha_{\mathbf{k}}}\right] = (\mathbf{k}^{\perp}\cdot\mathbf{m})\left\{\left(\frac{1}{|\mathbf{m}|^2}-\frac{1}{|\mathbf{k}|^2}\right)\frac{\partial}{\partial \beta_{\mathbf{m}+\mathbf{k}}} - \frac{1}{|\mathbf{m}|^2}\frac{\partial}{\partial \beta_{\mathbf{m}-\mathbf{k}}} - \frac{1}{|\mathbf{k}|^2}\frac{\partial}{\partial \beta_{\mathbf{k}-\mathbf{m}}}\right\}$$

Again the same summation convention as we discussed above applies.

Now fix any $\eta \in \mathbb{R}^{(N+1)^2-1}$. Take $\mathbf{m} = (1,1), \mathbf{k} = (0,1)$, then we have

$$\left[X_{\mathbf{m}}, \frac{\partial}{\partial \alpha_{\mathbf{k}}}\right] = \frac{1}{2}\left\{\frac{\partial}{\partial \alpha_{(1,2)}} + \frac{\partial}{\partial \alpha_{(1,0)}}\right\}$$

$$\left[Y_{\mathbf{m}}, \frac{\partial}{\partial \beta_{\mathbf{k}}}\right] = -\frac{1}{2}\left\{\frac{\partial}{\partial \alpha_{(1,2)}} - \frac{\partial}{\partial \alpha_{(1,0)}}\right\}$$

Therefore if $K_1 \subset K$, then $\frac{\partial}{\partial \alpha_{(1,2)}}, \frac{\partial}{\partial \alpha_{(1,0)}} \in \mathrm{SPAN}(\eta)$. Similarly take $\mathbf{k} = (1,0), \mathbf{m} = (1,2)$, we find $\frac{\partial}{\partial \alpha_{(0,2)}} \in \mathrm{SPAN}(\eta)$. Proceeding in this way, we find $\frac{\partial}{\partial \alpha_{(1,l)}}, \frac{\partial}{\partial \alpha_{(0,l)}} \in \mathrm{SPAN}(\eta)$, for $l \le N$. Take $\mathbf{k} = (1,0), \mathbf{m} = (1,l)$, we get $\frac{\partial}{\partial \alpha_{(2,l)}} \in \mathrm{SPAN}(\eta)$, for $l > 0$. Take $\mathbf{k} = (2,1)$ and $\mathbf{m} = (0,1)$ we get $\frac{\partial}{\partial \alpha_{(2,0)}} \in \mathrm{SPAN}(\eta)$. By induction, we obtain $\frac{\partial}{\partial \alpha_{(k,l)}}, \frac{\partial}{\partial \beta_{(k,l)}} \in \mathrm{SPAN}(\eta)$ for all $0 \le k, l \le N, (k,l) \ne (0,0)$.

This establishes hypoellipticity in the case when $K_1 \subset K$. The case when $K_2 \subset K$ can be dealt with in the same way.

5. Regularity and Dissipation

Formally, at the infinite Reynolds number limit, the Navier-Stokes equation reduce to Euler's equation which takes the form

$$\left\{\begin{array}{l} \mathbf{u}_t + (\mathbf{u}\cdot\nabla)\mathbf{u} + \nabla p = 0 \\ \nabla\cdot\mathbf{u} = 0 \end{array}\right. \tag{133}$$

in the absence of forcing. For solutions of (2) and (133), one can define the energy $\frac{1}{2}\int_D |\mathbf{u}|^2 d^3x$. For smooth solutions of (133), energy is conserved:

$$\frac{1}{2}\int_D |\mathbf{u}(\mathbf{x},t)|^2 d^3x = \frac{1}{2}\int_D |\mathbf{u}(\mathbf{x},0)|^2 d^3x \tag{134}$$

if there is no boundary effects. To prove (134), multiply (133) by $\mathbf{u}$ and integrate by parts. However, turbulence theory is concerned with solutions of (133) that dissipates energy. In fact, one of the basic assumptions in turbulence theory is that under a fixed external forcing, $\bar{\varepsilon} = \langle \nu |\nabla \mathbf{u}^{\nu}|^2\rangle$ the energy dissipation rate stays finite in the limit $\nu \to 0$. Hence the limiting invariant measure, if exists, supports singular Euler flows for which $\langle |\nabla \mathbf{u}|^2\rangle = +\infty$ and energy is decreased in the absence of forcing. This was pointed out by Onsager [**88**] in 1949. This is the striking property shared by all three problems that we discussed: the hydrodynamic turbulence in three dimension, the Burgers turbulence and the turbulent transport of passive scalar fields, namely that formally in the inviscid limit, smooth solutions conserve energy, but our interest is on solutions that dissipate energy. Hence they cannot be smooth and we are interested in their singularities. In the case of the inviscid Burgers equation (the Burgers-Hopf equation)

$$u_t + \frac{1}{2}(u^2)_x = 0 \tag{135}$$

the energy is $\frac{1}{2}\int u^2 dx$, which is conserved for smooth solutions, aside from boundary effects. However, it is well-known that solutions of (135) typically contain jump discontinuities called shocks. Physically relevant jump discontinuities have to be defined as the zero viscosity limit of the viscous problem

$$u_t + \frac{1}{2}(u^2)_x = \nu u_{xx} \tag{136}$$

as $\nu \to 0$. Such solutions satisfy the so-called entropy condition [**71**] which dissipates energy at the shocks. The statistics of the shocks give rise to the interesting behavior of the correlation functions.

Similarly, for the passive scalar transport equation in the "inviscid limit",

$$\theta_t + (\mathbf{u} \cdot \nabla)\theta = 0 \tag{137}$$

If $\mathbf{u}$ is incompressible $\nabla \cdot \mathbf{u} = 0$ then the energy, defined as $\frac{1}{2}\int \theta^2 d^3x$, is conserved when $\mathbf{u}$ is smooth. However, when $\mathbf{u}$ fails to be Lipschitz continuous, such as the case of the Kraichnan model, the generalized flow associated with (137) is in general non-degenerate and we have

$$\begin{aligned} \int \theta^2(\mathbf{x},t)dx &= \iiint \theta_0(\mathbf{y})\theta_0(\mathbf{z})g(\mathbf{x},t|\mathbf{y},0)g(\mathbf{x},t|\mathbf{z},0)dydzdx \\ &< \int \theta_0^2(\mathbf{x})dx \end{aligned}$$

We see in Section 3 that θ is in general only Hölder continuous.

In [**88**], Onsager conjectured that the solutions of the 3D incompressible Euler's equation conserve energy if they are spatially Hölder continuous with exponent larger than 1/3, and cease to conserve energy if the exponent is less than 1/3. The first half of this statement was proved in [**22**] in its sharp form formulated in terms of Besov spaces. For simplicity of presentation, we will assume periodic boundary condition on the domain D.

THEOREM 15. [**22**]. *Let* $\mathbf{u} = (u_1, u_2, u_3) \in L^3([0,T], B_3^{s,\infty}(D)) \cap C([0,T], L^2(D))$ *be a weak solution of the 3D incompressible Euler's equation, i.e.*

$$\begin{aligned} &-\int_0^T \int_D u_\alpha(\mathbf{x},t)\frac{\partial}{\partial t}\psi_\alpha(\mathbf{x},t)d^3x dt - \int_D u_\alpha(\mathbf{x},0)\psi_\alpha(\mathbf{x},0)d^3x \\ &-\int_0^T \int_D u_\alpha(\mathbf{x},t)u_\beta(\mathbf{x},t)\nabla_\alpha\psi_\beta(\mathbf{x},t)d^3x dt - \int_0^T \int_D p(\mathbf{x},t)\nabla_\alpha\psi_\alpha(\mathbf{x},t)d^3x dt = 0, \end{aligned}$$

(summation convention is used) for every test function $\psi = (\psi_1, \psi_2, \psi_3) \in C^\infty(D \times [0,T))$ *with compact support. If* $s > \frac{1}{3}$, *then*

$$\int_D |\mathbf{u}(\mathbf{x},t)|^2 d^3x = \int_D |\mathbf{u}(\mathbf{x},0)|^2 d^3x, \quad \textit{for} \ \ t \in [0,T).$$

Besov space is the natural setting for formulating this result since its definition closely resembles the definition of structure functions except that the ensemble average is replaced by the spatial average. In fact, in more physical terms, Theorem 15 states that if

$$\left(\int_D |u(\mathbf{x}+\mathbf{r},t)-u(\mathbf{x},t)|^3 d^3x\right)^{1/3} \le C|\mathbf{r}|^\alpha, \tag{138}$$

for $\alpha > 1/3$, then $\bar{\varepsilon} = 0$. This is the deterministic analog of Kolmogorov's 4/5 law. It is clear from this that Theorem 15 is sharp. Furthermore, Theorem (15) suggests that in the inviscid limit, turbulent velocity fields live in a space close to $B_3^{1/3,\infty}$.

This result further suggests that the typical "turbulent" velocity fields, the ones that Kolmogorov was interested in when he discussed the inertial range, are the smoothest velocity fields that dissipate energy. As far as regularity is concerned, they live at the boundary between conservative and dissipative weak solutions.

An exception to this scenario might be two dimensional turbulence. It is easy to prove, using the techniques in [**25**], that for the two-dimensional Euler's equation the enstrophy $\int \omega^2 dx$ is conserved as long as it is finite. This observation was made by E and Vanden-Eijnden (see [**41**]). It is likely that there are no intrinsic dissipation mechanism present in the two dimensional Euler equation, and dissipation has to be added in order for the forced system to have an invariant measure.

6. Conclusions

Stochastic PDEs provide the natural framework for studying the longtime behavior of complex systems. In this regard, we have only touched upon one type of behavior: the statistical stationary states. Another type that is often found in applications is the self-similar behavior. In particular, there has been quite some interest in recent years on the scaling behavior of Hamilton-Jacobi-like growth models [**64**] and Cahn-Hilliard-like models for phase ordering and phase separation [**14**]. The rigorous treatment of such scaling results often hinges on well-posedness of the governing differential equation with rough initial data. Such a theory is available for the Hamilton-Jacobi equation, thanks to the work of Crandall and Lions [**75**]. As a result, most of the results on the scaling behavior for Hamilton-Jacobi-like growth models can be rigorously proved [**85**]. The scaling exponents often depends on the statistics of the maximum value of a large number of identically distributed random variables, which has been classified in the literature of probability theory [**73**]. For phase ordering and phase separation type of problems, on the other hand, the mathematical difficulties involved in defining solutions for the relevant governing equations (here the various forms of curvature flows) with rough initial data seem rather severe and rigorous treatment of physical theories such as the Lifshitz-Slyozov theory [**14**] does not seem to be in sight.

In this review, we have concentrated on dissipative PDEs and hence omitted work on invariant measures for Hamiltonian PDEs such as the nonlinear Schrödinger equation. We refer to [**13, 84**] for details on these results. Also omitted is the work on stochastic quantization, which is very relevant to our discussions on stochastic Ginzburg-Landau equation [**26, 57**].

We end this review with a list of open problems on the topics that we discussed.

1. On stochastic Burgers and Hamilton-Jacobi equations, the most pressing open problems seem to be the existence or nonexistence of an invariant measure on $\mathbb{R}^n$, and the hyperbolicity of the TSM in high dimensions.

2. On the stochastic passive scalar equation, it is of great interest to describe geometrically the structure of the stationary passive scalar field. In particular, we are interested in characterizing the numerically observed "cliffs" that resulted in the saturation of the scaling exponents of the structure functions.
3. On the stochastic Navier-Stokes equations, from a technical point of view, it is highly desirable to fully develop the machinery of Malliavin calculus for PDEs. In connection with two dimensional turbulence, we would like to study the inviscid limit of the stationary solutions in the presence of large scale forcing and damping.

Also of interest is the problem of weak turbulence or wave turbulence. This concerns a large class of nonlinear dispersive wave equations such as the nonlinear Schrodinger equation. Even though the predictions are quite precise at a physical level, from a mathematical point of view, the problem is not very well formulated. It would be interesting even to define the rather framework under which the problem of weak turbulence can be studied in precise mathematical terms.

Acknowledgement: It is a pleasure to acknowledge the contributions from many people I have worked with, including K. Khanin, D. Liu, J. Mattingly, A. Mazel, Ya. Sinai and E. Vanden-Eijnden. My work is partially supported by a Presidential Faculty Fellowship from NSF.

References

[1] Batchelor, G.K., *Computation of the energy spectrum in homogeneous two-dimensional turbulence*, Phys. Fluids. Suppl. II, **12** 233–239 (1969).

[2] Balkovsky, E., Lebedev, V., *Instanton for the Kraichnan passive scalar problem*, Phys. Rev. E, **58**, 5776–5795 (1998).

[3] Barenblatt, G.I., Chorin, A.J., *Scaling laws and vanishing viscosity limits for wall-bounded shear flows and for local structures in developed turbulence*, Comm. Pure Appl. Math., **50**, 381–398 (1997).

[4] Barenblatt, G.I., Chorin, A.J., *Turbulence: an old challenge and new perspective*, Meccanica, **33**, 445–468 (1998).

[5] Barenblatt, G.I., Chorin, A.J., *New Perspectives in Turbulence: Scaling Laws, Asymptotics, and Intermittency*, SIAM Rev., **40**, 265–291 (1998).

[6] Bec, J., *Universality of velocity gradients in forced Burgers turbulence*, preprint, 2001.

[7] Bec, J., Frisch, U., Khanin, K., *Kicked Burgers turbulence*, J. Fluid Mech., **416** 239–267 (2000).

[8] Bernard, D., Gawędzki, K., and Kupiainen, A., *Anomalous scaling in the N-point function sof a passive scalar*, Phys. Rev. E, **54**, 2564–2572 (1996).

[9] Billingsley, P., *Convergence of Probability Measures*, John Wiley & Sons, Inc., New York (1968).

[10] Blatter, G., Feigelman, M. V., Geshkenbein, V. B., Larkin, A. I., and Vinokur, V. M., *Vortices in high-temperature superconductors*, Rev. Modern Phys., **66**, 1125–1388 (1994).

[11] Boldyrev, S.A., *Velocity-difference probability density functions for Burgers turbulence*, Phys. Rev. E, **55**, 6907–6910 (1997).

[12] Bouchaud, J.-P., and Mézard, M., *Velocity fluctuations in forced Burgers turbulence*, Phys. Rev. E, **54**, 5116–5121 (1996).

[13] Bourgain, J., *Nonlinear Schrödinger equations*, in *Hyperbolic Equations and Frequency Interactions*, IAS/Park City Mathematics Series, **5**, Caffarelli and E ed., American Mathematical Society (1999).

[14] Bray, A.J., *Theory of phase-ordering kinetics*, Advances in Physics, **43**(3), 357–459 (1994).

[15] Brenier, Y., *The Least Action Principle and the Related Concept of Generalized Flows for Incompressible Perfect Fluids*, J. Amer. Math. Soc., **2**, 225–255 (1989).

[16] Brenier, Y., *The dual least action problem for an ideal, incompressible fluid*, Arch. Rational Mech. Anal. **122**, 323–351 (1993).

[17] Bricmont, J., Kupiainen, A., and Lefevere, R., *Exponential mixing of the 2D stochastic Navier-Stokes dynamics*, preprint.

[18] Chertkov, M., Falkovich, G., Kolokolov, I., Lebedev, V., *Statistics of a passive scalar advected by a large-scale two-dimensional velocity field: Analytic solution*, Phys. Rev. E, **51**, 5609–5627 (1995).

[19] Chertkov, M., Falkovich, G., Kolokolov, I., Lebedev, V., *Normal and anomalous scaling of the fourth-order correlation function of a randomly advected passive scalar*, Phys. Rev. E, **52**(5), 4924–4941 (1995).

[20] Chorin, A. J., *Vorticity and Turbulence*, Springer, New York, 1994.

[21] Constantin, P., and Foiaş, C., *Navier-Stokes Equations*, University of Chicago Press, Chicago (1988).

[22] Constantin, P., E, W., and Titi, E., *Onsager's conjecture on the energy conservation for solutions of Euler's equation*, Comm. Math. Phys., **165**, 207–209 (1994).

[23] Dafermos, C., *Generalized characteristics and the structure of solutions of hyperbolic conservation laws*, Indiana Univ. Math. J., **26**, 1097–1119 (1977).

[24] DiPerna, R. and Majda, A.J., *Oscillations and concentrations in weak solutions of the incompressible fluid equations*, Comm. Math. Phys., **108**, 667–689 (1986).

[25] DiPerna, R. and Lions, P.L., *Ordinary differential equations, transport theory and Sobolev spaces*, Invent. Math. **98**, 511–547 (1989).

[26] Doering, C., Thesis, University of Texas, Austin.

[27] E, W., *Aubry-Mather theory and periodic solutions of forced Burgers equations*, Comm. Pure Appl. Math., **LII** 0811–0828 (1999).

[28] E, W., *A class of homogenization problems in the calculus of variations*, Comm. Pure Appl. Math, **44**(7), 733–759 (1991).

[29] E, W., Khanin, K., Mazel, A., and Sinai, Ya. G., *Probability distributions functions for the random forced Burgers equation*, Phys. Rev. Lett., **78**, 1904–1907 (1997).

[30] E, W., Khanin, K., Mazel, A., and Sinai, Ya. G., *Invariant measures for the random-forced Burgers equation*, Ann. Math., **151**, 877–960 (2000).

[31] E, W. and Liu, D., *Gibbsian dynamics and invariant measure for stochastic dissipative PDEs*, submitted to J. Stat. Phys.

[32] E, W. and Mattingly, J., *Ergodicity for the Navier-Stokes Equation with Degenerate Random Forcing: Finite Dimensional Approximation*, to appear in Comm. Pure. Appl. Math.

[33] E, W., Mattingly, J. and Sinai, Ya., *Gibbsian dynamics and ergodicity for the stochastically forced Navier-Stokes Equation*, to appear in Comm. Math. Phys.

[34] E, W., and Vanden-Eijnden, E., *Asymptotic theory for the probability density functions in Burgers turbulence*, Phys. Rev. Lett., **83**, 2572–2575 (1999).

[35] E, W., and Vanden-Eijnden, E., *Statistical theory of the stochastic Burgers equation in the inviscid limit*, Comm. Pure Appl. Math., **LIII**, 0852–0901 (2000).

[36] E, W., and Vanden-Eijnden, E., *Generalized flows, intrinsic stochasticity and turbulent transport*, Proc. Nat. Acad. Sci., **97**(15), 8200–8205 (2000).

[37] E, W., and Vanden-Eijnden, E., *Prandtl number effect in passive scalar advection*, Physica D (2000).

[38] E, W., and Vanden-Eijnden, E., *Generalized flows and transport equations*, preprint.

[39] Eckmann, J. P., and Hairer, M., *Uniqueness of the invariant measure for a stochastic PDE driven by degenerate noise*, preprint.

[40] Evans, L.C., *The perturbed test function method for viscosity solutions of nonlinear PDE*, Proc. Roy. Soc. Edinburgh, Ser A **III**(3-4), 359–375 (1989).

[41] Eyink, G., *Dissipation in Turbulent Solutions of 2D Euler equations*, Nonlinearity, **14**, 787–802 (2001).

[42] Fathi, A., *Théorème KAM faible et théorie de Mather sur les systémes lagrangivens*, C.R. Acad. Sci. Paris Sér. I Math. **324**(9), 1043–1046 (1997).

[43] Feigelman, M. V., *One-dimensional periodic structures in a weak random potential*, Sov. Phys. JETP, **52**, 555–561 (1980).

[44] Ferrario, B., *Ergodic results for stochastic Navier-Stokes equation*, Stochastics and Stochastics Reports, **60**(3-4), 271–288 (1997).

[45] Flandoli, F., *Dissipativity and invariant measures for stochastic Navier-Stokes equations*, NoDEA, **1**, 403–426 (1994).

[46] Flandoli, F., Maslowski, B., *Ergodicity of the 2-D Navier-Stokes equation under random perturbations*, Comm. in Math. Phys., **171**, 119–141 (1995).

[47] Foias, C., Manley, O., Rosa, R., Temam, R., *Navier-Stokes Equations and Turbulence*, to be published.

[48] Frisch, U., *Turbulence: The Legacy of A. N. Kolmogorov*, Cambridge University Press (1995).

[49] Frisch, U., Mazzino, A., Noullez, A., and Vergassola, M., *Lagrangian method for multiple correlations in the passive scalar advection*, cond-mat/9810074, submitted to Phys. Fluids (1998).

[50] Gawędzki, K., Kupiainen, A., *Anomalous scaling of a passive scalar*, Phys. Rev. Lett., **75**, 3834–3837 (1995).

[51] Gawędzki, K., Vergassola, M., *Phase transition in the passive scalar advection* Physica D, **138**, 63–90 (2000).

[52] Gotoh, T. and Kraichnan, R.H., *Steady-state Burgers turbulence with large-scale forcing*, Phys. Fluids **10**, 2859–2866 (1998).

[53] Harris, T.E., *The existence of stationary measures for certain Markov processes*, in *Proceedings of the Third Berkeley Symposium on Mathematical Statistics and Probability, 1954–1955*, **II**, 113–124, University of California Press (1956).

[54] Ichihara, K. and Kunita, H., *A classification of the second order degenerate elliptic operators and its probabilistic characterization*, Z. Wahrscheinlichkeitstheorie und Verw. Gebiete, **30**, 235–254 (1974).

[55] Iturriaga, R. and Khanin, K., *Burgers turbulence and random Lagrangian systems*, preprint.

[56] Jauslin, H. R., Kreiss, H. O., and Moser, J., *On the forced Burgers equation with periodic boundary conditions*, preprint (1997).

[57] Jona Lasinio, G. and Mitter, S.K., *On the stochastic quantization of field theory*, Comm. Math. Phys., **101**, 409–436 (1985).

[58] Kardar, M., Parisi, G., Zhang, Y.C., *Dynamic scaling of growing interfaces*, Phys. Rev. Lett., **56**, 889–892 (1986).

[59] Kolmogorov, A.N., *The local structure of turbulence in incompressible viscous fluid for very large Reynolds numbers*, C.R. (Doklady) Acad. Sci. URSS (N.S.), **30**, 301–305 (1941a).

[60] Kolmogorov, A.N., *On degeneration of isotropic turbulence in an incompressible viscous liquid*, C.R. (Doklady) Acad. Sci. URSS (N.S.), **31**, 538–540 (1941b).

[61] Kolmogorov, A.N., *A refinement of previous hypothese concerning the local structure of turbulence in a viscous incompressible fluid at high Reynolds number*, J. Fluid Mech., **13**, 82–85 (1962).

[62] Kraichnan, R.H., *Inertial ranges in two-dimensional turbulence*, Phys. Fluids, **10**, 1417–1423 (1967).

[63] Kraichnan, R.H., *Dispersion of particle pairs in homogeneous turbulence*, Phys. Fluids **9**, 1937–1943 (1966).

[64] Krug, J. and Spohn, H., *Kinetic roughening of growing surfaces*, in *Solids far from equilibrium* (Godrèche, G.C., ed.), 477–582, Cambridge University Press, England (1992).

[65] Kuksin, S., Shirikyan, A., *Stochastic Dissipative PDEs and Gibbs Measures*, Comm. Math. Phys., **213**, 291–330 (2000).

[66] Kuksin, S.B. and Shirikyan, A., *On Dissipative Systems Perturbed by Bounded Random Kick-forces*, preprint (2000).

[67] Kunita, H., *Supports of diffusion processes and controllability problems*, in *Proceedings of the International Symposium on Stochastic Differential Equations (Res. Inst. Math. Sci., Kyoto Univ., Kyoto, 1976)*, 163–185, New York, Wiley, (1978).

[68] Kunita, H., *Stochastic flows and stochastic differential equations* Cambridge University Press, Cambridge (1990).

[69] Kunita, H., *Stochastic differential equations*, Cambridge University Press, Cambridge (1990).

[70] Kusuoka, S. and Stroock, D., *Applications of the Malliavin calculus. II*, J. Fac. Sci. Univ. Tokyo Sect. IA Math., **32**(1), 1–76, (1985).

[71] Lax, P.D., *Hyperbolic Systems of Conservation Laws and the Mathematical Theory of Shock Waves*, SIAM, Philadelphia (1973).

[72] Le Jan, Y., Raimond, O., *Integration of Brownian vector fields*, preprint.

[73] Leadbetter, M., Lindgren, G. and Rootzen, H., *Extremes and Related Properties of Random Sequences and Processes*, Springer-Verlag, 1983.

[74] Lifshitz, I.M. and Slyozov, V.V., J. Phys. Chem. Solids, **19**, 35 (1961).

[75] Lions, P.L., *Generalized Solutions of Hamilton-Jacobi Equations*, Pitman Advanced Publishing Program (1982).
[76] Lions, P.L., Papanicolaou, G., Varadhan, S.R.S., *Homogenization of Hamilton-Jacobi equations*, unpublished.
[77] Majda, A.J., J. Stat. Phys., **73**, 515–542 (1993).
[78] Majda, A.J., Kramer, P.R., *Simplified models for turbulent diffusion: Theory, numerical modeling and physical phenomena*, Phys. Rep., **314**, 237–574 (1999).
[79] Masmoudi, N., and Young, L.-S., *Ergodicity Theory of Infinite Dimensional Systems with Application to Dissipative Parabolic PDEs*, preprint.
[80] Mather, J.N., *Existence of quasi-periodic orbits for twist homeomorphisms of the annulus*, Topology, **21**, 457–467 (1982).
[81] Mattingly, J.C., *The stochastically forced Navier-Stokes equations:energy estimates and phase space contraction*, Ph.D. thesis, Princeton University (1998).
[82] Mattingly, J.C., *Ergodicity of 2D Navier-Stokes equations with random forcing and large viscosity*, Comm. Math. Phys. **206**(2), 273–288 (1999).
[83] Mattingly, J.C., *Exponential convergence for the stochastically forced Navier-Stokes equations and other partially dissipative dynamics*, preprint.
[84] McKean, H. and Vaninski, K., *Statistical mechanics of nonlinear wave equations and Brownian motion with restoring drift: The petit and microcanonical ensembles*, Comm. Math. Phys., **160**, 615–630 (1994).
[85] Molchanov, S.A., Surgailis, D. and Woyczynski, W.A., *The large-scale structures in the universe and quasi-Voronoi tessellation of shock fronts in forced inviscid Burgers turbulence in* $\mathbb{R}^d$, Ann. Appl. Prob., **7**, 200–228 (1997).
[86] Monin, A.S. and Yaglom, A.M., *Statistical Fluid Mechanics: Mechanics of Turbulence*, MIT Press, Cambridge (1975).
[87] Norris, J., *Simplified Malliavin calculus*, in *Séminaire de Probabilités, XX, 1984/85*, 101–130. Springer, Berlin, (1986).
[88] Onsager, L., *Statistical hydrodynamics*, Nuovo Cimento (Supplemento), **6**, 279 (1949).
[89] Parisi, G., Frisch, U., *Turbulence and Predictability in Geophysical Fluid Dynamics* (Varenna, 1983) 83–87 (North-Holland, Amsterdam) (1985).
[90] Polyakov, A.M., *Turbulence without pressure*, Phys. Rev. E, **52**, 6183–6188 (1995).
[91] Shnirelman, A.I., Geom. Funct. Anal., **4**, 586–620 (1994).
[92] Shraiman, B., Siggia, E., Phys. Rev. E, **49**, 2912–2927 (1994).
[93] Sinai, Ya., *Two results concerning asymptotic behavior of solutions of the Burgers equation with force*, J. Stat. Phys., **64**, 1-12 (1992).
[94] Sinai, Ya., *Burgers system driven by a periodic stochastic flow*, in *Itô's Stochastic Calculus and Probability Theory*, Springer-Verlag, New York, Berlin 347–355 (1996).
[95] Smoller, J., *Shock Waves and Reaction Diffusion Equations*, Springer-Verlag, New York, Berlin (1983).
[96] Stroock, D. and Varadhan, S.R.S., *Multidimensional Diffusion Processes*, Springer-Verlag, Berlin, New York (1979).
[97] Tadmor, E., *L^1-stability and error estimates for approximate Hamilton-Jacobi solutions*, UCLA CAM Report 98-21.
[98] Tartar, L., *Systems of nonlinear partial differential equations*, NATO Adv. Sci. Inst. Ser. C: Math. Phys. Sci., (Oxford, 1982) 263–285, 111 (Reidel, Dordrecht–Boston) (1983).
[99] Woyczynski, W.A., *Burgers-KPZ Turbulence*, Lecture Notes in Mathematics, **1700**, Springer-Verlag (1998).
[100] Vishik, M. and Fursikov, A., *Mathematical problems of statistical hydrodynamics*, Kluwer Academic Publishers, (1988).
[101] Vol'pert, A., *The space BV and quasilinear equations*, Math. USSR, Sb., **2** 225–267 (1967).

Department of Mathematics and, Program in Applied and Computational Mathematics,, Princeton University, Princeton NJ 08544

Current Developments in Mathematics, 2000
Pages 149 – 196

Arithmetic Progressions in Sparse Sets.

W. T. Gowers

1. Introduction.

The following statement is a famous theorem of van der Waerden.

Theorem 1.1. *Let k and r be positive integers. Then there exists a positive integer N such that, no matter how the set $\{1, 2, \ldots, N\}$ is partitioned into r subsets $X_1 \cup \cdots \cup X_r$, at least one X_i contains an arithmetic progression of length k.*

It is customary to refer to the partition of $\{1, 2, \ldots, N\}$ as an *r-colouring* and to the sets $X_1, \ldots, X_r$ as *colour classes* or even simply *colours.* An arithmetic progression contained in one colour class is then known as *monochromatic.*

Over the years there has been considerable interest in the dependence of N on k and r. The proof of van der Waerden uses a double induction argument and the resulting bound is as bad as the Ackermann function, even when r is fixed at 2. The bound was not improved for several decades, and there was some speculation that it might even reflect the true state of affairs, until Shelah discovered a beautiful argument [Sh], surprisingly similar to the original one, which reduced the bound to a primitive recursive function. This function was still large, however: if you define T recursively by $T(1) = 1$ and $T(n+1) = 2^{T(n)}$, and define W recursively by $W(1) = 1$ and $W(n+1) = T(W(n))$, then W gives you some idea of the order of growth of Shelah's improved upper bound.

In 1936, Erdős and Turán conjectured the following dramatic strengthening of van der Waerden's theorem [ET], which was finally proved by Szemerédi almost four decades later [Sz1,2].

Theorem 1.2. *For every $\delta > 0$ and every positive integer k there exists a positive integer N such that every set $A \subset \{1, 2, \ldots, N\}$ of size at least δN contains an arithmetic progression of length k.*

This is a strengthening since one can deduce Theorem 1.1 by setting $\delta = 1/r$. In other words, Szemerédi's theorem implies that in van der Waerden's theorem one can find an arithmetic progression not just in *some* colour class, but in the *largest* one. For this reason, Szemerédi's theorem is known as the *density version* of van der Waerden's theorem.

Erdős and Turán had two main reasons for making their conjecture. The first was that there seems to be no way to adapt the proof of van der Waerden's theorem to prove the density version. Thus, the density version needs a fundamentally new idea, and one can hope that this new idea will lead to much better bounds for van der Waerden's theorem. The second is that if one could prove the result for $N = \exp(\delta^{-1})$ (when δ is sufficiently small) then it would follow from the prime number theorem that the primes contain arithmetic progressions of every length.

Unfortunately, Szemerédi's proof used van der Waerden's theorem. As a result, he gave no new information about van der Waerden's theorem itself (at least from the point of view of bounds) and certainly not about arithmetic progressions in the primes. A further breakthrough was made by Furstenberg a few years later, who discovered a statement in ergodic theory which was equivalent to Szemerédi's theorem and proved the equivalent statement using ergodic-theoretic methods [Fu] (see also [FKO] for a simpler proof along similar lines). Furstenberg's approach led to a whole industry of powerful generalizations which is still continuing. However, it was non-constructive, so gave no information at all about bounds.

Indeed, the question about the primes is still an open problem, so the hope of Erdős and Turán has not yet been fulfilled. The situation regarding van der Waerden's theorem is different, however. The main topic of these lectures is a new proof of Szemerédi's theorem which, for the first time, gives the best known bound for van der Waerden's theorem as well. Our precise result is as follows. We shall write $a \uparrow b$ for a^b, with the obvious bracketing convention: $a \uparrow b \uparrow c$ means $a \uparrow (b \uparrow c)$.

Theorem 1.3. *Let $0 < \delta \leq 1/2$, let k be a positive integer, let $N \geq 2 \uparrow 2 \uparrow \delta^{-1} \uparrow 2 \uparrow 2 \uparrow (k+9)$ and let A be a subset of the set $\{1, 2, \ldots, N\}$ of size at least δN. Then A contains an arithmetic progression of length k.* □

Corollary 1.4. *Let k be a positive integer and let $N \geq 2 \uparrow 2 \uparrow 2 \uparrow 2 \uparrow 2 \uparrow (k+9)$. Then however the set $\{1, 2, \ldots, N\}$ is coloured with two colours, there will be a monochromatic arithmetic progression of length k.* □

The bound given by Corollary 1.4 is large, but it is two levels below Shelah's bound in the Ackermann hierarchy, and could be regarded as the first 'reasonable' bound for van der Waerden's theorem.

An additional interesting feature of our proof is that it uses methods of additive number theory, such as estimating exponential sums. Since many additive theorems about the primes have been proved using similar methods, there is a possibility that some combination of all the ideas may prove that the primes contain arbitrarily long arithmetic progressions. That is, it may be possible to solve the primes problem using more about the primes than just the prime number theorem. However, this is still some way off, and even the following question is unsolved: do the primes contain infinitely many arithmetic progressions of length four?

It is not be possible even to sketch the entire proof in an hour and a half - as it is written at present [G] it takes 129 pages. My main aim in this article is to explain the ideas from [G] for progressions of length three and four. Even for progressions of length three the result is interesting, and was first proved by Roth in 1952 [R]. Progressions of length four turn out to be significantly harder. This is true for all known proofs: the usual experience seems to be that if you have proved the result for progressions of length four, then you will eventually prove it for all lengths, though it may take a lot of effort. After discussing progressions of length four, I shall give some idea of why progressions of length five are harder again, and of how to deal with the extra difficulties that arise. Beyond five, there are no interesting further difficulties: once again, this was true for Szemerédi and Furstenberg as well.

The next few sections of this paper are lifted directly from [G] and modified, sometimes a little and sometimes a lot. In general, this paper is more discursive and replaces some of the precise arguments of [G] by sketches which may be easier to read. In this way I have tried to cater for the reader who wishes to skip details, while providing enough precision for the more careful reader to get a good idea of the most important arguments. The final, quite long section is a survey of open problems related to the results and techniques of this paper.

2. Uniform Sets and Roth's Theorem.

This section begins with some notation and one or two concepts that are fundamental to the rest of the paper, and then uses these concepts to give Roth's proof (slightly rewritten) of his theorem (which is Szemerédi's theorem for progressions of length three). I have given this proof in full detail, and then followed it by a less detailed version of the same argument for those who prefer it.

It is not hard to prove that a random subset of the set $\{1, 2, \ldots, N\}$ of cardinality δN contains, with high probability, roughly the expected number of arithmetic progressions of length k, that is, δ^k times the number of such progressions in the whole of $\{1, 2, \ldots, N\}$. A natural approach to Szemerédi's theorem is therefore to try to show that random sets contain the fewest progressions of length k, which would then imply the theorem. In view of many other examples in combinatorics where random sets are extremal, this is a plausible statement, but unfortunately it is false. Indeed, if random sets were the worst, then the value of δ needed to ensure an arithmetic progression of length three would be of order of magnitude $N^{-2/3}$, whereas in fact it is known to be at least $\exp(-c(\log N)^{1/2})$ for some absolute constant $c > 0$ [Be]. We shall present this lower bound in the last section of the paper. (The random argument suggested above is to choose δ so that the expected number of arithmetic progressions is less than one. Using a standard trick in probabilistic combinatorics, we can instead ask for the expected number to be at most $\delta N/2$ and then delete one point from each one. This slightly better argument lifts the density significantly, but still only to $cN^{-1/2}$.)

Despite Behrend's example, it is tempting to try to exploit the fact that random sets contain long arithmetic progressions. Such a proof could be organized as follows.

(1) Define an appropriate notion of pseudorandomness.
(2) Prove that every pseudorandom subset of $\{1, 2, \ldots, N\}$ contains roughly the number of arithmetic progressions of length k that you would expect.
(3) Prove that if $A \subset \{1, 2, \ldots, N\}$ has size δN and is *not* pseudorandom, then there exists an arithmetic progression $P \subset \{1, 2, \ldots, N\}$ with length tending to

infinity with N, such that $|A \cap P| \geq (\delta + \epsilon)|P|$, for some $\epsilon > 0$ that depends on δ (and k) only.

If these three steps can be carried out, then a simple iteration proves Szemerédi's theorem. As we shall see, this is exactly the scheme of Roth's proof for progressions of length three.

First, we must introduce some notation. Throughout the paper we shall be considering subsets of $\mathbb{Z}_N$ rather than subsets of $\{1, 2, \ldots, N\}$. It will be convenient (although not essential) to take N to be a prime number. We shall write ω for the number $\exp(2\pi i/N)$. Given a function $f : \mathbb{Z}_N \to \mathbb{C}$ and $r \in \mathbb{Z}_N$ we set

$$\hat{f}(r) = \sum_{s \in \mathbb{Z}_N} f(s)\omega^{-rs} .$$

The function $\hat{f}$ is the discrete Fourier transform of f. (In most papers in analytic number theory, the above exponential sum is written $\sum_{s=1}^{N} e(-rs/N)$, or possibly $\sum_{s=1}^{N} e_N(-rs)$.) Let us write $f * g$ for the function

$$f * g(s) = \sum_{t \in \mathbb{Z}_N} f(t)\overline{g(t-s)} .$$

(This is not standard notation, but we shall have no use for the convolution $\sum f(t)g(s-t)$ in this paper, so it is very convenient.) From now on, all sums will be over $\mathbb{Z}_N$ unless it is specified otherwise. We shall use the following basic identities over and over again in the paper.

$$(f * g)^\wedge(r) = \hat{f}(r)\overline{\hat{g}(r)} \tag{1}$$

$$\sum_r \hat{f}(r)\overline{\hat{g}(r)} = N \sum_s f(s)\overline{g(s)} \tag{2}$$

$$\sum_r |\hat{f}(r)|^2 = N \sum_s |f(s)|^2 \tag{3}$$

$$f(s) = N^{-1} \sum_r \hat{f}(r)\omega^{rs} \tag{4}$$

Of these, the first tells us that convolutions transform to pointwise products, the second and third are Parseval's identities and the last is the inversion formula. To check them directly, note that

$$\begin{aligned} (f * g)(r) &= \textstyle\sum_s (f * g)(s)\omega^{-rs} \\ &= \textstyle\sum_{s,t} f(t)\overline{g(t-s)}\omega^{-rt}\omega^{r(t-s)} \\ &= \textstyle\sum_{t,u} f(t)\omega^{-rt}\overline{g(u)\omega^{-ru}} \\ &= \hat{f}(r)\overline{\hat{g}(r)} \end{aligned}$$

which proves (1). We may deduce (2), since

$$\sum_r \hat{f}(r)\overline{\hat{g}(r)} = \sum_r \sum_s f * g(s)\omega^{-rs} = Nf * g(0) = N \sum_s f(s)\overline{g(s)} ,$$

where for the second equality we used the fact that $\sum_s \omega^{-rs}$ is N if $r = 0$ and zero otherwise. Identity (3) is a special case of (2). Noting that the function $r \mapsto \omega^{-rs}$ is the Fourier transform of the characteristic function of the singleton $\{s\}$, we can deduce (4) from (2) as well (though it is perhaps more natural just to expand the right hand side and give a direct proof).

There is one further identity, sufficiently important to be worth stating as a lemma.

Lemma 2.1. *Let f and g be functions from $\mathbb{Z}_N$ to $\mathbb{C}$. Then*

$$\sum_r |\hat{f}(r)|^2 |\overline{\hat{g}(r)}|^2 = N \sum_t \Big| \sum_s f(s)\overline{g(s-t)} \Big|^2 . \tag{5}$$

Proof. By identities (1) and (2),

$$\begin{aligned}\sum_r |\hat{f}(r)|^2 |\overline{\hat{g}(r)}|^2 &= \textstyle\sum_r |(f * g)^\wedge(r)|^2 \\ &= N \textstyle\sum_t |f * g(t)|^2 \\ &= N \textstyle\sum_{t,s,u} f(s)\overline{g(s-t)f(u)}g(u-t) \\ &= N \textstyle\sum_t \Big| \sum_s f(s)\overline{g(s-t)} \Big|^2\end{aligned}$$

as required. □

Setting $f = g$ and expanding the right hand side of (5), one obtains another identity which shows that sums of fourth powers of Fourier coefficients have an interesting interpretation.

$$\sum_r |\hat{f}(r)|^4 = N \sum_{a-b=c-d} f(a)\overline{f(b)f(c)}f(d) . \tag{6}$$

It is of course easy to check this identity directly.

Nearly all the functions in this paper will take values with modulus at most one. In such a case, one can think of Lemma 2.1 as saying that if f has a large inner product with a large number of rotations of g, then f and g must have large Fourier coefficients in common, where large means of size proportional to N. We shall be particularly interested in the Fourier coefficients of characteristic functions of sets $A \subset \mathbb{Z}_N$ of cardinality δN, which we shall denote by the same letter as the set itself. Notice that identity (6), when applied to (the characteristic function of) a set A, tells us that the sum $\sum_r |\hat{A}(r)|^4$ is N times the number of quadruples $(a, b, c, d) \in A^4$ such that $a - b = c - d$.

For technical reasons it is also useful to consider functions of mean zero. Given a set A of cardinality δN, let us define the *balanced* function of A to be $f_A : \mathbb{Z}^N \to [-1, 1]$ where

$$f_A(s) = \begin{cases} 1-\delta & \text{s} \in \text{A} \\ -\delta & \text{s} \notin \text{A} \end{cases} .$$

This is the characteristic function of A minus the constant function $\delta\mathbf{1}$. Note that $\sum_{s \in \mathbb{Z}_N} f_A(s) = \hat{f}_A(0) = 0$ and that $\hat{f}_A(r) = \hat{A}(r)$ for $r \neq 0$.

We are now in a position to define a useful notion of pseudorandomness. The next lemma (which is not new) gives several equivalent definitions involving constants c_i. When we say that one property involving c_i implies another involving c_j, we mean that if the first holds, then so does the second for a constant c_j that tends to zero as c_i tends to zero. (Thus, if one moves from one property to another and then back again, one does not necessarily recover the original constant.) From the point of view of the eventual bounds obtained, it is important that the dependence is no worse than a fixed power. This is always true below.

In this paper we shall use the letter D to denote the closed unit disc in $\mathbb{C}$ (unless it obviously means something else).

Lemma 2.2. *Let f be a function from $\mathbb{Z}_N$ to D. The following are equivalent.*
(i) $\sum_k \left|\sum_s f(s)\overline{f(s-k)}\right|^2 \le c_1 N^3$.
(ii) $\sum_{a-b=c-d} f(a)\overline{f(b)}\overline{f(c)}f(d) \le c_1 N^3$.
(iii) $\sum_r |\hat{f}(r)|^4 \le c_1 N^4$.
(iv) $\max_r |\hat{f}(r)| \le c_2 N$.
(v) $\sum_k \left|\sum_s f(s)\overline{g(s-k)}\right|^2 \le c_3 N^2 \|g\|_2^2$ *for every function* $g : \mathbb{Z}_N \to \mathbb{C}$.

Proof. The equivalence of (i) and (ii) comes from expanding the left hand side of (i), and the equivalence of (i) and (iii) follows from identity (6) above. It is obvious that (iii) implies (iv) if $c_2 \ge c_1^{1/4}$. Since

$$\sum_r |\hat{f}(r)|^4 \le \max_r |\hat{f}(r)|^2 \sum_r |\hat{f}(r)|^2 \le N^2 \max_r |\hat{f}(r)|^2 ,$$

we find that (iv) implies (iii) if $c_1 \ge c_2^2$. It is obvious that (v) implies (i) if $c_1 \ge c_3$. By Lemma 2.1, the left hand side of (v) is

$$N^{-1} \sum_r |\hat{f}(r)|^2 |\hat{g}(r)|^2 \le N^{-1} \Big(\sum_r |\hat{f}(r)|^4\Big)^{1/2} \Big(\sum_r |\hat{g}(r)|^4\Big)^{1/2}$$

by the Cauchy-Schwarz inequality. Using the additional inequality

$$\Big(\sum_r |\hat{g}(r)|^4\Big)^{1/2} \le \sum_r |\hat{g}(r)|^2 ,$$

we see that (iii) implies (v) if $c_3 \ge c_1^{1/2}$. □

A function $f : \mathbb{Z}_N \to D$ satisfying condition (i) above, with $c_1 = \alpha$, will be called *α-uniform*. If f is the balanced function f_A of some set $A \subset \mathbb{Z}_N$, then we shall also say that A is α-uniform. If $A \subset \mathbb{Z}_N$ is an α-uniform set of cardinality δN, and f is its balanced function, then

$$\sum_r |\hat{A}(r)|^4 = |A|^4 + \sum_r |\hat{f}(r)|^4 \le |A|^4 + \alpha N^4 .$$

We noted earlier that $\sum_r |\hat{A}(r)|^4$ is N times the number of quadruples $(a,b,c,d) \in A^4$ such that $a-b=c-d$. If A were a random set of size δN, then we would expect about $\delta^4 N^3 = N^{-1}|A|^4$ such quadruples (which from the above is clearly a lower bound). Therefore, the number α is measuring how close A is to being random in this particular sense. Notice that quadruples (a,b,c,d) with $a-b=c-d$ are the same as quadruples of the form $(x, x+s, x+t, x+s+t)$.

We remark that our definition of an α-uniform set coincides with the definition of *quasirandom* subsets of $\mathbb{Z}_N$, due to Chung and Graham [CG]. They prove that several formulations of the definition (including those of this paper) are equivalent. They do not mention the connection with Roth's theorem, which we shall now explain. We need a very standard lemma, which we prove in slightly greater generality than is immediately necessary, so that it can be used again later. Let us define the *diameter* of a subset $X \subset \mathbb{Z}_N$ to be the smallest integer s such that $X \subset \{n, n+1, \ldots, n+s\}$ for some $n \in \mathbb{Z}_N$.

Lemma 2.3. *Let r, s and N be positive integers with $r, s \leq N$ and $rs \geq N$, and let $\phi : \{0, 1, \ldots, r-1\} \to \mathbb{Z}_N$ be linear (i.e., of the form $\phi(x) = ax + b$). Then the set $\{0, 1, \ldots, r-1\}$ can be partitioned into arithmetic progressions $P_1, \ldots, P_M$ such that for each j the diameter of $\phi(P_j)$ is at most s and the length of P_j lies between $(rs/4N)^{1/2}$ and $(rs/N)^{1/2}$.*

Proof. Let $t = \lceil (rN/4s)^{1/2} \rceil$. Of the numbers $\phi(0), \phi(1), \ldots, \phi(t)$, at least two must be within N/t. Therefore, by the linearity of ϕ, we can find a non-zero $u \leq t$ such that $|\phi(u) - \phi(0)| \leq N/t$. Split $\{0, 1, \ldots, r-1\}$ into congruence classes mod u. Each congruence class is an arithmetic progression of cardinality either $\lfloor r/u \rfloor$ or $\lceil r/u \rceil$. If P is any set of at most st/N consecutive elements of a congruence class, then $\mathrm{diam}\phi(P) \leq s$. It is easy to check first that $st/N \leq r/3t \leq (1/2)\lfloor r/u \rfloor$, next that this implies that the congruence classes can be partitioned into sets P_j of consecutive elements with every P_j of cardinality between $\lceil st/2N \rceil$ and $\lfloor st/N \rfloor$, and finally that this proves the lemma. □

Corollary 2.4. *Let f be a function from the set $\{0, 1, \ldots, r-1\}$ to the closed unit disc in $\mathbb{C}$, let $\phi : \mathbb{Z}_N \to \mathbb{Z}_N$ be linear and let $\alpha > 0$. If*

$$\Big|\sum_{x=0}^{r-1} f(x)\omega^{-\phi(s)}\Big| \geq \alpha r$$

then there is a partition of $\{0, 1, \ldots, r-1\}$ into $m \leq (8\pi r/\alpha)^{1/2}$ arithmetic progressions $P_1, \ldots, P_m$ such that

$$\sum_{j=1}^{m}\Big|\sum_{x \in P_j} f(x)\Big| \geq (\alpha/2)r$$

and such that the lengths of the P_j all lie between $(\alpha r/\pi)^{1/2}/4$ and $(\alpha r/\pi)^{1/2}/2$.

Proof. Let $s \leq \alpha N/4\pi$ and let $m = (16\pi r/\alpha)^{1/2}$. By Lemma 2.3 we can find a partition of $\{0, 1, \ldots, r-1\}$ into arithmetic progressions $P_1, \ldots, P_m$ such that the diameter of $\phi(P_j)$ is at most s for every j and the length of each P_j lies between r/m and $2r/m$. By the triangle inequality,

$$\sum_{j=1}^{m}\Big|\sum_{x \in P_j} f(x)\omega^{-\phi(x)}\Big| \geq \alpha r\ .$$

Let $x_j \in P_j$. The estimate on the diameter of $\phi(P_j)$ implies that $|\omega^{-\phi(x)} - \omega^{-\phi(x_j)}|$ is at most $\alpha/2$ for every $x \in P_j$. Therefore

$$\begin{aligned}\sum_{j=1}^{m}\Big|\sum_{x \in P_j} f(x)\Big| &= \textstyle\sum_{j=1}^{m}\Big|\sum_{x \in P_j} f(x)\omega^{-\phi(x_j)}\Big| \\ &\geq \textstyle\sum_{j=1}^{m}\Big|\sum_{x \in P_j} f(x)\omega^{-\phi(x)}\Big| - \sum_{j=1}^{m}(\alpha/2)|P_j| \\ &\geq \alpha r/2\end{aligned}$$

as claimed. □

Corollary 2.5. *Let $A \subset \mathbb{Z}_N$ and suppose that $|\hat{A}(r)| \geq \alpha N$ for some $r \neq 0$. Then there exists an arithmetic progression $P \subset \{0,1,\ldots,N-1\}$ of length at least $(\alpha^3 N/128\pi)^{1/2}$ such that $|A \cap P| \geq (\delta + \alpha/8)|P|$.*

Proof. Define $\phi(x) = rx$ and let f be the balanced function of A (regarded as a function on $\{0,1,\ldots,N-1\}$). By Corollary 2.4 we can partition the set $\{0,1,\ldots,N-1\}$ into $m \leq (16\pi N/\alpha)^{1/2}$ arithmetic progressions $P_1,\ldots,P_m$ of lengths between N/m and $2N/m$ such that

$$\sum_{j=1}^{m} \Big|\sum_{x\in P_j} f(x)\Big| \geq \alpha N/2\ .$$

Since $\sum_{x\in P_j} f(x)$ is real for all j, and since $\sum_{j=1}^{m}\sum_{x\in P_j} f(x) = 0$, if we define J to be the set of j with $\sum_{x\in P_j} f(x) \geq 0$, we have

$$\sum_{j\in J}\sum_{x\in P_j} f(x) \geq \alpha N/4\ .$$

Therefore, we can find j such that $\sum_{x\in P_j} f(x) \geq \alpha N/4m$. But $|P_j| \leq 2N/m$, so $\sum_{x\in P_j} f(x) \geq \alpha|P_j|/8$, which implies that $|A \cap P_j| \geq (\delta + \alpha/8)|P_j|$. □

We can now give Roth's proof of his theorem on arithmetic progressions of length three.

Theorem 2.6. *Let $\delta > 0$, let $N \geq \exp\exp(C\delta^{-1})$ (where C is an absolute constant) and let $A \subset \{1,2,\ldots,N\}$ be a set of size at least δN. Then A contains an arithmetic progression of length three.*

Proof. Since we are passing to smaller progressions and iterating, we cannot simply assume that N is prime, so we shall begin by dealing with this small technicality. Let N_0 be a positive integer and let A_0 be a subset of $\{1,2,\ldots,N_0\}$ of size at least $\delta_0 N_0$.

By Bertrand's postulate (which is elementary - it would be a pity to use the full strength of the prime number theorem in a proof of Roth's theorem) there is a prime p between $N/3$ and $2N/3$. Write q for $N-p$. If $|A_0 \cap \{1,2,\ldots,p\}| \leq \delta_0(1-\delta_0/160)p$, then we know that

$$|A_0 \cap \{p+1,\ldots,N\}| \geq \delta_0(N-(1-\delta_0/160)p) = \delta_0(q+\delta_0 p/160) \geq \delta_0(1+\delta_0/320)q\ .$$

Let us call this situation case 0.

If case 0 does not hold, then let N be the prime p obtained above, let $A = A_0 \cap \{1,\ldots,N\}$ and let $\delta = \delta_0(1-\delta_0/160)$. Let $B = A\cap[N/3,2N/3)$. If $|B| \leq \delta N/5$, then either $A\cap[0,N/3)$ or $A\cap[2N/3,N)$ has cardinality at least $2\delta N/5 = (6\delta/5)(N/3)$. This situation we shall call case 1.

Next, let $\alpha = \delta^2/10$ and suppose that $|\hat{A}(r)| > \alpha N$ for some non-zero r. In this case, by Corollary 2.5 there is an arithmetic progression P of cardinality at least $(\alpha^3 N/128\pi)^{1/2}$ such that $|A \cap P| \geq (\delta + \delta^2/80)|P|$. This situation will be case 2.

If case 2 does not hold, then $|\hat{A}(r)| \leq \alpha N$ for every non-zero r, which says that A satisfies condition (iv) of Lemma 2.2. The number of triples $(x, y, z) \in A \times B^2$ such that $x + z = 2y$ is then

$$N^{-1} \sum_{x \in A} \sum_{y \in B} \sum_{z \in B} \sum_r \omega^{r(2y-x-z)} = N^{-1} \sum_r \hat{A}(r)\hat{B}(-2r)\hat{C}(r)$$
$$\geq N^{-1}|A||B|^2 - N^{-1} \max_{r \neq 0} |\hat{A}(r)| \Big(\sum_{r \neq 0} |\hat{B}(-2r)|^2\Big)^{1/2} \Big(\sum_{r \neq 0} |\hat{B}(r)|^2\Big)^{1/2}$$
$$\geq \delta |B|^2 - \alpha |B| N \ .$$

If in addition case 1 does not hold, then this quantity is minimized when $|B| = \delta N/5$, and the minimum value is $\delta^3 N^2/50$, implying the existence of at least this number of triples $(x, y, z) \in A \times B^2$ in arithmetic progression mod N. Since B lives in the middle third, these are genuine progressions in $\{1, 2, \ldots, N\}$ and since there are only N degenerate progressions (i.e., with difference zero) we can conclude that A contains an arithmetic progression of length three as long as $N \geq 50\delta^{-3}$. This we shall call case 3.

To summarize, if case 3 holds and $N \geq 50\delta^{-3}$, then A contains an arithmetic progression of length three. In case 2, we can find a subprogression P of $\{1, \ldots, N\}$ of cardinality at least $(\alpha^3 N/128\pi)^{1/2}$ such that $|A \cap P| \geq \delta(1 + \delta/80)|P|$. Since $\{1, \ldots, N\}$ is a subprogression of $\{1, \ldots, N_0\}$, $A = A_0 \cap \{1, \ldots, N\}$ and one can easily check that $\delta(1 + \delta/80) \geq \delta_0(1 + \delta_0/320)$, we may conclude that in case 2 there is a subprogression P of $\{1, \ldots, N_0\}$ of cardinality at least $(\alpha^3 N_0/384\pi)^{1/2}$ such that $|A_0 \cap P| \geq \delta_0(1 + \delta_0/320)|P|$. As for cases 0 and 1, it is easy to see that the same conclusion also holds, and indeed a much stronger one as P has a length which is linear in N_0.

This gives us the basis for an iteration argument. If A_0 does not contain an arithmetic progression of length three, then we drop down to a progression P where the density of A is larger and repeat. If the density at step m of the iteration is δ_m, then at each subsequent iteration the density increases by at least $\delta_m^2/320$. It follows that the density reaches $2\delta_m$ after at most $320\delta_m^{-1}$ further steps. It follows that the total number of steps cannot be more than $320(\delta^{-1} + (2\delta)^{-1} + (4\delta)^{-1} + \ldots) = 640\delta^{-1}$. At each step, the size of the progression in which A lives is around the square root of what it was at the previous step. The result now follows from a simple calculation (left to the reader). □

Before we move to the next section, here is a sketch of the above argument, for the benefit of those who do not have the time or inclination to follow the details.

A sketched version of the above argument.

Let A be a subset of $\{1,2,\ldots,N\}$ of size at least δN. It is convenient to identify $\{1,2,\ldots,N\}$ with $\mathbb{Z}_N$ because the group structure of $\mathbb{Z}_N$ allows us to define discrete Fourier coefficients (see the precise definition at the beginning of this section). This is very useful, because the number of triples (x,y,z) in A^3 such that $x+z=2y$ mod N turns out to be $N^{-1}\sum_r \hat{A}(r)\hat{A}(-2r)\hat{A}(r)$. To see this, expand out the Fourier coefficients to obtain

$$N^{-1}\sum_r\sum_{x,y,z} A(x)\omega^{-rx}A(y)\omega^{2ry}A(z)\omega^{-rz} = N^{-1}\sum_{x,y,z} A(x)A(y)A(z)\sum_r \omega^{-r(x-2y+z)} .$$

The sum over r on the right hand side gives N if $x+z=2y$ mod N and otherwise vanishes, which proves the assertion.

Unfortunately, this raises a technical problem, since not every triple (x,y,z) with $x+z=2y$ mod N corresponds to an arithmetic progression in $\{1,2,\ldots,N\}$. To get round this, it is convenient to look at three different sets A, B and C and count the number of triples $(x,y,z)\in A\times B\times C$ such that $x+z=2y$ mod N. If B and C are subsets of the interval $[N/3,2N/3]$, then any such triple must correspond to a genuine arithmetic progression (apart from the degenerate triples (x,x,x) but there are few enough of these that they will not cause us any trouble).

Exactly as above, the number of triples in question can be interpreted on the Fourier side as $N^{-1}\sum_r \hat{A}(r)\hat{B}(-2r)\hat{C}(r)$. We then divide this sum into two parts, the contribution from $r=0$ and the rest. Now $N^{-1}\hat{A}(0)\hat{B}(0)\hat{C}(0) = |A||B||C|$. Writing $|A|=\alpha N$, $|B|=\beta N$ and $|C|=\gamma N$, we see that this is $\alpha\beta\gamma N^2$, which can be thought of as the expected number of progressions (mod N) if the elements of A are chosen randomly and independently with probability α, and similarly for B and C.

Standard probabilistic arguments show that with very high probability the number of mod-N progressions in $A\times B\times C$ will be very close to $\alpha\beta\gamma N^2$, so we can think of the rest of the sum as the *deviation from randomness* of the triple (A,B,C). In particular, if we can show that $N^{-1}\sum_{r\neq 0}\hat{A}(r)\hat{B}(-2r)\hat{C}(r)$ is at most $\alpha\beta\gamma N/2$, then we will have shown that $A\times B\times C$ contains many mod-N arithmetic progressions.

Of course, it is easy to think of examples of triples (A,B,C) such that $A\times B\times C$ contains no mod-N arithmetic progressions, so we cannot prove this unconditionally. However, we can prove it under the assumption that $\max_{r\neq 0}\hat{A}(r)\leq cN$ for a sufficiently small constant c (depending on α, β and γ). To do this, one simply applies the Cauchy-Schwarz inequality and Parseval's identity to say

$$\begin{aligned} N^{-1}\sum_{r\neq 0}\hat{A}(r)\hat{B}(-2r)\hat{C}(r) &\leq c\sum_r \hat{B}(-2r)\hat{C}(r) \\ &\leq c\Big(\sum_r |\hat{B}(-2r)|^2\Big)^{1/2}\Big(\sum_r |\hat{C}(r)|^2\Big)^{1/2} \\ &= c(N|B|)^{1/2}(N|C|)^{1/2} = c(\beta\gamma)^{1/2}N^2 . \end{aligned}$$

Hence, for the argument to work, one needs $c(\beta\gamma)^{1/2}\leq \alpha\beta\gamma$, or $c\leq\alpha(\beta\gamma)^{1/2}$.

What happens if there exists r such that $|\hat{A}(r)|>cN$ for this c? Then we rely on a different argument. The statement that A has a non-trivial large Fourier coefficient at r is the statement that the sum $\sum_{x\in A}\omega^{-rx}$ is large, which means that the values rx for $x \in A$ are not evenly distributed in $\mathbb{Z}_N$. This implies that there is a long interval $I = \{k, k+1, \ldots, l\}$ such that $|rA \cap I|$ is significantly larger than $\alpha|I|$, which implies that $|A \cap r^{-1}I|$ is significantly larger than $|r^{-1}I| = |I|$. (The length of I is proportional to N.)

The set $r^{-1}I$ is an arithmetic progression mod N with common difference r^{-1}. A standard application of the pigeonhole principle (see Lemma 2.3 for the method of proof) allows us to partition I into about $\sqrt{N}$ subsets corresponding to genuine subprogressions of $\{1, 2, \ldots, N\}$ and then, by an easy averaging argument, to find one of these, P say, such that $|P|$ is about $\sqrt{N}$ and $|A \cap P|$ is significantly larger than $\alpha|P|$. In fact, the proof gives $|A \cap P| \geq \alpha(1 + c_1\alpha)|P|$ for an absolute constant $c_1 > 0$.

This gives us the basis for an iteration argument along the lines sketched at the beginning of this section. Either A is random-like, meaning that it has no unexpectedly large Fourier coefficients, in which case it contains plenty of arithmetic progressions of length three, or it isn't, in which case we can use a large Fourier coefficient to find a longish subprogression where A has increased density. We can then restrict our attention to this subprogression and repeat the argument. Eventually, the density reaches 1 and the iteration cannot continue. It can be checked that this argument implies that any subset of $\{1, 2, \ldots, N\}$ of size at least $CN/\log\log N$ contains an arithmetic progression of length three.

I have ignored two technicalities: N may not be prime (this becomes important when we iterate) and B and C, which we set to be $A \cap [N/3, 2N/3]$, may be very small. However, these are easily dealt with, as can be seen in the detailed proof of Theorem 2.6.

3. Higher-Degree Uniformity.

As I have said, the scheme of proof of Roth's theorem can be summarized as follows: if A is random-like, then it contains many progressions of length three. If, on the other hand, it is not random-like, then there is a reasonably long subprogression inside which A has increased density, making possible an iterative argument. A set A counts as random-like for this argument if it is α-uniform for some small α (the definition appears just after Lemma 2.2), which means, amongst other things, that $\hat{A}(r)$ is small for every non-zero r.

As will be shown in the next section, α-uniformity is not an appropriate definition of pseudorandomness when we come to look at progressions of length greater than three, because it turns out not to imply all that much about the number of such progressions. The main purpose of this section is therefore to introduce a stronger definition which can be used instead. I shall do so informally to begin with, and state the main result that is needed. The rest of the section contains a complete proof of the result.

Recall from the remarks following Lemma 2.2 that a set A of cardinality δN is α-uniform for a small α if and only if the number of quadruples $(x, x+a, x+b, x+a+b) \in A^4$ is not much larger than the lower bound of $\delta^4 N^4$, which holds for all such sets A and is more or less attained by random ones. The quadruples $(x, x+a, x+b, x+a+b)$ can be thought of as two-dimensional Hilbert cubes, and to produce a definition of pseudorandomness suitable for progressions of length four we simply increase this dimension to three. Accordingly, we shall say that a set A of size δN is pseudorandom (the term we shall actually use is *quadratically uniform*) if the number of octuples

$$(x, x+a, x+b, x+c, x+a+b, x+a+c, x+b+c, x+a+b+c) \in A^8$$

is not much more than $\delta^8 N^4$, which, once again, can be shown quite easily to be a lower bound, almost attained by random sets.

The result needed later will be that if A is quadratically uniform in this sense (of course, to make this precise it is necessary to make the definition quantitative, but this is easily done) then A contains roughly the expected number of progressions of length four that will be contained in a random set of the same cardinality. In other words, when we talk about progressions of length four, quadratic uniformity is an appropriate definition of pseudorandomness. Moreover, the obvious generalization of this idea is the correct one: if we define a set A to be uniform of degree k when it contains roughly $\delta^{2^{k+1}} N^{k+2}$ cubes of dimension $k+1$, then, as one might hope, a set that is sufficiently uniform of degree k will contain roughly the expected number of progressions of length $k+2$.

Of course, these definitions will not help us unless we can go on to say something useful about sets that *fail* to be quadratically uniform or uniform of higher degree. Since quadratic uniformity is a stronger condition than uniformity, one would expect this to be harder than the corresponding step in Roth's theorem - proving that a set with a large Fourier coefficient can be restricted to a subprogression where it has increased density. And indeed it is, much harder. This is where most of the work is needed, and where the main interest in the proof lies.

It is worth mentioning an equivalent definition of quadratic uniformity, which relates it to Fourier coefficients. A set A is α-uniform if $\hat{A}(r)$ is at most αN whenever r is non-zero. It is quadratically uniform if, for almost every $k \in \mathbb{Z}_N$, the set $A \cap (A+k) = \{x : x \in A, x-k \in A\}$ is uniform. More precisely, one could say that A is α-quadratically uniform if the number of k for which $A \cap (A+k)$ fails to be α-uniform is at most αN. (Similarly, a set is cubically uniform if almost all the sets $A \cap (A+k)$ are quadratically uniform, or equivalently if almost all the sets $A \cap (A+k) \cap (A+l) \cap (A+k+l)$ are uniform.)

The rest of this section makes the above ideas precise and contains a proof that appropriately uniform sets contain appropriately many arithmetic progressions. For technical reasons we shall actually work with *functions* rather than sets, so the definitions below do not quite coincide with the one I have given. However, the reader who wishes to skip technical details ought to be able to jump to the next section and still understand the rest of the paper reasonably well.

In order to simplify the presentation for the case of progressions of length four, we shall now prove two lemmas, even though the second implies the first. Given a function $f : \mathbb{Z}_N \to \mathbb{Z}_N$, we shall define, for any k, the *difference function* $\Delta(f;k)$ by $\Delta(f;k)(s) = f(s)\overline{f(s-k)}$. The reason for the terminology is that if, as will often be the case, $f(s) = \omega^{\phi(s)}$ for some function $\phi : \mathbb{Z}_N \to \mathbb{Z}_N$, then $\Delta(f;k)(s) = \omega^{\phi(k)-\phi(s-k)}$.

Now let us define iterated difference functions in two different ways as follows. The first is inductive, setting $\Delta(f;a_1,\ldots,a_d)(s)$ to be $\Delta(\Delta(f;a_1,\ldots,a_{d-1});a_d)(s)$. The second makes explicit the result of the inductive process. Let C stand for the map from $\mathbb{C}^N$ to $\mathbb{C}^N$ which takes a function to its pointwise complex conjugate.

Given a function $f : \mathbb{Z}_N \to \mathbb{C}$, we define

$$\Delta(f; a_1, \ldots, a_d)(s) = \prod_{\epsilon_1,\ldots,\epsilon_d} (C^{\epsilon_1+\cdots+\epsilon_d} f)\Big(s - \sum_{i=1}^{d} a_i\epsilon_i\Big)$$

where the product is over all sequences $\epsilon_1, \ldots, \epsilon_d$ with $\epsilon_i \in \{0,1\}$. When $d = 3$, for example, this definition becomes

$$\begin{aligned}\Delta(f; a, b, c)(s) \qquad &= f(s)\overline{f(s-a)f(s-b)f(s-c)}\times \\ &\times f(s-a-b)f(s-a-c)f(s-b-c)\overline{f(s-a-b-c)}\ .\end{aligned}$$

We now define a function f from $\mathbb{Z}_N$ to the closed unit disc $D \subset \mathbb{C}$ to be *α-uniform of degree d* if

$$\sum_{a_1,\ldots,a_d} \Big|\sum_s \Delta(f; a_1, \ldots, a_d)(s)\Big|^2 \leq \alpha N^{d+2}\ .$$

When d equals two or three, we say that f is quadratically or cubically α-uniform respectively. As with the definition of α-uniformity (which is the same as α-uniformity of degree one) this definition has several useful reformulations.

Lemma 3.1. *Let f be a function from $\mathbb{Z}_N$ to D. The following are equivalent.*
(i) f is c_1-uniform of degree d.
(ii) $\sum_s \sum_{a_1,\ldots,a_{d+1}} \Delta(f; a_1, \ldots, a_{d+1})(s) \leq c_1 N^{d+2}$.
(iii) There is a function $\alpha : \mathbb{Z}_N^{d-1} \to [0,1]$ such that $\sum_{a_1,\ldots,a_{d-1}} \alpha(a_1, \ldots, a_{d-1}) \leq c_1 N^{d-1}$ and $\Delta(f; a_1, \ldots, a_{d-1})$ is $\alpha(a_1, \ldots, a_{d-1})$-uniform for every $(a_1, \ldots, a_{d-1})$.
(iv) There is a function $\alpha : \mathbb{Z}_N \to [0,1]$ such that $\sum_r \alpha(r) = c_1 N$ and $\Delta(f; r)$ is $\alpha(r)$-uniform of degree $d-1$ for every r.
(v) $\sum_{a_1,\ldots,a_{d-1}} \sum_r \big|\Delta(f; a_1, \ldots, a_{d-1})^\wedge(r)\big|^4 \leq c_1 N^{d+3}$.
(vi) For all but $c_2 N^{d-1}$ choices of $(a_1, \ldots, a_{d-1})$ the function $\Delta(f; a_1, \ldots, a_{d-1})$ is c_2-uniform.
(vii) There are at most $c_3 N^{d-1}$ values of $(a_1, \ldots, a_{d-1})$ for which there exists some $r \in \mathbb{Z}_N$ with $\big|\Delta(f; a_1, \ldots, a_{d-1})^\wedge(r)\big| \geq c_3 N$.

Proof. The equivalence of (i) and (ii) is easy, as the left hand sides of the relevant expressions are equal. It is also obvious that (ii) and (iii) are equivalent. A very simple inductive argument shows that (ii) is equivalent to (iv). The equivalence of (i) and (v) follows, as in the proof of the equivalence of (i) and (iii) in Lemma 2.1, by expanding the left hand side of (v). Alternatively, it can be deduced from Lemma 2.1 by applying that equivalence to each function $\Delta(f; a_1, \ldots, a_{d-1})$ and adding.

Averaging arguments show that (iii) implies (vi) as long as $c_1 \leq c_2^2$, and that (vi) implies (iii) as long as $c_1 \geq 2c_2$. Finally, the equivalence of (i) and (ii) in Lemma 2.1 shows that in this lemma (vi) implies (vii) if $c_3 \geq c_2^{1/4}$ and (vii) implies (vi) if $c_2 \geq c_3$. □

Notice that properties (i) and (ii) above make sense even when $d = 0$. Therefore, we shall define a function $f : \mathbb{Z}_N \to D$ to be α-uniform of degree zero if

$\left|\sum_s f(s)\right|^2 \leq \alpha N^2$. Property (iv) now makes sense when $d = 1$. This definition will allow us to begin an inductive argument at an earlier and thus easier place.

The next result is the main one of this section. Although it will not be applied directly, it easily implies the results that are needed for later.

Theorem 3.2. *Let $k \geq 2$ and let $f_1, \ldots, f_k$ be functions from $\mathbb{Z}_N$ to D such that f_k is α-uniform of degree $k-2$. Then*

$$\left|\sum_r \sum_s f_1(s) f_2(s-r) \ldots f_k(s-(k-1)r)\right| \leq \alpha^{1/2^{k-1}} N^2 .$$

Proof. When $k = 2$, we know that

$$\left|\sum_r \sum_s f_1(s) f_2(s-r)\right| = \left|\left(\sum_s f_1(s)\right)\left(\sum_t f_2(t)\right)\right| \leq \alpha^{1/2} N^2 ,$$

since $\left|\sum_s f_1(s)\right| \leq N$ and $\left|\sum_t f_2(t)\right| \leq \alpha^{1/2} N$.

When $k > 2$, assume the result for $k-1$, let f_k be α-uniform of degree $k-2$ and let $\alpha : \mathbb{Z}_N \to [0,1]$ be a function with the property that $\Delta(f_k; r)$ is $\alpha(r)$-uniform of degree $k-3$ for every $r \in \mathbb{Z}_N$. Then

$$\begin{aligned}
\Big|\sum_r & \sum_s f_1(s) \ldots f_k(s-(k-1)r)\Big|^2 \\
&\leq N \sum_s \Big|\sum_r f_1(s) f_2(s-r) \ldots f_k(s-(k-1)r)\Big|^2 \\
&\leq N \sum_s \Big|\sum_r f_2(s-r) f_3(s-2r) \ldots f_k(s-(k-1)r)\Big|^2 \\
&= N \sum_s \sum_r \sum_t f_2(s-r)\overline{f_2(s-t)} \ldots f_k(s-(k-1)r)\overline{f_k(s-(k-1)t)} \\
&= N \sum_s \sum_r \sum_u f_2(s)\overline{f_2(s-u)} \ldots f_k(s-(k-2)r)\overline{f_k(s-(k-2)r-(k-1)u)} \\
&= N \sum_s \sum_r \sum_u \Delta(f_2; u)(s) \Delta(f_3; 2u)(s-r) \ldots \Delta(f_k; (k-1)u)(s-(k-2)r) .
\end{aligned}$$

Since $\Delta(f_k; (k-1)u)$ is $\alpha((k-1)u)$-uniform of degree $k-3$, our inductive hypothesis implies that this is at most $N \sum_u \alpha((k-1)u)^{1/2^{k-2}} N^2$, and since $\sum_u \alpha((k-1)u) \leq \alpha N$, this is at most $\alpha^{1/2^{k-2}} N^2$, which proves the result for k. □

The interest in Theorem 3.2 is of course that the expression on the left hand side can be used to count arithmetic progressions. Let us now define a set $A \subset \mathbb{Z}_N$ to be α-uniform of degree d if its balanced function is. (This definition makes sense when $d = 0$, but only because it applies to all sets.) The next result implies that a set A which is α-uniform of degree $d-2$ for some small α contains about the number of arithmetic progressions of length d that a random set of the same cardinality would have, where this means arithmetic progressions mod N. We shall then show how to obtain genuine progressions, which turns out to be a minor technicality, similar to the corresponding technicality in the proof of Roth's theorem.

Corollary 3.3. *Let $A_1, \ldots, A_k$ be subsets of $\mathbb{Z}_N$, such that A_i has cardinality $\delta_i N$ for every i, and is $\alpha^{2^{i-1}}$-uniform of degree $i-2$ for every $i \geq 3$. Then*

$$\left|\sum_r |(A_1 + r) \cap \cdots \cap (A_k + kr)| - \delta_1 \ldots \delta_k N^2\right| \leq 2^k \alpha N^2 .$$

Proof. For each i, let f_i be the balanced function of A_i. Then

$$|(A_1 + r) \cap \cdots \cap (A_k + kr)| = \sum_s (\delta_1 + f_1(s - r)) \ldots (\delta_k + f_k(s - kr)) \ ,$$

so we can rewrite $|(A_1 + r) \cap \cdots \cap (A_k + kr)| - \delta_1 \ldots \delta_k N$ as

$$\sum_{B \subset [k], B \neq \emptyset} \prod_{i \notin B} \delta_i \sum_s \prod_{i \in B} f_i(s - ir) \ .$$

Now if $j = \max B$, then $\sum_r \sum_s \prod_{i \in B} f_i(s - ir)$ is at most $\alpha^{2^{j-1}/2^{j-1}} N^2$, by Theorem 3.2. It follows that

$$\begin{aligned} \Big|\sum_r |(A_1 + r) \cap \cdots \cap (A_k + kr)| - \delta_1 \ldots \delta_k N^2\Big| &\leq \textstyle\sum_{B \subset [k], B \neq \emptyset} \prod_{i \notin B} \delta_i . \alpha N^2 \\ &= \alpha N^2 \Big(\textstyle\prod_{i=1}^k (1 + \delta_i) - 1\Big) \end{aligned}$$

which is at most $2^k \alpha N^2$, as required. □

We now prove two simple technical lemmas.

Lemma 3.4. *Let $d \geq 1$ and let $f : \mathbb{Z}_N \to D$ be α-uniform of degree d. Then f is $\alpha^{1/2}$-uniform of degree $d - 1$.*

Proof. Our assumption is that

$$\sum_{a_1, \ldots, a_d} \Big|\sum_s \Delta(f; a_1, \ldots, a_d)(s)\Big|^2 \leq \alpha N^{k+2} \ .$$

By the Cauchy-Schwarz inequality, this implies that

$$\Big|\sum_{a_1, \ldots, a_d} \sum_s \Delta(f; a_1, \ldots, a_d)(s)\Big| \leq \alpha^{1/2} N^{k+1} \ ,$$

which, by the equivalence of properties (i) and (ii) in Lemma 3.1, proves the lemma. □

Lemma 3.5. *Let A be an α-uniform subset of $\mathbb{Z}_N$ of cardinality δN, and let P be an interval of the form $\{a + 1, \ldots, a + M\}$, where $M = \beta N$. Then $\big||A \cap P| - \beta \delta N\big| \leq \alpha^{1/4} N$.*

Proof. First, we can easily estimate the Fourier coefficients of the set P. Indeed,

$$\begin{aligned} |\tilde{P}(r)| &= \big|\textstyle\sum_{s=1}^M \omega^{-r(a+s)}\big| \\ &= |(1 - \omega^{rM})/(1 - \omega^r)| \leq N/2r \ . \end{aligned}$$

(We also know that it is at most M, but will not need to use this fact.) This estimate implies that $\sum_{r \neq 0} |\tilde{P}(r)|^{4/3} \leq N^{4/3}$. Therefore,

$$\big||A \cap P| - \beta \delta N\big| = N^{-1} \Big|\textstyle\sum_{r \neq 0} \hat{A}(r) \tilde{P}(r)\Big|$$

$$\leq N^{-1}\Big(\sum_{r\neq 0}|\hat{A}(r)|^4\Big)^{1/4}\Big(\sum_{r\neq 0}|\tilde{P}(r)|^{4/3}\Big)^{3/4}$$
$$\leq \Big(\sum_{r\neq 0}|\hat{A}(r)|^4\Big)^{1/4}\leq \alpha^{1/4}N$$

using property (iv) of Lemma 3.1.

Corollary 3.6. *Let $A\subset\mathbb{Z}_N$ be α-uniform of degree $k-2$ and have cardinality δN. If $\alpha\leq(\delta/2)^{k2^k}$ and $N\geq 32k^2\delta^{-k}$ then A contains an arithmetic progression of length k.*

Proof. Let $A_1=A_2=A\cap\big[(k-2)N/(2k-3),(k-1)N/(2k-3)\big]$, and let $A_3=\cdots=A_k=A$. By Lemma 3.4 A is $\alpha^{1/2^{k-3}}$-uniform (of degree one), so by Lemma 3.5 the sets A_1 and A_2 both have cardinality at least $\delta N/4k$ since, by the first inequality we have assumed, we know that $\alpha^{1/2^{k-1}}\leq\delta/4k$.

Therefore, by Corollary 3.3, A contains at least $\big((\delta^k/16k^2)-2^k\alpha^{1/2^{k-1}}\big)N^2$ arithmetic progressions modulo N with the first two terms belonging to the interval $\big[(k-2)N/(2k-3),(k-1)N/(2k-3)\big]$. The only way such a progression can fail to be genuine is if the common difference is zero, and there are at most δN such degenerate progressions. Thus the corollary is proved, since the two inequalities we have assumed imply that $(\delta^k/16k^2)-2^k\alpha^{1/2^{k-1}}\geq\delta^k/32k^2$ and $\delta^kN^2/32k^2>\delta N$. □

Remark. Notice that the proof of Corollary 3.6 did not use Fourier coefficients. This shows that in the proof of Theorem 2.6, the Fourier analysis was not really needed for the analysis of case 3. However, it was used in a more essential way for case 2.

In order to prove Szemerédi's theorem, it is now enough to prove that if $A\subset\mathbb{Z}_N$ is a set of size δN which is not $(\delta/2)^{k2^k}$-uniform of degree $d-2$, then there is an arithmetic progression $P\subset\mathbb{Z}_N$ of length tending to infinity with N, such that $|A\cap P|\geq(\delta+\epsilon)|P|$, where $\epsilon>0$ depends on δ and d only. Thus, we wish to deduce a structural property of A from information about its differences. We do not quite have an inverse problem, as usually defined, of additive number theory, but it is certainly in the same spirit, and we shall relate it to a well-known inverse problem, Freiman's theorem, later in the paper. For the rest of this section we shall give a combinatorial characterization of α-uniform sets of degree d. The result will not be needed for Szemerédi's theorem but gives a little more insight into what is being proved. Also, Lemma 3.7 below will be used near the end of the paper.

Let A be a subset of $\mathbb{Z}_N$ and let $d\geq 0$. By a *d-dimensional cube in A* we shall mean a function $\phi:\{0,1\}^d\to A$ of the form

$$\phi:(\epsilon_1,\ldots,\epsilon_d)\mapsto a_0+\epsilon_1a_1+\cdots+\epsilon_da_d\ ,$$

where $a_0,a_1,\ldots,a_d$ all belong to $\mathbb{Z}_N$. We shall say that such a cube is contained in A, even though it is strictly speaking contained in $A^{\{0,1\}^d}$.

Let $A\subset\mathbb{Z}_N$ have cardinality δN. Then A obviously contains exactly δN cubes of dimension zero and δ^2N^2 cubes of dimension one. As remarked after Lemma

2.2, the number of two-dimensional cubes in A can be written as $N^{-1}\sum_r |\hat{A}(r)|^4$, so A is α-uniform if and only if there are at most $(\delta^4+\alpha)N^3$ of them. We shall now show that A contains at least $\delta^{2^d}N^{d+1}$ cubes of dimension d, and that equality is nearly attained if A is α-uniform of degree $d-1$ for some small α. The remarks we have just made prove this result for $d=1$. Notice that equality is also nearly attained (with high probability) if A is a random set of cardinality δN. This is why we regard higher-degree uniformity as a form of pseudorandomness.

Lemma 3.7. *Let A be a subset of $\mathbb{Z}_N$ of cardinality δN and let $d \geq 0$. Then A contains at least $\delta^{2^d}N^{d+1}$ cubes of dimension d.*

Proof. We know the result for $d=0$ or 1 so let $d>1$ and assume that the result is known for $d-1$. The number of d-dimensional cubes in A is the sum over all r of the number of $(d-1)$-dimensional cubes in $A\cap(A+r)$. Write $\delta(r)N$ for the cardinality of $A\cap(A+r)$. Then by induction the number of d-dimensional cubes in A is at least $\sum_r \delta(r)^{2^{d-1}}N^d$. Since the average value of $\delta(r)$ is exactly δ^2, this is at least $\delta^{2^d}N^{d+1}$ as required. □

The next lemma is little more than the Cauchy-Schwarz inequality and some notation. It will be convenient to use abbreviations such as x for $(x_1,\dots,x_k)$ and $x.y$ for $\sum_{i=1}^k x_iy_i$. If $\epsilon\in\{0,1\}^k$ then we shall write $|\epsilon|$ for $\sum_{i=1}^k \epsilon_i$. Once again, C is the operation of complex conjugation.

Lemma 3.8. *For every $\epsilon\in\{0,1\}^k$ let f_ϵ be a function from $\mathbb{Z}_N$ to D. Then*

$$\Big|\sum_{x\in\mathbb{Z}_N^d}\sum_s\prod_{\epsilon\in\{0,1\}^d}C^{|\epsilon|}f_\epsilon(s-\epsilon.x)\Big| \leq \prod_{\epsilon\in\{0,1\}^d}\Big|\sum_{x\in\mathbb{Z}_N^d}\sum_s\prod_{\eta\in\{0,1\}^d}C^{|\eta|}f_\epsilon(s-\eta.x)\Big|^{1/2^d}.$$

Proof.

$$\begin{aligned}&\Big|\textstyle\sum_{x\in\mathbb{Z}_N^d}\sum_s\prod_{\epsilon\in\{0,1\}^d}C^{|\epsilon|}f_\epsilon(s-\epsilon.x)\Big|\\ =&\Big|\textstyle\sum_{x\in\mathbb{Z}_N^{d-1}}\Big(\sum_s\prod_{\epsilon\in\{0,1\}^{d-1}}C^{|\epsilon|}f_{\epsilon,0}(s-\epsilon.x)\Big)\Big(\sum_t\prod_{\epsilon\in\{0,1\}^{d-1}}C^{|\epsilon|}f_{\epsilon,1}(t-\epsilon.x)\Big)\Big|\\ \leq&\Big(\textstyle\sum_{x\in\mathbb{Z}_N^{d-1}}\Big|\sum_s\prod_{\epsilon\in\{0,1\}^{d-1}}C^{|\epsilon|}f_{\epsilon,0}(s-\epsilon.x)\Big|^2\Big)^{\frac12}\\ &\Big(\textstyle\sum_{x\in\mathbb{Z}_N^d}\Big|\sum_s\prod_{\epsilon\in\{0,1\}^{d-1}}C^{|\epsilon|}f_{\epsilon,1}(s-\epsilon.x)\Big|^2\Big)^{\frac12}\end{aligned}$$

Let us write $P_d(\epsilon)$ and $Q_d(\epsilon)$ for the sequences $(\epsilon_1,\dots,\epsilon_{d-1},0)$ and $(\epsilon_1,\dots,\epsilon_{d-1},1)$. Then

$$\sum_{x\in\mathbb{Z}_N^{d-1}}\Big|\sum_s\prod_{\epsilon\in\{0,1\}^{d-1}}C^{|\epsilon|}f_{\epsilon,0}(s-\epsilon.x)\Big|^2=\sum_{x\in\mathbb{Z}_N^d}\sum_s\prod_{\epsilon\in\{0,1\}^d}C^{|\epsilon|}f_{P_d(\epsilon)}(s-\epsilon.x)$$

and similarly for the second bracket with Q_d, so the two parts are square roots of expressions of the form we started with, except that the function f_ϵ no longer depends on ϵ_d. Repeating this argument for the other coordinates, we obtain the result. □

If we regard Lemma 3.8 as a modification of the Cauchy-Schwarz inequality, then the next lemma is the corresponding modification of Minkowski's inequality.

Lemma 3.9. *Given any function $f : \mathbb{Z}_N \to \mathbb{C}$ and any $d \geq 2$, define $\|f\|_d$ by the formula*

$$\|f\|_d = \Big|\sum_{x\in\mathbb{Z}_N^d}\sum_s \prod_{\epsilon\in\{0,1\}^d} C^{|\epsilon|} f(s-\epsilon.x)\Big|^{1/2^d} .$$

Then $\|f+g\|_d \leq \|f\|_d + \|g\|_d$ for any pair of functions $f, g : \mathbb{Z}_N \to \mathbb{C}$. In other words, $\|.\|_d$ is a norm.

Proof. If we expand $\|f+g\|^{2^d}$, we obtain the sum

$$\sum_{x\in\mathbb{Z}_N^d}\sum_s \prod_{\epsilon\in\{0,1\}^d} C^{|\epsilon|}(f+g)(s-\epsilon.x) .$$

If we expand the product we obtain 2^{2^d} terms of the form $\prod_{\epsilon\in\{0,1\}^d} C^{|\epsilon|} f_\epsilon(s-\epsilon.x)$, where each function f_ϵ is either f or g. For each one of these terms, if we take the sum over $x_1, \dots, x_d$ and s and apply Lemma 3.8, we have an upper estimate of $\|f\|_d^k \|g\|_d^l$, where k and l are the number of times that f_ϵ equals f and g respectively. From this it follows that

$$\|f+g\|^{2^d} \leq \sum_{k=0}^{2^d} \binom{2^d}{k} \|f\|_d^k \|g\|_d^{2^d-k} = (\|f\|_d + \|g\|_d)^{2^d}$$

which proves the lemma. □

It is now very easy to show that equality is almost attained in Lemma 3.7 for sets that are sufficiently uniform.

Lemma 3.10. *Let A be α-uniform of degree $d-1$. Then A contains at most $(\delta + \alpha^{1/2^d})^{2^d} N^{d+1}$ cubes of dimension d.*

Proof. Write $A = \delta + f$, where $|A| = \delta N$ and f is the balanced function of A. Then $\|A\|_d \leq \|\delta\|_d + \|f\|_d$. It is easy to see that $\|A\|_d^{2^d}$ is the number of d-dimensional cubes in A and that $\|\delta\|_d^{2^d} = \delta^{2^d} N^{d+1}$. Moreover, the statement that A is α-uniform of degree $d-1$ is equivalent to the statement that $\|f\|_d^{2^d} \leq \alpha N^{d+1}$. Therefore, Lemma 3.9 tells us that A contains at most $(\delta + \alpha^{1/2^d})^{2^d} N^{d+1}$ cubes of dimension d. □

Remark. In a sense, the normed spaces just defined encapsulate all the information we need about the arithmetical properties of the functions we consider. In their definitions they bear some resemblance to Sobolev spaces. Although I cannot think of any potential applications, I still feel that it would be interesting to investigate them further.

4. Some Motivating Examples.

We now know that Szemerédi's theorem would follow from an adequate understanding of higher-degree uniformity. A natural question to ask is whether degree-one uniformity *implies* higher-degree uniformity (for which it would be enough to show that it implied quadratic uniformity). To make the question precise, if A has density δ and is α-uniform, does it follow that A is quadratically β-uniform, for some β depending on α and δ only? If so, then the same result for higher-degree uniformity can be deduced, and Szemerédi's theorem follows easily, by the method of §2.

I have already said that α-uniformity is not an appropriate definition of pseudorandomness for progressions of length greater than three, so the answer to this question is no. The first result of this section is a simple counterexample demonstrating this. Let A be the set $\{s \in \mathbb{Z}_N : |s^2| \leq N/100\}$. If $s \in A \cap (A+k)$, then $|s^2| \leq N/100$ and $|(s-k)^2| \leq N/100$ as well, which implies that $|2sk - k^2| \leq N/50$, or equivalently that s lies inside the set $(2k)^{-1}\{s : |s - k/2| \leq N/50\}$. It follows that $A \cap (A+k)$ is not uniform for any $k \neq 0$.

Let us also think about the number of mod-N arithmetic progressions of length four contained in A. Such a progression can be written in the form $(x-d, x, x+d, x+2d)$. Now $s \in A$ if and only if s^2 is 'small', in the sense that s^2 is close to zero. Suppose we know that $(x-d)^2$, x^2 and $(x+d)^2$ are all small in this sense. Then, by taking differences, we know that $2xd - d^2$ and $2xd + d^2$ are also small, from which it follows that $4xd$ and $2d^2$ are both small, from which it follows that $x^2 + 4xd + 4d^2 = (x+2d)^2$ is small. This means that it is more likely than it should be that $x + 2d$ also belongs to A, and therefore that A contains *more* arithmetic progressions of length four than would be contained in a random set of the same cardinality. (I am certain, but have not actually checked, that a similar example could be constructed with too few such progressions.)

It is possible, but not completely straightforward, to show that A itself is highly uniform. Rather than go into the details, we prove a closely related fact which is in some ways more natural. Let $f(s) = \omega^{s^2}$. We shall show that f is a very uniform function, while $\Delta(f;k)$ fails badly to be uniform for any $k \neq 0$. For the uniformity of f, notice that

$$|\hat{f}(r)| = \Big|\sum_s \omega^{s^2 - rs}\Big| = \Big|\sum_s \omega^{(s-r/2)^2}\Big| = \Big|\sum_s \omega^{s^2}\Big|$$

for every r. Therefore, $|\hat{f}(r)| = N^{1/2}$ for every $r \in \mathbb{Z}_N$, so f is as uniform as a function into the unit circle can possibly be. On the other hand, $\Delta(f;k)(s) = \omega^{2ks-k^2}$, so that

$$\Delta(f;k)^\wedge(r) = \begin{cases} N & \text{r=2k} \\ 0 & \textit{otherwise} \end{cases}$$

which shows that $\Delta(f;k)$ is, for $k \neq 0$, as non-uniform as possible.

More generally, if $\phi : \mathbb{Z}_N \to \mathbb{Z}_N$ is a quadratic polynomial and $f(s) = \omega^{\phi(s)}$, then f is highly uniform, but there is some $\lambda \in \mathbb{Z}_N$ such that, for every k,

$$\Delta(f;k)^\wedge(r) = \begin{cases} N & r = \lambda k \\ 0 & otherwise. \end{cases}$$

This suggests an attractive conjecture, which could perhaps replace the false idea that if A is uniform then so are almost all $A \cap (A+k)$. Perhaps if there are many values of k for which $A \cap (A+k)$ fails to be uniform, then there must be a quadratic function $\phi : \mathbb{Z}_N \to \mathbb{Z}_N$ such that $\left|\sum_{s\in A} \omega^{-\phi(s)}\right|$ is large. We shall see in the next section that such "quadratic bias" would actually imply the existence of a long arithmetic progression P_j such that $|A \cap P_j|/|P_j|$ was significantly larger than $|A|/N$. This would give a proof of Szemerédi's theorem for progressions of length four, and one can see how the above ideas might be generalized to higher-degree polynomials and longer arithmetic progressions.

The second example of this section shows that such conjectures are still too optimistic. As with the first example, we shall consider functions that are more general than characteristic functions of subsets of $\mathbb{Z}_N$. However, this should be enough to convince the reader not to try to prove the conjectures.

Let r be about $\sqrt{N}$ and for $0 \le a, b < r/2$ define $\phi(ar+b)$ to be a^2+b^2. Now define

$$f(s) = \begin{cases} \omega^{\phi(s)} & s = ar + b for some 0 \le a, b < r/2 \\ 0 & otherwise. \end{cases}$$

The function f is not quadratic, but it resembles a quadratic form in two variables (with the numbers 1 and r behaving like a basis of a two-dimensional space).

Suppose $s = ar + b$ and $k = cr + d$ are two numbers in $\mathbb{Z}_N$, where all of a, b, $a-c$ and $b-d$ lie in the interval $[0, r/2)$. Then

$$f(s)\overline{f(s-k)} = \omega^{2ac-c^2+2bd-d^2} = \omega^{\phi_k(ar+b)}$$

where ϕ_k depends linearly on the pair (a,b). The property that will interest us about ϕ_k is that, at least when c and d are not too close to $r/2$, there are several pairs (a,b) such that the condition on (a,b,c,d) applies, and therefore several quadruples $\big((a_i,b_i)\big)_{i=1}^4$ such that

$$(a_1,b_1) + (a_2,b_2) = (a_3,b_3) + (a_4,b_4)$$

and

$$\phi_k(a_1r+b_1) + \phi_k(a_2r+b_2) = \phi_k(a_3r+b_3) + \phi_k(a_4r+b_4) .$$

Here, "several" means a number proportional to N^3, which is the maximum it could be.

Let B be the set of all $s = ar + b$ for which a, b, c and d satisfy the conditions above. (Of course, B depends on k.) Then

$$\sum_q \left|\sum_{s\in B} \omega^{\phi_k(s)-qs}\right|^4 = N\sum\{\omega^{\phi_k(s)+\phi_k(t)-\phi_k(u)-\phi_k(v)} : s,t,u,v \in B,\ s+t = u+v\} .$$

Now the set B has been chosen so that if $s,t,u,v \in B$ and $s+t = u+v$, then $\phi_k(s) + \phi_k(t) = \phi_k(u) + \phi_k(v)$. Therefore, the right hand side above is N times the number of quadruples $(s,t,u,v) \in B^4$ such that $s+t = u+v$. It is not hard to check that if c and d are smaller than $r/4$, say, then B has cardinality proportional to N^3, and therefore that the right hand side above is proportional to N^4. Lemma

2.1 now tells us that ϕ_k has a large Fourier coefficient. Thus, at the very least, we have shown that, for many values of k, $\Delta(f;k)$ fails to be uniform.

If we could find a *genuinely* quadratic function $\phi(s) = as^2 + bs + c$ such that $\left|\sum_s f(s)\omega^{-\phi(s)}\right|^2$ was proportional to N^2, then, expanding, we would have

$$\sum_{s,k} f(s)\overline{f(s-k)}\omega^{-\phi(s)+\phi(s-k)} = \sum_{s,k} f(s)\overline{f(s-k)}\omega^{-2ask-bk}$$

proportional to N^2, which would imply that the number of k for which $\Delta(f;k)^\wedge(2ak)$ was proportional to N was proportional to N. A direct calculation (left to the interested reader) shows that such a phenomenon does not occur. That is, there is no value of λ such that $\Delta(f;k)^\wedge(\lambda k)$ is large for many values of k.

There are of course many examples like the second one above. One can define functions that resemble d-dimensional quadratic forms, and provided that d is small the same sort of behaviour occurs. Thus, we must accept that the ideas of this paper so far do not lead directly to a proof of Szemerédi's theorem, and begin to come to terms with these "multi-dimensional" examples. It is for this purpose that our major tool, an adaptation of Freiman's theorem, is used, as will be explained later in the paper.

5. Consequences of Weyl's inequality.

In the proof of Roth's theorem an important role was played by Lemma 2.3, which said that if $\phi : \{1, 2, \ldots, r\} \to \mathbb{Z}_N$ is a linear function (in the sense that it can be defined by a formula $\phi(x) = ax + b$) then the set $\{1, 2, \ldots, r\}$ can be partitioned into not too many arithmetic progressions, on each of which ϕ is roughly constant. The linear function arose because we were using the hypothesis that A had a large Fourier coefficient - that is, that $\sum_x A(x)\omega^{-rx}$ was large for some r. The linear function in question was then $x \mapsto rx$.

In a certain sense, then, a set that failed to be α-uniform was *linearly biased.* We shall see later that a set that fails to be α-quadratically uniform is *quadratically biased.* In a tidy world, this would mean that there was a quadratic function q with the property that $\sum_x A(x)\omega^{-q(x)}$ was large. Actually, all we can show (necessarily, in the light of the second example of the previous section) is that $\mathbb{Z}_N$ can be partitioned into a smallish number of subprogressions P_i in each of which there is a function q_i such that $\left|\sum_{x\in P_i} \omega^{-q_i(x)}\right|$ is, on average, an appreciable fraction of $|P|$.

This result explains our use of the term 'quadratically uniform'. A set that fails to be quadratically uniform exhibits a certain sort of quadratic bias. For this to be useful to us, we need to show that, given such a biased set, we can pass to a subprogression inside which it has increased density. Then, just as in Roth's theorem, we will be able to iterate. Rather than give the details of this step, I shall sketch the argument.

The main lemma of this part of the proof states that if $q : \{1, 2, \ldots, r\} \to \mathbb{Z}$ is a quadratic function, then $\{1, 2, \ldots, r\}$ can be partitioned into subprogressions P_i on each of which q is roughly constant. The average size of the P_i turns out to be r^c for some absolute constant $c > 0$.

An interesting result of Furstenberg states that the sequence of squares is recurrent, in the sense that, given any measure-preserving dynamical system (X, μ, T) and any set A of positive measure in X, there must exist a positive integer n such that $A \cap T^{-n^2}A$ has positive measure. (There is a similar statement for topological dynamical systems: every point x must return arbitrarily close to its starting point after a square number of iterations.) From this it is easy to deduce that, given any real number α and any $\epsilon > 0$ there exists a positive integer n such that $n^2\alpha$ is within ϵ of an integer. Equivalently, if one wishes to approximate α by a fraction of the form p/q^2, then one can do so, and can beat the trivial bound of $|\alpha - p/q^2| \leq c/q^2$.

This corollary was in fact first proved by Weyl, and from Weyl's argument it follows that one can find p and q with $|\alpha - p/q^2| \leq c/q^{2+\gamma}$ for some fixed $\gamma > 0$. Let us first see roughly how Weyl's argument works and then how the result is used for the purpose of partitioning the progression $\{1, 2, \ldots, r\}$ above.

The first step is closely related to Weyl's equidistribution theorem. If A is a random subset of $\mathbb{Z}_N$ of cardinality r, then the expected size of the intersection $A \cap \{-M, -(M-1), \ldots, (M-1), M\}$ is roughly $2Mr/N$. If rM is significantly larger than N and A has *empty* intersection with $\{-M, -(M-1), \ldots, (M-1), M\}$, then A is in some sense not random. Moreover, it fails to be random in a way that ought to be detected by the Fourier coefficients of the characteristic function of A, since in a certain sense A is not evenly distributed round the circle.

Here is a precise result along these lines.

Lemma 5.1. *Let A be a subset of $\mathbb{Z}_N$ of cardinality r, let M be an even integer and suppose that $A \cap [-M, M) = \emptyset$. Then there exists u with $0 < |u| \leq N^2M^{-2}$ such that $|\hat{A}(u)| \geq kM/2N$.*

If we apply this lemma to the set $\{q(1), q(2), \ldots, q(r)\}$ and assume that for no value of $x \leq r$ does $q(x)$ lie in the interval $\{-M, -(M-1), \ldots, (M-1), M\}$ for some suitably chosen M (significantly larger than N/r but significantly smaller than N) then we discover from Lemma 5.1 that there is a small u such that $\sum_{x=1}^{r} \omega^{uq(x)}$ is large.

If we write α for the number u/N, then this sum can be rewritten $\sum_{x=1}^{r} e(\alpha q(x))$, where $e(t)$ stands for $\exp(2\pi i t)$. A great deal is known about such sums, and one fundamental result, due to Weyl (it has since been surpassed but is sufficient for our purposes) is that such a sum can be large only if α is close to a rational number with a small denominator. (To see that some condition like this is necessary, one has only to consider a sum like $\sum_{x=1}^{r} e(x^2/3)$. The summand equals 1 a third of the time and $\exp(2\pi i/3)$ for the other two thirds, and therefore the sum has magnitude proportional to r, which counts as very large indeed since it is close to the trivial upper bound of r.) For a detailed statement and proof, see [V]. The result originally appeared in [We].

Now consider a function from $\{1, 2, \ldots, r\}$ to $\mathbb{Z}_N$ of the form $x \mapsto ax^2$. We know that if a/N is not close to a rational of small denominator then the image of this function is well distributed round the circle, and in particular contains values close to zero mod N. On the other hand, if a/N *is* close to a rational p/q with q small, then aq^2/N is close to the integer pq, from which it follows that aq^2 is close to pqN, and therefore that aq^2 is close to zero mod N. In other words, in this case the result holds for a very simple reason. Needless to say, one has to check that the numbers work out in the above argument, but, happily, they do.

Now let $q : \{1, 2, \ldots, r\} \to \mathbb{Z}_N$ be defined by the formula $x \mapsto ax^2 + bx$ and let us try to partition $\{1, 2, \ldots, r\}$ into subprogressions on which q is almost constant. (There is no loss of generality in assuming that q has a zero constant term.) We begin by finding some t significantly smaller than r such that at^2 is close to 0. (This we do by applying our result above to a small subinterval $\{1, 2, \ldots, s\}$, say.) If at^2 is close to 0, then $a(x + ht)^2 - ax^2 = 2axht + ah^2t^2$ which, if h is small, is close to $2axht$. Since $2axht$ depends linearly on h, for fixed x, it follows that if we partition $\{1, 2, \ldots, r\}$ into sufficiently short arithmetic progressions, each with common difference t, the function $x \mapsto ax^2$ is approximately linear on each of these progressions (though the 'gradients' will differ from progression to progression).

This reduces the problem to one we already know how to solve: given a linear function from an arithmetic progression to $\mathbb{Z}_N$, partition that progression into smaller ones on each of which the function is approximately constant. We proved exactly such a result in Lemma 2.3.

By similar techniques, one can prove more or less any result of this kind that one might need. For the general case of Szemerédi's theorem, it is necessary to prove generalizations to polynomials of higher degree, and several of them simultaneously. Another result that is needed is that if P is a progression and $\mu : P \to \mathbb{Z}_N$ is defined multilinearly, then P^d can be partitioned into subsets of the form $Q_1 \times \cdots \times Q_d$, where the Q_i are subprogressions of P with the same common difference and of roughly the same size, on each of which μ is roughly constant. And in fact this is needed for several multilinear functions simultaneously.

6. Somewhat Additive Functions.

We saw in §4 that it is possible for a set A to have small Fourier coefficients, but for $A \cap (A + k)$ to have at least one non-trivial large Fourier coefficient for every k. Moreover, the obvious conjecture concerning such sets, that they correlate with some function of the kind $\omega^{q(s)}$ where q is a quadratic polynomial, is false. The aim of the next three sections is to show that such a set A must nevertheless exhibit quadratic bias of some sort. We will then be able to use the results of the last section to find linear bias, which will complete the proof for progressions of length four. The generalization to longer progressions will use similar ideas, but involves one extra important difficulty.

Notice that what we are trying to prove is very natural. If we replace A by a function on $\mathbb{Z}_N$ of the form $f(s) = \omega^{\phi(s)}$, where $\phi : \mathbb{Z}_N \to \mathbb{Z}_N$, then we are trying to prove that if, for many k, the function $\phi_k(s) = \phi(s) - \phi(s - k)$ has some sort of linearity property, resulting in a large Fourier coefficient for the difference function $\Delta(f; k) = \omega^{\phi(s) - \phi(s-k)}$, then ϕ itself must in some way be quadratic. Many arguments in additive number theory (in particular Weyl's inequality) use the fact that taking difference functions reduces the degree of, and hence simplifies, a polynomial. We are trying to do something like the reverse process, "integrating" rather than "differentiating" and showing that the degree goes up by one. This is another sense in which we are engaged in an inverse problem.

This section contains a simple but crucial observation, which greatly restricts the possibilities for the Fourier coefficients of $A \cap (A + k)$ that are large. Let A be a set which is not quadratically α-uniform and let f be the balanced function of A.

Then by the definition of quadratic α-uniformity there are at least αN values of k such that we can find r for which

$$\Big|\sum_s f(s)f(s-k)\omega^{-rs}\Big| \geq \alpha N \ .$$

Letting B be the set of k for which such an r exists, we can find a function $\phi : B \to \mathbb{Z}_N$ such that

$$\sum_{k\in B}\Big|\sum_s f(s)f(s-k)\omega^{-\phi(k)s}\Big|^2 \geq \alpha^3 N^3 \ .$$

We shall show that the function ϕ has a weak-seeming property which we shall call γ-additivity, for a certain constant $\gamma > 0$ to be defined later. Using a variant of Freiman's theorem proved in the next section, we shall show that this property gives surprisingly precise information about ϕ.

Proposition 6.1. *Let $\alpha > 0$, let $f : \mathbb{Z}_N \to D$, let $B \subset \mathbb{Z}_N$ and let $\phi : B \to \mathbb{Z}_N$ be a function such that*

$$\sum_{k\in B}\big|\Delta(f;k)^\wedge(\phi(k))\big|^2 \geq \alpha N^3 \ .$$

Then there are at least $\alpha^4 N^3$ quadruples $(a,b,c,d) \in B^4$ such that $a+b = c+d$ and $\phi(a)+\phi(b) = \phi(c)+\phi(d)$.

Proof. Expanding the left hand side of the inequality we are assuming gives us the inequality

$$\sum_{k\in B}\sum_{s,t} f(s)\overline{f(s-k)f(t)}f(t-k)\omega^{-\phi(k)(s-t)} \geq \alpha N^3 \ .$$

If we now introduce the variable $u = s-t$ we can rewrite this as

$$\sum_k\sum_{s,u} f(s)\overline{f(s-k)f(s-u)}f(s-k-u)\omega^{-\phi(k)u} \geq \alpha N^3 \ .$$

Since $|f(x)| \leq 1$ for every x, it follows that

$$\sum_u\sum_s\Big|\sum_{k\in B}\overline{f(s-k)}f(s-k-u)\omega^{-\phi(k)u}\Big| \geq \alpha N^3$$

which implies that

$$\sum_u\sum_s\Big|\sum_{k\in B}\overline{f(s-k)}f(s-k-u)\omega^{-\phi(k)u}\Big|^2 \geq \alpha^2 N^4 \ .$$

For each u and x let $f_u(x) = \overline{f(-x)}f(-x-u)$ and let $g_u(x) = B(x)\omega^{\phi(x)u}$. The above inequality can be rewritten

$$\sum_u\sum_s\Big|\sum_k f(k-s)\overline{g(k)}\Big|^2 \geq \alpha^2 N^4 \ .$$

By Lemma 2.1, we can rewrite it again as

$$\sum_u\sum_r |\hat{f}_u(r)|^2|\hat{g}_u(r)|^2 \geq \alpha^2 N^5 \ .$$

Since $\sum_r |\hat{f}(r)|^4 \leq N^4$, the Cauchy-Schwarz inequality now implies that

$$\sum_u \Big(\sum_r |\hat{g}_u(r)|^4\Big)^{1/2} \geq \alpha^2 N^3 \ .$$

Applying the Cauchy-Schwarz inequality again, we can deduce that

$$\sum_{u,r} |\hat{g}(r)|^4 = \sum_{u,r} \Big|\sum_{k\in B} \omega^{\phi(s)u-rs}\Big|^4 \geq \alpha^4 N^5 \ .$$

Expanding the left hand side of this inequality we find that

$$\sum_{u,r} \sum_{a,b,c,d\in B} \omega^{u(\phi(a)+\phi(b)-\phi(c)-\phi(d))}\omega^{-r(a+b-c-d)} \geq \alpha^4 N^5 \ .$$

But now the left hand side is exactly N^2 times the number of quadruples $(a,b,c,d) \in B^4$ for which $a+b = c+d$ and $\phi(a)+\phi(b) = \phi(c)+\phi(d)$. This proves the proposition. □

If G is an Abelian group and a, b, c, d are elements of G such that $a+b = c+d$, we shall say that (a,b,c,d) is an *additive quadruple.* Given a subset $B \subset \mathbb{Z}_N$ and a function $\phi : B \to \mathbb{Z}_N$, let us say that a quadruple $(a,b,c,d) \in B^4$ is *ϕ-additive* if it is additive and in addition $\phi(a) + \phi(b) = \phi(c) + \phi(d)$. Let us say also that ϕ is *γ-additive* if there are at least γN^3 ϕ-additive quadruples. It is an easy exercise to show that if $\gamma = 1$ then B must be the whole of $\mathbb{Z}_N$ and $\phi : \mathbb{Z}_N \to \mathbb{Z}_N$ must be of the form $\phi(x) = \lambda x + \mu$, i.e., linear. Notice that the property of γ-additivity appeared, undefined, in §4 during the discussion of the function ϕ_k. Let us now give a simple but useful reformulation of the concept of γ-additivity.

Lemma 6.2. *Let $\gamma > 0$, let $B \subset \mathbb{Z}_N$, let $\phi : B \to \mathbb{Z}_N$ be a γ-additive function and let $\Gamma \subset \mathbb{Z}_N^2$ be the graph of ϕ. Then Γ contains at least γN^3 additive quadruples (in the group $\mathbb{Z}_N^2$).* □

As we have just remarked, a 1-additive function must be a linear. We finish this section with an important (and, in the light of the second example of §4, natural) example of a γ-additive function which cannot be approximated by a linear function even though γ is reasonably large. Let $x_1, \dots, x_d \in Z_N$ and $r_1, \dots, r_d \in \mathbb{N}$ be such that all the numbers $\sum_{i=1}^d a_i x_i$ with $0 \leq a_i < r_i$ are distinct. Let $y_1, \dots, y_d \in \mathbb{Z}_N$ be arbitrary, and define

$$\phi\Big(\sum_{i=1}^d a_i x_i\Big) = \sum_{i=1}^d a_i y_i \ .$$

Let $\phi(s)$ be arbitrary for the other values of s. Then a simple calculation shows that the number of additive quadruples is at least $(2/3)^d r_1^3 \dots r_d^3$. If $r_1 \dots r_d = \beta N$, then ϕ is $(2/3)^d \beta^3$-additive.

The function ϕ resembles a linear map between vector spaces, and the number d can be thought of as the dimension of the domain of the ϕ. In the next two sections we shall show that all γ-additive functions have, at least in part, something like the above form, with d not too large and $r_1 \dots r_d$ an appreciable fraction of N (both depending, of course, on γ).

7. Variations on a Theorem of Freiman.

The results of the previous section tell us that if f is a function from $\mathbb{Z}_N$ to the unit disc D, B is a large subset of $\mathbb{Z}_N$ and $\phi : B \to \mathbb{Z}_N$ is a function such that $\Delta(f;k)^\wedge(\phi(k))$ is large for every $k \in B$, then the graph Γ of ϕ contains many quadruples (x, y, z, w) such that $x+y = z+w$ (in $\mathbb{Z}_N^2$). The purpose of this section is to introduce a theorem of Freiman, and to explain how it helps to tell us about the structure of a function with such a graph.

It will not be possible to apply Freiman's theorem to Γ straight away, since Γ does not satisfy the necessary hypotheses. One of the main results of this section is that Γ has a large subset that does satisfy them. This result, a quantitative version of a theorem of Balog and Szemerédi, has recently been used by Bourgain, in a modified form, to obtain the best known lower bounds for the Hausdorff dimensions of Kakeya sets in $\mathbb{R}^n$ when n is large. Before we discuss it, let us see what Freiman's theorem says.

Let A be a subset of $\mathbb{Z}$ of cardinality m. It is easy to see that $A + A = \{x + y : x, y \in A\}$ has cardinality between $2m - 1$ and $m(m + 1)/2$. Suppose that $|A + A| \leq Cm$ for some constant C. What information does this give about the set A? This problem is called an *inverse* problem of additive number theory, since it involves deducing the structure of A from the behaviour of $A + A$ - in contrast to a *direct* problem where properties of A give information about $A + A$.

It is clear that $A + A$ will be small when A is a subset of an arithmetic progression of length not much greater than m. After a moment's thought, one realises that there are other examples. For instance, one can take a "progression of progressions" such as $\{aM + b : 0 \leq a < h, 0 \leq b < k\}$ where $M >> k$ and $hk = m$. This example can then be generalized to a large subset of a "d-dimensional" arithmetic progression, provided that d is reasonably small. Freiman's theorem is the beautiful result that these simple examples exhaust all possibilities [F1,2]. A precise statement of the theorem is as follows.

Theorem 7.1. *Let C be a constant. Then there exist constants d_0 and K depending only on C such that whenever A is a subset of $\mathbb{Z}$ with $|A| = m$ and $|A + A| \leq Cm$, there exist $d \leq d_0$, an integer x_0 and positive integers $x_1, \ldots, x_d$ and $k_1, \ldots, k_d$ such that $k_1 k_2 \ldots k_d \leq Km$ and*

$$A \subset \left\{x_0 + \sum_{i=1}^{d} a_i x_i : 0 \leq a_i < k_i \ (i = 1, 2, \ldots, d)\right\}.$$

The same is true if $|A - A| \leq Cm$.

It is an easy exercise to deduce from Theorem 7.1 the same result for subsets of $\mathbb{Z}^n$, where $x_0, x_1, \ldots, x_d$ are now points in $\mathbb{Z}^n$. We shall in fact be interested in the case $n = 2$, since we shall be applying Freiman's theorem to a graph coming from Proposition 6.1 and Lemma 6.2.

The number $k_1 k_2 \ldots k_d$ is called the *size* of the d-dimensional arithmetic progression. Note that this is not necessarily the same as the cardinality of the set since there may be numbers (or more generally points of $\mathbb{Z}^D$) which can be written in more than one way as $x_0 + \sum_{i=1}^d a_i x_i$. When every such representation is unique, we shall call the set a *proper* d-dimensional arithmetic progression. (This terminology is all standard.)

Freiman's original proof of Theorem 7.1 was long and very difficult to understand. Although a simplified version of his argument now exists [Bi], an extremely important breakthrough came a few years ago with a new and much easier proof by Ruzsa, which also provided a reasonable bound. This improved bound is very important for the purposes of our bound for Szemerédi's theorem. Full details of Ruzsa's proof can be found in [Ru1,2,3] or in a book by Nathanson [N], which also contains all necessary background material.

We shall in fact need a modification of Freiman's theorem, in which the hypothesis and the conclusion are weakened. In its qualitative form, the modification is a result of Balog and Szemerédi. However, they use Szemerédi's uniformity lemma, which for us is too expensive. Our argument will avoid the use of the uniformity lemma and thereby produce a much better bound than the bound of Balog and Szemerédi. It will be convenient (though not essential) to consider the version of Freiman's theorem where $A - A$, rather than $A + A$ is assumed to be small. Our weaker hypothesis concerns another parameter associated with a set A, which has several descriptions, and which we know applies to the graph Γ discussed above. It is

$$\|A * A\|_2^2 = \sum_{k\in\mathbb{Z}} |A \cap (A+k)|^2 = |\{(a,b,c,d) \in A^4 : a - b = c - d\}| \ .$$

(Freiman calls this invariant M' in his book [F2 p.41].) It is a straightforward exercise to show that

$$\|A * A\|_2^2 \le m^2 + 2(1^2 + \cdots + (m-1)^2)$$

with equality if and only if A is an arithmetic progression of length m. The Balog-Szemerédi theorem [BS] is the following result.

Theorem 7.2. *Let A be a subset of $\mathbb{Z}^D$ of cardinality m and suppose that $\|A * A\|_2^2 \ge c_0 m^3$. Then there are constants c, K and d_0 depending only on c_0 and an arithmetic progression P of dimension $d \le d_0$ and size at most Km such that $|A \cap P| \ge cm$.*

This result states that if $\|A * A\|_2^2$ is, to within a constant, as big as possible, then A has a proportional subset satisfying the conclusion of Freiman's theorem. Notice that, qualitatively at least, the conclusion of Theorem 7.2 cannot be strengthened, since if A has a proportional subset B with $\|B * B\|_2^2$ large, then $\|A * A\|_2^2$ is large whatever $A \setminus B$ is. To see that the new hypothesis is weaker, notice that if $|A - A| \le Cm$, then $A \cap (A+k)$ is empty except for at most Cm values of k, while $\sum_{k\in\mathbb{Z}} |A \cap (A+k)| = m^2$. It follows from the Cauchy-Schwarz inequality that $\sum_{k\in\mathbb{Z}} |A \cap (A+k)|^2 \ge m^3/C$.

The most obvious approach to deducing Theorem 7.2 from Theorem 7.1 is to show that a set satisfying the hypothesis of Theorem 7.2 has a large subset satisfying the hypothesis of Theorem 7.1. This is exactly what Balog and Szemerédi did and we shall do as well.

Proposition 7.3. *Let A be a subset of $\mathbb{Z}^n$ of cardinality m such that $\|A * A\|_2^2 \ge c_0 m^3$. There are constants c and C depending only on c_0 and a subset $A'' \subset A$ of* *cardinality at least cm such that $|A'' - A''| \leq Cm$. Moreover, c and C can be taken as $2^{-20}c_0^{12}$ and $2^{38}c_0^{-24}$ respectively.*

We shall need the following lemma for the proof.

Lemma 7.4. *Let V be a set of size m, let $\delta > 0$ and let $A_1, \ldots, A_n$ be subsets of V such that $\sum_{x=1}^n \sum_{y=1}^n |A_x \cap A_y| \geq \delta^2 mn^2$. Then there is a subset $K \subset [n]$ of cardinality at least $2^{-1/2}\delta^5 n$ such that for at least 90% of the pairs $(x, y) \in K^2$ the intersection $A_x \cap A_y$ has cardinality at least $\delta^2 m/2$. In particular, the result holds if $|A_x| \geq \delta m$ for every x.*

Proof. For every $j \leq m$ let $B_j = \{i : j \in A_i\}$ and let $E_j = B_j^2$. Choose five numbers $j_1, \ldots, j_5 \leq m$ at random (uniformly and independently), and let $X = E_{j_1} \cap \cdots \cap E_{j_5}$. The probability p_{xy} that a given pair $(x, y) \in [n]^2$ belongs to E_{j_r} is $m^{-1}|A_x \cap A_y|$, so the probability that it belongs to X is p_{xy}^5. By our assumption we have that $\sum_{x,y=1}^n p_{xy} \geq \delta^2 n^2$, which implies (by Hölder's inequality) that $\sum_{x,y=1}^n p_{xy}^5 \geq \delta^{10} n^2$. In other words, the expected size of X is at least $\delta^{10} n^2$.

Let Y be the set of pairs $(x, y) \in X$ such that $|A_x \cap A_y| < \delta^2 m/2$, or equivalently $p_{xy} < \delta^2/2$. Because of the bound on p_{xy}, the probability that $(x, y) \in Y$ is at most $(\delta^2/2)^5$, so the expected size of Y is at most $\delta^{10} n^2/32$.

It follows that the expectation of $|X| - 16|Y|$ is at least $\delta^{10} n^2/2$. Hence, there exist $j_1, \ldots, j_5$ such that $|X| \geq 16|Y|$ and $|X| \geq \delta^{10} n^2/2$. It follows that the set $K = B_{j_1} \cap \cdots \cap B_{j_5}$ satisfies the conclusion of the lemma. □

Proof of Proposition 7.3. The function $f(x) = A * A(x)$ (from $\mathbb{Z}^n$ to $\mathbb{Z}$)is non-negative and satisfies $\|f\|_\infty \leq m$, $\|f\|_2^2 \geq c_0 m^3$ and $\|f\|_1 = m^2$. This implies that $f(x) \geq c_0 m/2$ for at least $c_0 m/2$ values of x, since otherwise we could write $f = g + h$ with g and h disjointly supported, g supported on fewer than $c_0 m/2$ points and $\|h\|_\infty \leq c_0 m/2$, which would tell us that

$$\|f\|_2^2 \leq \|g\|_2^2 + \|h\|_\infty \|h\|_1 < (c_0 m/2)m^2 + (c_0 m/2).m^2 = c_0 m^3 .$$

Let us call a value of x for which $f(x) \geq c_0 m/2$ a *popular difference* and let us define a graph G with vertex set A by joining a to b if $b - a$ (and hence $a - b$) is a popular difference. The average degree in G is at least $c_0^2 m/4$, so there must be at least $c_0^2 m/8$ vertices of degree at least $c_0^2 m/8$. Let $\delta = c_0^2/8$, let $a_1, \ldots, a_n$ be vertices of degree at least $c_0^2 m/8$, with $n \geq \delta m$, and let $A_1, \ldots, A_n$ be the neighbourhoods of the vertices $a_1, \ldots, a_n$. By Lemma 7.4 we can find a subset $A' \subset \{a_1, \ldots, a_n\}$ of cardinality at least $\delta^5 n/\sqrt{2}$ such that at least 90% of the intersections $A_i \cap A_j$ with $a_i, a_j \in A'$ are of size at least $\delta^2 m/2$. Set $\alpha = \delta^6/\sqrt{2}$ so that $|A'| \geq \alpha m$.

Now define a graph H with vertex set A', joining a_i to a_j if and only if $|A_i \cap A_j| \geq \delta^2 m/2$. The average degree of the vertices in H is at least $(9/10)|A'|$, so at least $|A'|/2$ vertices have degree at least $4|A'|/5$. Define A'' to be the set of all such vertices.

We claim now that A'' has a small difference set. To see this, consider any two elements $a_i, a_j \in A''$. Since the degrees of a_i and a_j are at least $(4/5)|A'|$ in H, there are at least $(3/5)|A'|$ points $a_k \in A'$ joined to both a_i and a_j. For every such k we have $|A_i \cap A_k|$ and $|A_j \cap A_k|$ both of size at least $\delta^2 m/2$. If $b \in A_i \cap A_k$, then both $a_i - b$ and $a_k - b$ are popular differences. It follows that there are at least $c_0^2 m^2/4$ ways of writing $a_i - a_k$ as $(p-q)-(r-s)$, where $p,q,r,s \in A$, $p-q = a_i - b$ and $r - s = a_k - b$. Summing over all $b \in A_i \cap A_k$, we find that there are at least $\delta^2 c_0^2 m^3/8$ ways of writing $a_i - a_k$ as $(p-q)-(r-s)$ with $p,q,r,s \in A$. The same is true of $a_j - a_k$. Finally, summing over all k such that a_k is joined in H to both a_i and a_j, we find that there are at least $(3/5)|A'|\delta^4 c_0^4 m^6/64 \geq \alpha\delta^4 c_0^4 m^7/120$ ways of writing $a_i - a_j$ in the form $(p-q)-(r-s)-\big((t-u)-(v-w)\big)$ with $p,q,\ldots,w \in A$.

Since there are at most m^8 elements in A^8, the number of differences of elements of A'' is at most $120m/\alpha\delta^4 c_0^4 \leq 2^{38}m/c_0^{24}$. Note also that the cardinality of A'' is at least $(1/2)\alpha m \geq c_0^{12}m/2^{20}$. The proposition is proved. □

It is possible to apply Theorem 7.2 as it stands in order to prove Szemerédi's theorem for progressions of length four (and quite possibly in general). However, it is better to combine Proposition 7.3 with a weaker version of Freiman's theorem that gives less information about the structure of a set A with small difference set. There are three advantages in doing this. The first is that with the weaker version one can get a much better bound. The second is that using the weaker version is cleaner, particularly when we come to the general case. The third is that the weaker version is easier to prove than Freiman's theorem itself, as it avoids certain arguments from the geometry of numbers.

Rather than give the details, let me state a precise result and then explain in broad terms why Theorem 7.2 should lead us to expect such a result. I have given explicit constants just to stress that the result is completely effective, but some readers might prefer to replace expressions like $N^{2^{-3770}\gamma^{2328}\alpha^2}$ by ones like $N^{c\gamma^K\alpha^2}$.

Corollary 7.5. *Let N be sufficiently large, let $B_0 \subset \mathbb{Z}_N$ have cardinality αN and let $\phi : B_0 \to \mathbb{Z}_N$ have $\gamma(\alpha N)^3$ additive quadruples. Then there exist a mod-N arithmetic progression P of length at least $N^{2^{-3770}\gamma^{2328}\alpha^2}$, a subset $H \subset P$ of cardinality at least $2^{-1849}\gamma^{1164}\alpha|P|$ and constants $\lambda, \mu \in \mathbb{Z}_N$ such that $\phi(s) = \lambda s + \mu$ for every $s \in H$.*

What this corollary says is that if there are many ϕ-additive quadruples, then there must be a reasonably long arithmetic progression P and a linear function ψ such that $\phi(x) = \psi(x)$ for many values of $x \in P$. (Here, as in several other places, the word 'many' means 'as many as possible, up to a constant which depends, ultimately, only on the density of the original set A'.)

Why does this follow from Theorem 7.2? Well, that theorem implies that a large subset of Γ is contained in a smallish multidimensional (but low-dimensional) arithmetic progression $P \subset \mathbb{Z}_N^2$. Such a progression will be a set of the form

$$\Big\{(x_0,y_0) + \sum_{i=1}^{d} a_i(x_i,y_i) : 0 \leq a_i < k_i \ (i = 1,2,\ldots,d)\Big\}\,,$$

which is the graph of a linear-like function ψ that takes $x_0 + \sum_{i=1}^d a_i x_i$ to $y_0 + \sum_{i=1}^d a_i y_i$. Now the size of P is proportional to N, so at least one of the k_i will be of size $m \geq N^{1/d}$, where d is the dimension of P. Now P can be partitioned into one-dimensional progressions obtained by fixing all the (x_j, y_j) apart from (x_i, y_i), and these all have size m. These progressions are of the form $\{(u,v)+a_i(x_i,y_i): 0\le a_i<m\}$, which, provided x_i is non-zero (which can easily be shown), is the graph of a linear function defined on a one-dimensional arithmetic progression in $\mathbb{Z}_N$ of length m. By averaging, at least one of these small graphs has a substantial intersection with Γ, which is the same as saying that ϕ has a small part which is linear.

8. Progressions of Length Four.

We have now shown that if $A\cap(A+k)^{\sim}(\phi(k))$ is large for many values of k then ϕ resembles a linear function. To simplify the exposition, let us now assume that ϕ *is* linear, and see what we can deduce. The constant term is unimportant, so let us suppose, for convenience, that $\phi(k)=2ck$ for every k, for some constant $c\in\mathbb{Z}_N$. Recall from the beginning of the proof of Proposition 6.1 that this implies (indeed, is equivalent to) the inequality

$$\sum_k\sum_{s,u} f(s)\overline{f(s-k)f(s-u)}f(s-k-u)\omega^{-\phi(k)u}\ge\alpha N^3\ ,$$

which in our case becomes

$$\sum_k\sum_{s,u} f(s)f(s-k)f(s-u)f(s-k-u)\omega^{-2cku}\ge\alpha^3N^3\ .$$

Using the identity

$$2ku=s^2-(s-k)^2-(s-u)^2+(s-k-u)^2$$

we can deduce that

$$\sum_r\sum_{a,b,c,d} f(a)f(b)f(c)f(d)\omega^{-r(a-b-c+d)}\omega^{-c(a^2-b^2-c^2+d^2)}\ge\alpha^3N^4\ ,$$

or in other words that

$$\sum_r\Big|\sum_s f(s)\omega^{-cs^2}\omega^{-rs}\Big|^4\ge\alpha^3N^4\ .$$

Now $\sum_s f(s)\omega^{-cs^2}\omega^{-rs}$ is just $\hat{g}(r)$, where h is the function $s\mapsto f(s)\omega^{-cs^2}$. By the implication of (iii) from (iv) in Lemma 2.2, we therefore deduce that for some value of r we have the lower bound

$$\Big|\sum_s f(s)\omega^{-cs^2}\omega^{-rs}\Big|\ge\alpha^{3/2}N$$

or in other words that A exhibits quadratic bias of a particularly strong kind.

If ϕ is not linear but does at least have a small linear part, such as is given by Corollary 7.10, then a similar result can be proved, but the conclusion is weaker. The proof is similar as well, but messier. Since the main idea is contained in the above argument, I shall simply state the result that comes out.

Proposition 8.1. *Let $A\subset\mathbb{Z}_N$ have balanced function f. Let P be an arithmetic progression (in $\mathbb{Z}_N$) of cardinality T. Suppose that there exist λ and μ such* *that* $\sum_{k\in P}|\Delta(f;k)^{\sim}(\lambda k+\mu)|^2 \geq \beta N^2 T$. *Then there exist quadratic polynomials* $\psi_0,\psi_1,\ldots,\psi_{N-1}$ *such that*

$$\sum_s \Big|\sum_{z\in P+s} f(z)\omega^{-\psi_s(z)}\Big| \geq \beta NT/\sqrt{2}\ .$$

Since our aim is to iterate by passing to subprogressions, the fact that f exhibits only this weak form of quadratic bias is not particularly damaging.

We are now ready to give a proof, with a couple of details sketched, of Szemerédi's theorem for progressions of length four. Once again, all numbers have been put in explicitly - I hope the reader can see through them to the main idea of the argument.

Theorem 8.2. *There is an absolute constant C with the following property. Let A be a subset of $\mathbb{Z}_N$ with cardinality δN. If $N \geq \exp\exp\big((1/\delta)^C\big)$, then A contains an arithmetic progression of length four.*

Proof. Our assumption certainly implies that $N \geq 32k^2\delta^{-k}$. Suppose now that the result is false. Then Corollary 3.6 implies that A is not α-quadratically uniform, where $\alpha = (\delta/2)^{64}$. By Lemma 3.1 (in particular the implication of (i) from (v)) there is a set $B \subset \mathbb{Z}_N$ of cardinality at least $\alpha N/2$ together with a function $\phi : B \to \mathbb{Z}_N$, such that $|\Delta(f;k)^{\sim}(\phi(k))| \geq \alpha N/2$ for every $k \in B$. In particular,

$$\sum_{k\in B} |\Delta(f;k)^{\sim}(\phi(k))|^2 \geq (\alpha/2)^3 N^3\ .$$

Hence, by Proposition 6.1, B contains at least $(\alpha/2)^{12}N^3$ ϕ-additive quadruples.

By Corollary 7.10, we can find a mod-N arithmetic progression P of size at least $N^{2^{-32000}\alpha^{30000}}$ and constants $\lambda,\mu \in \mathbb{Z}_N$ such that

$$\sum_{k\in P} |\Delta(f;k)^{\sim}(\lambda k+\mu)|^2 \geq 2^{-16000}\alpha^{15000}|P|N^2\ .$$

Therefore, by Proposition 8.1, we have quadratic polynomials $\psi_0,\psi_1,\ldots,\psi_{N-1}$ such that

$$\sum_s \Big|\sum_{z\in P+s} f(z)\omega^{-\psi_s(z)}\Big| \geq \beta N|P|/\sqrt{2}$$

where $\beta = 2^{-16000}\alpha^{15000}$.

By a simple averaging argument we can find a partition of $\mathbb{Z}_N$ into mod-N arithmetic progressions $P_1,\ldots,P_M$ of length $|P|$ or $|P|+1$ and also a sequence $\psi_1,\ldots,\psi_M$ (after renaming) of quadratic polynomials such that

$$\sum_{j=1}^{M} \Big|\sum_{z\in P_j} f(z)\omega^{-\psi_j(z)}\Big| \geq \beta N/2\ .$$

(Each P_j is either a translate of P or a translate of P extended by one point. Because of the small extensions we have changed $\sqrt{2}$ to 2.)

Using Lemma 2.3, one can partition each P_j into genuine arithmetic progressions (rather than just mod-N ones), obtaining a refinement $Q_1,\ldots,Q_L$, which automatically satisfies an inequality of the form

$$\sum_{j=1}^{M}\Big|\sum_{z\in Q_j} f(z)\omega^{-\psi_j(z)}\Big| \geq \beta N/2\ .$$

Once again, we have renamed the functions ψ_j. It turns out that one can take $L \leq N^{1-2^{-32002}\alpha^{30000}}$. Next, the results of Section 5 give us a further refinement of $Q_1, \ldots, Q_L$ into arithmetic progressions $R_1, \ldots, R_H$ such that the functions $\psi(j)$ are approximately constant on each R_h, and can therefore be ignored. More precisely, one obtains the inequality

$$\sum_{i=1}^{H}\Big|\sum_{s\in R_i} f(s)\Big| \geq \beta N/4$$

where H is at most $N^{1-2^{-32010}\alpha^{30000}}$. Finally, an averaging argument gives us an arithmetic progression R of cardinality at least $\beta N^{2^{-32010}\alpha^{30000}}$ such that $\sum_{s\in R} f(s) \geq \beta|R|/16$. This implies that the cardinality of $A \cap R$ is at least $|R|(\delta + 2^{-16004}\alpha^{15000})$. Recalling that $\alpha = (\delta/2)^{64}$, we find that the density of A has gone up from δ in $\mathbb{Z}_N$ to at least $\delta(1 + (\delta/2)^{980000})$ inside the arithmetic progression R.

We now iterate this argument as in the proof of Roth's theorem. The iteration can be performed at most $(\delta/2)^{-1000000}$ times, and at each step the value of N is raised to a power which exceeds $(\delta/2)^{2000000}$. It is not hard to check that N will always remain sufficiently large for the argument to work, as long as the initial value of N is at least $\exp\exp(\delta^{-C})$, where C can be taken to be 2000000. □

An alternative formulation of the condition on N and δ is that δ should be at least $(\log\log N)^{-c}$ for some absolute constant $c > 0$. We have the following immediate corollary.

Corollary 8.3. *There is an absolute constant $c > 0$ with the following property. If the set $\{1, 2, \ldots, N\}$ is coloured with at most $(\log\log N)^c$ colours, then there is a monochromatic arithmetic progression of length four.* □

9. Progressions of length greater than four.

As I said in the introduction, dealing with arithmetic progressions of length five or more is considerably more difficult than dealing with progressions of length four, even though similar ideas are used. In this section I shall outline the main steps of the proof for progressions of length five, and in the next one I shall say more about the step that causes by far the most extra difficulty.

An outline of the proof.

A. By Corollary 3.6, if a set A fails to contain an arithmetic progression of length five, then it fails to be α-cubically uniform for some α that depends only on the density of A.

B. By the definition of α-cubically uniform (or at least, by one of the equivalent definitions), this means that there exists a set $B \subset \mathbb{Z}_N^2$ of size at least αN^2 and a function $\phi : B \to \mathbb{Z}_N^2$ such that $\Delta(f;k,l)^\wedge(\phi(k,l)) \geq \alpha N$ for every $(k,l) \in B$. Here, f is the balanced function of A (which is defined shortly after Lemma 2.1).

C. Just as the function ϕ that occurred in the proof for progressions of length four turned out to have a small linear piece, so this function ϕ has a small bilinear piece. To be precise, there exist arithmetic progressions P and Q of size $N^{c(\alpha)}$ with the same common difference, and a bilinear function $\beta : P \times Q \to \mathbb{Z}_N$, such that $\phi(x,y) = \beta(x,y)$ for at least $c'(\alpha)|P||Q|$ values of $(x,y) \in P \times Q$.

D. To simplify the exposition, let us now suppose that $P = Q = \mathbb{Z}_N$ and β is the function $(x,y) \mapsto 6cxy$ for some constant $c \in \mathbb{Z}_N$. Since $\Delta(f;k,l)^\wedge(\phi(k,l)) \geq \alpha N$ for every $(k,l) \in \mathbb{Z}_N^2$, we have the inequality

$$\sum_{k,l} |\Delta(f;k,l)^\wedge(6ckl)|^2 \geq \alpha N^4$$

Expanding out the modulus squared as

$$\sum_{s,t} \Delta(f;k,l)(s)\overline{\Delta(f;k,l)(t)}\omega^{-6ckl(s-t)}$$

and then making the substitution $m = s - t$, we find that

$$\sum_s \sum_{k,l,m} \Delta(f;k,l,m)(s)\omega^{-6cklm} \geq \alpha N^4 .$$

If we now use the identity

$$6klm = \sum_{\epsilon_1,\epsilon_2,\epsilon_3} (s - \epsilon_1 k - \epsilon_2 l - \epsilon_3 m)^3 ,$$

where the sum is over the eight triples $(\epsilon_1,\epsilon_2,\epsilon_3)$ with $\epsilon_i = 0$ or 1, then, writing C for the operation of complex conjugation, we can deduce that

$$\sum_s \sum_{k,l,m} \prod_{\epsilon_1,\epsilon_2,\epsilon_3} C^{\epsilon_1+\epsilon_2+\epsilon_3}\Big(f(s-\epsilon_1 k-\epsilon_2 l-\epsilon_3 m)\omega^{-c(s-\epsilon_1 k-\epsilon_2 l-\epsilon_3 m)^3}\Big) \geq \alpha N^4 .$$

This does not immediately tell us that f is cubically biased, as would be the most obvious generalization of the proof for progressions of length four. However, what it does tell us is that the function $x \mapsto f(x)\omega^{-cx^3}$ fails to be *quadratically* uniform. Indeed, it is precisely the negation of equivalence (ii) of Lemma 3.1 for this function.

E. If P and Q are not equal to $\mathbb{Z}_N$, then a similar statement can be proved, but this time it involves partitioning $\mathbb{Z}_N$ into subprogressions P_i and finding cubic functions κ_i such that $x \mapsto f(x)\omega^{-\kappa_i(x)}$ is not quadratically uniform inside P_i (in a sense that can, with a little effort, be made precise).

F. We can now apply the results proved earlier about quadratic non-uniformity (in particular, Proposition 6.1, Corollary 7.5 and Proposition 8.1) to find a partition of $\mathbb{Z}_N$ into subprogressions P_{ij} for each of which there is a quadratic function q_{ij} such that $\sum_{x\in P_{ij}} f(x)\omega^{-\kappa_i(x)-q_{ij}(x)}$ is, on average, of magnitude comparable to $|P_{ij}|$.

G. As in the proof of Theorem 8.2, one can then use the results of Section 5 to pass to further subprogressions in which the cubic functions $\kappa_i + q_{ij}$ are approximately constant, and then ignore these functions. Finally, by an averaging argument one finds one of these subprogressions, R say, with the property that $\sum_{x\in R} f(x) \geq c''(\alpha)|R|$, which implies that $|A\cap R| \geq (\delta + c''(\alpha))$.

H. It turns out that $c''(\alpha)$ is a power of α (albeit a very large one) and therefore that the above argument can be iterated to prove that every subset of $\{1, 2, \ldots, N\}$ of size at least $N/(\log\log N)^\gamma$ contains an arithmetic progression of length at least five. Here, γ is a large power of α.

10. Finding a small bilinear piece.

Of the steps outlined in the previous section, most are straightforward generalizations of the corresponding steps for progressions of length four. Step E is annoyingly messy, though not fundamentally difficult. The genuinely troublesome step is C. Why should this be so hard? The answer is that we have left behind the comfortable world of linearity, and in fact the proof involves quadratic functions in a fundamental way.

One might try to reason as follows. If one considers the function $\Delta(f; k, l)$ for fixed k, then it is simply the function $\Delta(g_k; l)$, where g_k is the function $D(f; k)$. If $\Delta(f; k, l)(\phi(k, l))$ is often large, then for many values of k we must find that $\Delta(g_k; l)(\phi(k, l))$ is often large. But then, by the results of Sections 6 and 7, it follows that the function $l \mapsto \phi(k, l)$ is linear-like. In other words, fixing k often gives a linear-like function in l and, by symmetry, fixing l often gives a linear-like function in k. This suggests that ϕ is bilinear-like.

Unfortunately, it is difficult to get the linear behaviours of ϕ in the individual variables to interact with one another. The following example illustrates one sort of problem that can occur. Let λ be an arbitrary function from $\mathbb{Z}_N$ to $\mathbb{Z}_N$, and define

$$\phi(x, y) = \begin{cases} \lambda(x)y & 0\leq x \leq y < N \\ x\lambda(y) & 0\leq y < x < N\ . \end{cases}$$

There are certainly many additive quadruples in each variable, and plenty of resulting linearity, but if λ does not have special additivity properties, then the quadruples with x fixed do not mix with those with y fixed and there is nothing more to say about ϕ, and in particular no restriction of ϕ that looks bilinear.

This example can be dealt with by the observation that not only ϕ but any restriction of ϕ to a reasonably large set should be linear-like in each variable. However, there is another phenomenon that causes far more difficulty. Let us informally call a function *quasilinear* if it resembles a low-dimensional linear function (such as, for example, the function defined at the end of §6). A serious complication arises even if we know for every x that $\phi(x, y)$ is quasilinear in y for every x and vice-versa.

It is tempting to suppose that one might be able to find a large subset $B' \subset B$, and numbers $x_0, x_1, \ldots, x_d$, $r_1, \ldots, r_d$, $y_0, y_1, \ldots, y_d$, $s_1, \ldots, s_d$ and $(c_{ij})_{i,j=0}^d$ such that the restriction of ϕ to B' was of the form

$$\phi\Big(x_0 + \sum_{i=1}^d a_i x_i, y_0 + \sum_{i=1}^d b_j y_j\Big) = \sum_{i,j=0}^d c_{ij} a_i b_j$$

for $0 \leq x_i < r_i$ and $0 \leq y_j < s_j$.

However, this would imply that one could find a small "common basis" for all the functions $y \mapsto \phi(x, y)$ (and similarly the other way round) and a simple example shows that such a statement is too strong. Indeed, let ψ be a non-trivial (i.e., non-linear) quasilinear function from $\mathbb{Z}_N$ to $\mathbb{Z}_N$. (For definiteness one could let $\psi(z) = z \pmod m$ for some m near $\sqrt{N}$.) Define $\phi(x, y)$ to be $\psi(xy)$. The natural bases for the functions $y \mapsto \psi(xy)$ are all completely different, and there is no small basis that can be used for all (or even a large proportion) of them. We shall not prove this here, but it rules out simple proofs, or even definitions, of the bilinearity of ϕ.

It is not easy to say much about how we do actually go about the task. Here is a very vague sketch.

(1) Given any $h \in \mathbb{Z}_N$, define $\phi_h(x, y)$ to be $\phi(x, y + h) - \phi(x, y)$ (with the obvious convention that this is undefined unless both $\phi(x, y + h)$ and $\phi(x, y)$ are defined).

(2) Prove a lemma, along similar lines to Proposition 6.1, to the effect that many of the ϕ_h have many additive quadruples. Notice that since the ϕ_h are defined in terms of more than one row of $\mathbb{Z}_N$, knowledge about the ϕ_h collectively has the effect of 'linking' these rows.

(3) Use another lemma to pass to a subset B' of B with the property that, for almost every ϕ_h, almost every additive quadruple where ϕ_h is defined (now defining ϕ_h only at those (x, y) such that both (x, y) and $(x, y + h)$ belong to B') is ϕ_h-additive. That is, if $x_1 + x_2 = x_3 + x_4$ then we tend to expect

$$\phi_h(x_1, y) + \phi_h(x_2, y) = \phi_h(x_3, y) + \phi_h(x_4, y)$$

(4) Now define functions g_h by setting $g_h(x)$ to be the number of y such that both (x, y) and $(x, y + h)$ belong to B'. Another proposition similar to Proposition 6.1 can be used to show that the Fourier coefficients of the functions g_h are related. Specifically, if σ is any function such that $\hat{g}_h(\sigma(h))$ is large for many values of h, then there are many σ-additive quadruples, and hence, by the results of Section 7, σ is linear-like.

(5) The functions ϕ_h are also linear-like. Moreover, the Fourier coefficients of the functions g_h are closely related to the multidimensional arithmetic progressions that serve as the 'domains' of the ϕ_h

(6) Using the results of Section 5, the above facts and many averaging arguments, one can find arithmetic progressions P and Q with the same common difference (this detail is important and turns out to require quadratic methods) such that, for many values of $h \in P$, ϕ_h behaves linearly in y for many $y \in Q$.

(7) By further averaging arguments, we can find P' and Q' such that for many $x \in P'$, $\phi(x, y)$ behaves linearly for many $y \in Q'$.

(8) Because ϕ is linear-like in x as well as in y, we can deduce that the 'gradients' of the functions $x \mapsto \phi(x, y)$ are often related in a linear way. This gives us the required bilinear piece of ϕ.

I do not expect the above sketch to be fully comprehensible, but I hope that it gives some flavour of the argument, and some idea of its complexity.

11. Some unsolved problems in additive/combinatorial number theory.

There are many interesting open problems that it may now be possible to tackle, though they all seem to be challenging. In the remainder of this paper, we shall discuss some of these, and in this way give a brief survey of the more general area of number theory of which Szemerédi's theorem forms a part. There is no completely satisfactory name for this area: it lies at the interface between additive number theory, harmonic analysis and combinatorics. Perhaps one could characterize it negatively as that corner of number theory where neither algebraic methods nor the Riemann zeta function and its generalizations play a central role. Alternatively, one could describe it as the part of number theory that immediately appeals to combinatorialists, even if they cannot rely exclusively on combinatorics to solve its problems.

Problem 11.1.

It is known by Furstenberg's methods that the following multidimensional version of Szemerédi's theorem holds: for every δ, k and d and any finite subset $K \subset \mathbb{Z}^d$ there exists N such that every subset $A \subset \{1, 2, \ldots, N\}^d$ of size at least δN^d contains a homothetic copy $a + bK$ of K. However, there is no proof known that gives any bound for N. In fact, even when $d = 2$ and A is the set $\{(0,0), (0,1), (1,0)\}$ there is no good bound known.

In fact, the best bound so far was discovered very recently by Solymosi [So]. His proof relies on a curious lemma of Ruzsa and Szemerédi. In order to explain it, we must first recall some terminology from graph theory. Let G be a bipartite graph whose edges join the two (disjoint) vertex sets X and Y. A *matching* in G is defined to be a set of edges (x_i, y_i) with $x_i \in X$, $y_i \in Y$ and all the x_i and y_i distinct. In other words, thinking of each edge as the set consisting of its two end-vertices, all edges in a matching are required to be disjoint. Given any graph G, an *induced subgraph* of G is any graph H whose vertex set W is a subset of the vertex set V of G and whose edges consist of all pairs $\{x, y\} \subset W$ such that x and y are joined in G. (This is very different from the more general notion of a subgraph, which is simply any graph formed by a subset of the edges of G.)

Returning to the bipartite case, an *induced matching* is, of course, a matching that happens also to be an induced subgraph of G. To find an induced matching, one must choose subsets $\{x_1, \ldots, x_k\} \subset X$ and $\{y_1, \ldots, y_k\} \subset Y$ such that x_i is joined to y_j in G if *and only if* $i = j$. If G has many edges, then one naturally expects induced matchings to be hard to come by since they have very few edges. The Ruzsa-Szemerédi lemma provides some confirmation of this.

Lemma 11.2. *Let C be a constant and let G be a bipartite graph with vertex sets X and Y of size n. Suppose that the edges of G can be expressed as a union of Cn induced matchings. Then G has $o(n^2)$ edges.*

What is curious about this lemma is that its conclusion is so weak. If G had cn^2 edges for some constant $c > 0$, and could be written as a union of only Cn induced matchings, then the average size of a matching would be cn/C. The number of vertices of a typical matching would therefore be within a constant of maximal, and there would be almost no edges between these vertices (because the matching is induced). Thus, G would be full of enormous holes wherever its edges concentrated.

One might expect such thoughts to lead to a fairly easy proof that the number of edges in G was at most n^α for some $\alpha < 2$. However, not only do they not do so, but this conclusion is not even true, as it would imply an upper bound for Roth's theorem which is better than the best known lower bound. See the discussion of problem 11.6 below for more details on this point.

The proof of Lemma 11.2 is not too hard, but it relies on Szemerédi's famous regularity lemma, which is an extremely useful graph-theoretic tool but which gives rise to bounds of tower type. This is the reason that the Ruzsa-Szemerédi lemma is usually not stated in a more quantitative form. What their proof actually shows is that the number of edges can be at most $n^2/f(n)$ where f is a function that grows roughly as slowly as $\log^*(n)$. (This is defined as the number of times you must take logarithms in order to get n down to 1.)

Now let A be a dense subset of $\{1, 2, \ldots, N\}^2$. Armed Solymosi defines a bipartite graph G with vertex sets X and Y both copies of $\{1, 2, \ldots, N\}$ and joins x to y if and only if $(x, y) \in A$. In addition, for each $d \in \{-(N-1), -(N-2) \ldots, N-1\}$ he defines a matching M_d to consist of all edges (x, y) with $x - y = d$. Since G has more than $o(n^2)$ edges and these have been written as a union of fewer than $2N$ matchings, the Ruzsa-Szemerédi lemma implies that not all these matchings are induced. This, it can be checked, implies that A contains a configuration of the form $\{(x, y), (x + d, y), (x, y - d)\}$. If we therefore apply the argument to a suitable reflection of A, we obtain a triangle of the form $\{(x, y), (x + d, y), (x, y - d)\}$ as required by the theorem. (Irritatingly, there seems to be no obvious way to force d to be positive, so the rightangle may be the bottom left corner or it may be the top right.)

Problem 11.3

Although a power-type bound for the Ruzsa-Szemerédi lemma does not hold, it is extremely unlikely that a function like $n^2/\log^*(n)$ gives the correct order of magnitude. Given that the lemma has such interesting direct consequences, there is strong motivation for the following problem: find a proof of the Ruzsa-Szemerédi lemma which does not use Szemerédi's regularity lemma, and which (consequently, one imagines) gives a significantly better bound.

Problem 11.4

Recently, Bergelson and Leibman proved the following beautiful 'polynomial Szemerédi theorem' [BL]. For any $\delta > 0$ and any collection $p_1, \ldots, p_k$ of polynomials that have integer coefficients and vanish at zero, there exists N such that every set $A \subset \{1, 2, \ldots, N\}$ of size at least δN contains a subset of the form $\{a + p_1(d), a + p_2(d), \ldots, a + p_k(d)\}$. Letting p_i be the polynomial id, one immediately recovers Szemerédi's theorem. Once again, this theorem is known only by the ergodic theory method and hence no bound for N is known.

Even guaranteeing the existence of a subset of the form $\{a, a+d^2\}$ is not trivial, but for simple examples like this there are explicit bounds, due to Sárközy [S] and others [PSS]. (In its qualitative form, this result was discovered independently of Sárközy by Furstenberg.) The analytic proof of Szemerédi's theorem outlined in this paper suggests that it ought to be possible to prove a quantitative version of the Bergelson-Leibman theorem as well. This, one might hope, could be developed from the proof of a simple case such as $\{a, a+d^2\}$ rather as the proof of Szemerédi's theorem has its roots in Roth's much simpler argument for progressions of length three.

One difficulty with this project is that Sárközy's argument is *not* all that simple. Very recently, however, Green [Gre1] has discovered a proof of the Furstenberg-Sárközy theorem which, though giving a worse bound than Sárközy obtains, has the merit of being simpler and, more importantly, closely analogous to the proof of Roth's theorem. If a quantitative version of the Bergelson-Leibman theorem is ever discovered, it will probably begin with Green's argument.

Not too surprisingly, progress has so far been modest. Green's paper contains a highly ingenious quantitative proof that if $\delta > 0$, N is sufficiently large and A is a subset of $\{1, 2, \ldots, N\}$ of size at least δN, then A must contain an arithmetic progression of length three whose common difference is a sum of two squares. This restriction on the common difference forces him to use quadratic methods similar to those needed in this paper to deal with progressions of length four.

Just as Szemerédi's theorem can be thought of as the density version of van der Waerden's theorem, so the Bergelson-Leibman theorem is the density version of the following colouring statement.

Theorem 11.5. *Let $p_1, \ldots, p_k$ be polynomials that vanish at zero and have integer coefficients. Then for every positive integer r there exists N (depending only on r and $p_1, \ldots, p_k$) such that, however the set $\{1, 2, \ldots, N\}$ is coloured with r colours there exist a and $d \neq 0$ such that all the numbers $a + p_i(d)$ have the same colour.*
Bergelson and Leibman proved this theorem first and then applied Furstenberg's methods to obtain the stronger density statement. Even their proof of the colouring statement used ergodic theory, so it was of considerable interest when Walters [W] found a purely combinatorial proof of Theorem 11.5 which was very much in the spirit of van der Waerden's original arguments. It seems not to be possible to find a 'Shelah-ization' of Walters's proof, so it is still an open problem whether the bounds for Theorem 11.5 can be made primitive recursive.

Problem 11.6

The following famous question was discussed in the introduction: do the primes contain arbitrarily long arithmetic progressions? There are two obvious approaches to it. The first, and more ambitious, is to improve the density bound in Szemerédi's theorem enough to show that a density of $(\log n)^{-1}$ is sufficient. (In fact, it is an amusing exercise to show that even a density of $C \log\log n/\log n$, for a certain absolute constant C, will do.) The second approach, which at the moment seems more realistic, is to start with Vinogradov's methods, which can be used to prove that the primes contain infinitely many progressions of length three, and try to generalize them in the way that the methods of this paper generalize the proof of Roth's theorem.

There are at least two major obstacles to carrying out this very natural programme. The first, which seems to be more fundamental, is that any use of Freiman's theorem will be for a constant C which is comparable to $\log n$. Since the best known estimate for the dimension of the arithmetic progression given by the theorem is itself comparable to C (see Problem 11.9 for further discussion of this), and since not much can be said about a $(\log n)$-dimensional arithmetic progression, it appears that progress with the primes will have to wait until there has been progress in our understanding of Freiman's theorem.

The second obstacle is related to the way we *used* Freiman's theorem. When proving Szemerédi's theorem, one can afford to pass to a small subprogression and start again, as long as the set is reasonably dense. However, if one wishes to use the structure of the set of primes, then this move is ruled out: next to nothing is known about the structure of the primes when they are restricted to an arithmetic progression of length, say, $N^{1/100}$. One can imagine ways round this difficulty: it ought to be possible to strengthen the argument of Section 8 to deduce from the quadratic non-uniformity of a function f not just that $\sum_{x\in P} f(x)\omega^{q(x)}$ is unexpectedly large for some quadratic function q and some smallish arithmetic progression P, but that $\sum_x f(x)\omega^{\psi(x)}$ is large, where now the sum is over the whole of $\mathbb{Z}_N$ and ψ is some sort of 'multidimensional quadratic form'. A more global statement such as this should be easier to disprove in the case of the primes using standard methods. Unfortunately, it is not easy even to formulate an appropriate statement, let alone prove it.

Problem 11.7

The following is probably the most famous of all the unsolved problems of Erdős. Let X be a subset of $\mathbb{N}$ with the property that $\sum_{x\in X} x^{-1} = \infty$. Does X necessarily contain arithmetic progressions of every length? It is not known even whether X must contain an arithmetic progression of length three. If the problem has a positive answer, then it implies that the primes contain arbitrarily long progressions.

Although the form of the conjecture is amusingly neat, one should not be misled into thinking that there is anything particularly natural about the sum of reciprocals. It is an easy exercise to show that if $\sum_{x\in X} x^{-1} = \infty$ then for any $\epsilon > 0$ the size of $X \cap \{1, 2, \ldots, N\}$ is at least $N/(\log N)^{1+\epsilon}$ infinitely often. Thus, Erdős's conjecture would follow if one could show that a density of $1/(\log N)^{1+\epsilon}$ was enough in Szemerédi's theorem. Conversely, if a sequence of sets $A_1, A_2, \ldots$ can be found, where each A_m is a subset of $\{1, 2, 3, \ldots, 2^m\}$ of size at least $2^m/m$ not containing an arithmetic progression of length k, then the union $X = (A_1 + 2) \cup (A_3 + 2^3) \cup (A_5 + 2^5) \cup \ldots$ still contains no arithmetic progression of length k even though $\sum_{n\in X} n^{-1} = \infty$. Thus, Erdős's conjecture follows from, and is roughly equivalent to, the statement that Szemerédi's theorem is true with a density significantly better than $1/\log n$. I say 'roughly equivalent' because it is conceivable that, although a density of $1/\log n$ is insufficient for Szemerédi's theorem, the counterexamples are so few and far between that it is not possible to put them together to obtain a set X such that $\sum_{n\in X} n^{-1} = \infty$. For example, this would be true if a density of $\log n$ was *usually* sufficient, but not quite always, owing to a strange construction that worked only inside intervals of length of the form 2^{m^2}. Of course, not only is such a scenario highly unlikely, it would also not matter in the case of sets such as the primes, which have density $1/\log n$ not just sporadically, but all the time.

Problem 11.8

These observations show that the prettiness of Erdős's conjecture is somewhat artificial, and that the real question is the more prosaic (but still fascinating) one about the correct density in Szemerédi's theorem. Given that progressions of length three are much easier to handle than longer ones, it is very frustrating that the following special case of the problem is still wide open: what is the correct bound for Roth's theorem?

In Section 2 we saw that a density of $C/\log\log N$ is enough to guarantee an arithmetic progression of length three. This bound was improved by Szemerédi [Sz3] and Heath-Brown [H-B] to $(\log N)^{-c}$ for an absolute constant $c > 0$. The best known result in this direction was obtained recently by Bourgain [Bou], who showed that a density of $C\log\log N/(\log N)^{1/2}$ was enough. The reason Bourgain obtains a much stronger bound than Roth is, very roughly, as follows. The main source of inefficiency in Roth's argument is the fact that one passes many times to a subprogression of size the square root of what one had before. This means that the iteration argument is very costly. Moreover, at each stage of the iteration, one obtains increased density on a mod-N arithmetic progression of *linear* size and simply discards almost all of this information in the process of restricting to a 'genuine' arithmetic progression.

Bourgain does not throw away information in this way. Instead, he tries to find increased density not on arithmetic progressions but on translates of *Bohr neighbourhoods*, which are sets of the form $\{x \in \mathbb{Z}_N : r_i x \in [-\delta_i N, \delta_i N]\}$. Note that these sets are just intersections of a few mod-N arithmetic progressions. Roughly speaking, if a set A is not evenly distributed inside a Bohr neighbourhood B, then, using a large Fourier coefficient of $A \cap B$, one can pick out a new mod-N arithmetic progression P such that the density of A inside $B \cap P$, which is still a Bohr neighbourhood, is larger. The reason this approach can be expected to work is that Bohr neighbourhoods have a great deal of arithmetic structure: indeed, they are rather similar to multidimensional arithmetic progressions. I should make clear that this sketch of Bourgain's method, although it conveys the basic idea, is not quite an accurate portrayal of what he actually does. The technicalities involved in getting something like this idea to work are formidable and Bourgain's paper is a tour de force.

As I have said, the discrepancy between this bound and the best known lower bound is very large. The lower bound comes from a construction of Behrend [Be]. It was published in 1946, and nobody has found even the smallest improvement. Since the construction is beautiful, simple and gives an important insight into why the problem is difficult, it is worth giving in full.

To begin with, let us construct a different object. Let m and d be parameters to be chosen later, and let us search for a subset A of the grid $\{0,1,2,\dots,m-1\}^d$ containing no arithmetic progression of length three. (This means a set of three points x, y and z in the grid such that $x+z=2y$.) A simple way to do this is to choose a positive integer t and let A_t be the set of all x such that $x_1^2+\cdots+x_d^2=t$. Since all points in A_t then lie on the surface of a sphere, it is clear that A_t contains no arithmetic progression (or even a set of three collinear points). Furthermore, since A_t is only ever non-empty for $d \le t \le m^2 d$, and every point in the grid lies in some A_t, averaging tells us that there exists a t such that A_t has cardinality at least $m^d/m^2 d$.

The next observation is that the grid can be embedded into $\mathbb{N}$ in such a way that arithmetic progressions of length three are preserved and no new ones are created. More precisely, given a point x in $\{0,1,2,\dots,m-1\}^d$, let $\phi(x)$ be the positive integer obtained by thinking of x as a number written backwards in base $2m$. (In other words, $\phi(x)=\sum_{i=1}^d x_i(2m)^{i-1}$.) Then it is not hard to check that $\phi(x)+\phi(z)=2\phi(y)$ if and only if $x+z=2y$. (It is to obtain the 'only if' that we use base $2m$ rather than the more obvious base m.)

Furthermore, the range of ϕ is contained in an interval of length $(2m)^d$. Therefore, we can use the map ϕ to take the set A_t to a subset A of such an interval, where A has size at least $m^d/m^2 d$ and contains no arithmetic progression of length three. All that remains is to optimize the choice of m and d given that $(2m)^d=N$. It turns out that a good choice is to set $d=\sqrt{\log N}$, which results in a subset A of $\{1,2,\dots,N\}$ with no arithmetic progression of length three and with cardinality $N\exp(-c\sqrt{\log N})$.

It is perhaps easier to see how far this bound is from Bourgain's upper bound if we state the bounds in the following equivalent way. For a fixed $\delta>0$, let $D=\delta^{-1}$. Then the N needed by Bourgain to guarantee that every subset of $\{1,2,\dots,N\}$ of size at least δN contains an arithmetic progression of length three is $\exp(cD^2\log D)$, whereas Behrend's construction gives a counterexample when $N=\exp(C(\log D)^2)$.

In view of the apparent weakness of the Behrend bound, why is it regarded as so interesting? The main reason is that, as mentioned at the beginning of Section 2, it disproves a very natural conjecture (which at one time was even made by Erdős and Turán). This conjecture is that a density of $CN^{-\alpha}$ is sufficient to guarantee a progression of length k, for some $\alpha>0$ depending only on k. This is the sort of bound one would expect from the general heuristic principle that probabilistic arguments always do best. This principle is simply wrong in the case of Szemerédi's theorem.

This fact is interesting in itself, but it also has interesting metamathematical consequences. If one is trying to improve the upper bound, one can immediately rule out several potential arguments on the grounds that, if they worked, they would give rise to power-type bounds. When planning an approach to the upper bound, it is very important that there should be some foreseeable 'unpleasantness', some difficulty that would give rise to a bound expressed by a less neat function. This shows, for example, that you will not be able to prove Roth's theorem using only a little formal manipulation of Fourier coefficients. (A different way to see this is to note that the Fourier expression that counts the arithmetic progressions of length three in A does not have to be non-negative if A, a characteristic function of a set, is replaced by a more general function.) Also, there seems little hope of a compression-type proof that successively modifies an AP-free set without decreasing its size until eventually it is forced to have some extremal structure - simply because it is hard to imagine forcing a set to have the very particular and not wholly natural quadratic structure of Behrend's example.

Problem 11.9

We saw in the discussion of Problem 11.6 some of the motivation for the following question: what are the correct bounds for Freiman's theorem? In fact, this is a question of major importance, with potential applications to all sorts of different problems. In order to discuss bounds, it is helpful to summarize very briefly Ruzsa's proof of the theorem.

Ruzsa starts with a set A_0 such that $|A_0 + A_0| \leq C|A_0|$. Then, by a highly ingenious argument, he finds a subset $A_1 \subset A$ of proportional size such that A_1 is 'isomorphic' to a subset $A \subset \mathbb{Z}_N$, where N is also proportional to $|A_0|$ (that is, not too large). Rather than say precisely what 'isomorphic' means, let me give instead the main relevant consequence, which is that if $2A - 2A$ contains a d-dimensional arithmetic progression of a certain size, then so does $2A_0 - 2A_0$.

Further arguments of a combinatorial nature can be used to show that if $2A_0 - 2A_0$ contains a large and small-dimensional arithmetic progression, then A_0 is contained in one. (Of course, this also uses the assumption that $|A_0 + A_0| \leq C|A_0|$.)

As a result, Ruzsa shows that Freiman's theorem is (non-trivially) equivalent to the following statement: if A is a subset of $\mathbb{Z}_N$ of size δN, then $2A - 2A$ contains an arithmetic progression P of size at least $c(\delta)N$ and dimension at most $d(\delta)$.

It turns out that, from the point of view of applications, the quantity for which one would most like a good estimate is $d(\delta)$. A fairly straightforward argument, based on a technique of Bogolyubov [Bo], shows that d can be taken to be at most δ^{-2}. Very recently this was improved by Chang [C], who added some interesting refinements to Ruzsa's approach and obtained a bound of $\delta^{-1}\log(\delta^{-1})$. Almost certainly, however, this bound, which is the best known, is a long way from being best possible. This may be as low as $C\log(\delta^{-1})$, which would have very significant consequences. Even an estimate of $(\log(\delta^{-1}))^C$ would be extremely interesting - for example, it would be good enough to use for Problem 11.6.

Chang also found a much more efficient way than Ruzsa's of passing from the progression inside $2A_0 - 2A_0$ to the one containing A_0, so she obtains the following bounds for Freiman's theorem. If $|A + A| \leq C|A|$ then A is contained in a progression P of dimension at most $a(C\log C)^2$ and cardinality at most $\exp(aC^2(\log C)^3)$, where a is an absolute constant. A simple example shows that these bounds are almost best possible. (Just to make this statement clear: the bounds for the progression *containing* A are close to best possible, but it would be very interesting to improve the bounds for the progression *contained in* $2A - 2A$, which are far from best possible.)

The example is the following. Let m be a large integer and let A be the geometric progression $\{1, m, \ldots, m^{k-1}\}$. (As will be clear, any set with no small additive relations would do just as well.) Then $|A| = k$ and $|A + A| = k(k+1)/2$, so we have $|A + A| \leq C|A|$ for a constant C proportional to k. Now the elements of A are independent in the following sense: if $a_1, \ldots, a_k$ are integers such that $\sum_{i=1}^{k} a_i m^{i-1} = 0$, then at least one a_i has modulus at least $m/2$. Using this fact, it is easy to see that any arithmetic progression of dimension less than k containing

A must have cardinality at least m. Since m is not bounded by any function of k, it is impossible to prove a bound for the dimension that is better than linear in C.

Note that this example can be 'fattened up': simply replace A by $A+\{1,2,\ldots,t\}$. With appropriate choices of t, k and m one can find similar examples for C and $|A|$ of any desired size. Note also that such examples have no bearing on the 'inner' progression. If you are looking for a low-dimensional progression inside $2B-2B$, where $B = A+\{1,2,\ldots,t\}$ as above, then all you have to do is consider one of the k 'pieces' of B, which is an interval of length t, which shows that $2B-2B$ contains a one-dimensional progression of length comparable to $C^{-1}|A|$. In general, it seems that the weakness in the known arguments for Freiman's theorem is that they do not take into account the possibility that a set with small sumset may well have a subset with much better structure.

As for further potential applications of Freiman's theorem, here are a few problems that seem to be related. Others are listed in [F3] and [C].

Problem 11.10

Yet another beautiful question of Erdős is the following. Let A be a set of n integers and let $\epsilon > 0$. Is it true that either $A+A$ or $A.A$ (the set of all products ab with $a, b \in A$) has cardinality at least $n^{2-\epsilon}$?

The idea behind this problem is, of course, that if you try to make $A+A$ small, say by making A into an arithmetic progression, then $A.A$ will be almost maximal, whereas if you try to minimize $A.A$, say by making A a geometric progession, then $A+A$ will be almost maximal. In general, whatever you do to pull the sums together seems to drive the products apart, and vice-versa.

It will probably never be possible to use Freiman's theorem directly to solve this problem: to say anything about a set A with $|A+A| \leq |A|^{1+\alpha}$ for *any* fixed $\alpha > 0$ is way beyond what is possible at the moment, and it seems unlikely that there is a useful structural statement when, say, $\alpha = 0.99$. Nevertheless, it might be possible, and would be interesting, to show that if $|A+A| \leq |A|^{1.01}$ then $|A.A| \geq |A|^{2-\epsilon}$. The best known bound for the problem as stated is due to Elekes [E], who proved that one of $A+A$ and $A.A$ must have cardinality at least $|A|^{5/4}$. His proof used the Szemerédi-Trotter theorem [ST], which is a very useful tool in combinatorial geometry.

Problem 11.11

A similar question for which Freiman's theorem may eventually be useful is the so-called Erdős ring problem. It asks whether $\mathbb{R}$ contains a subring of dimension 1/2 - that is, a subset A of Hausdorff dimension 1/2 which is closed under addition and multiplication. Very roughly, this corresponds to asking about the structure of sets A of integers such that $|A+A| \leq |A|^{1+\epsilon}$. An interesting discrete version of the problem is the following. Does there exist a subset A of the field $\mathbb{F}_p$ of cardinality about $p^{1/2}$ such that both $A+A$ and $A.A$ have cardinality at most $p^{o(1)}|A|$? Such a set would be 'approximately closed' under addition and multiplication. Note that if one replaces $\mathbb{F}_p$ by $\mathbb{F}_{p^2}$ then the answer is trivially yes, so any proof would have somehow to distinguish between different kinds of finite fields. (A similar remark applies to the original ring problem - it is not hard to find a subring of $\mathbb{C}$ of half the dimension of $\mathbb{C}$, so a proof in $\mathbb{R}$ would have to distinguish $\mathbb{R}$ from $\mathbb{C}$.)

Problem 11.12

Freiman has suggested that a good enough bound for his theorem would have a bearing on a famous problem in additive number theory: what is the correct order of magnitude of the Waring number $G(k)$? Recall that this is the smallest integer m such that every sufficiently large integer can be written as a sum of m k^{th} powers. The best known upper bound is $(1 + o(1))k \log k$, due to Wooley [Wo], but it is conjectured that the correct bound is linear in k - or even, more ambitiously, that it is linear with constant 1. Note that k is a trivial lower bound.

Let N be very large and let K be the set of all k^{th} powers less than N. If m is significantly larger than k and if it is not possible to write every integer between, say, $kN/2$ and $kN/4$ as a sum of m elements of K, then the cardinalities of the sets K, $K + K$, $K + K + K$, and so on, eventually cease to be close to their maximum possible values of $N^{1/k}$, $N^{2/k}$, $N^{3/k}$ and so on. Indeed, at some point there must be an r such that $rK + K$ has significantly smaller cardinality than $|rK||K|$. With a very good bound for Freiman's theorem, one might possibly be able to exploit this information to obtain a contradiction - though nobody has come up with a theorem to this effect.

Problem 11.13

A *Sidon set* of integers is a set A with the property that all its sums are distinct. That is, the only solutions of the equation $x + y = z + w$ with x, y, z and w in A are the trivial ones $x + y = x + y$ or $x + y = y + x$. There are several interesting open problems connected with such sets, of a very similar flavour to the questions discussed in this paper. The most obvious three are the following.

1. How large is the largest possible Sidon subset of $\mathbb{Z}_N$?

2. How large is the largest possible Sidon subset of $\{1, 2, \ldots, N\}$?

3. Suppose that A is a Sidon subset of $\mathbb{N}$ such that $A \cap \{1, 2, \ldots, N\}$ has cardinality at least N^α for all sufficiently large N. How large can α be?

Easy counting arguments provide upper bounds for all three problems. For example, if A is a Sidon subset of $\mathbb{Z}_N$ and has cardinality m, then $A+A$ has cardinality at least $m(m + 1)/2$, from which it follows that $m \leq (2N)^{1/2}$. If instead $A \subset \{1, 2, \ldots, N\}$ then $A + A \subset \{2, 3, \ldots, 2N\}$ and the same argument implies that $m \leq 2N^{1/2}$. This fact, in turn, gives an upper bound of $1/2$ for α in the third question.

It is interesting to reflect on why it is that the absence of non-degenerate solutions to the equation $x + y = z + w$ gives rise, by an easy argument, to a power-type upper bound, while the absence of non-degenerate solutions to the superficially similar equation $x + z = 2y$ leads to a difficult open problem for which a power-type bound is known not to hold. One reason is simply that counting argument we have just given relies on the symmetry in the first equation, and the second does not have this symmetry. A second, which is really the first reason in a different guise, becomes clear when we look at the problems on the Fourier side. Let $A \subset \mathbb{Z}_N$. As we saw in §2, the number of solutions in A to the equation $x + z = 2y$ is $N^{-1} \sum_r \hat{A}(r)^2 \hat{A}(-2r)$, while the number of solutions to $x + y = z + w$ is

$N^{-1}\sum_r |\hat{A}(r)|^4$. A big difference between these two expressions is that the second is automatically positive. Of course, the first is positive as well, since it counts the number of solutions to $x+z=2y$, but we know it is positive only because we know that $\hat{A}$ is the Fourier transform of a non-negative function. If f is an arbitrary real-valued function, then it is perfectly possible for $N^{-1}\sum_r \hat{f}(r)^2\hat{f}(-2r)$ to be negative, but obviously not possible for $N^{-1}\sum_r |\hat{f}(r)|^4$ to be. (See the remark towards the end of the discussion of Problem 11.8.)

Let us now consider what is known about the three questions above. Two very simple arguments show that a Sidon subset of $\mathbb{Z}_N$ can have cardinality $cN^{1/3}$. The first is simply to choose *any* maximal Sidon set $A=\{x_1,\dots,x_m\}$. If no further element can be added to A without its losing the Sidon property, then every $x\in\mathbb{Z}_N$ can be expressed in some way as $z+w-y$ with $y,z,w\in A$. Since there are at most $|A|^3$ such numbers (clearly a more careful argument will improve this estimate by a constant) it follows that $m\geq N^{1/3}$. The same argument obviously works for subsets of $\{1,2,\dots,N\}$. A similar argument shows also that $\alpha=1/3$ is possible for infinite Sidon sets: if one greedily chooses an infinite sequence $x_1,x_2,x_3,\dots$ such that each x_k is as small as possible, given that $\{x_1,\dots,x_k\}$ remains a Sidon set, then the order of magnitude of x_k is at most k^3.

Similar bounds can be obtained by the most basic form of the probabilistic argument. Suppose, for example, that A is a subset of $\mathbb{Z}_N$ with each element chosen randomly and independently with probability p. The expected size of A is pN and the expected number of non-degenerate solutions to $x+y=z+w$ is at most p^4N^3 (actually smaller by a constant because each solution is counted more than once). If $pN\geq 2p^4N^3$, then the expected value of the size of A minus the number of non-degenerate solutions to $x+y=z+w$ is at least $pN/2$. Since $p=N^{-2/3}$ satisfies this condition, we can find a set A of size at least $N^{1/3}/2$ such that, throwing away one point from each non-degenerate solution, we end up with a Sidon set of sze at least $N^{1/3}/4$. Again, with a bit more care one can improve the constant in this bound.

The best known bounds for the first two questions were obtained by Singer in 1938 using a fairly simple algebraic construction [Si]. If N happens to be of the form p^2-1 for a prime p, then he obtains a Sidon subset of $\{1,2,\dots,N\}$ of size p. By the prime number theorem, this implies a bound of $(1+o(1))N^{1/2}$ in general, which is the same order of magnitude as the trivial upper bound. In fact, even the correct constant is known, since Erdős and Turán [ET2] improved the trivial upper bound to $N^{1/2}+N^{1/4}+1$.

Despite the close agreement between these two bounds, there is considerable interest in improving them, and especially in answering the following unsolved problem: is the correct upper bound of the form $N^{1/2}+C$ for an absolute constant C? This problem, or indeed the weaker problem of finding *any* improvement of the Erdős-Turán bound, has remained open for sixty years.

The infinite problem is interestingly different from the finite one, because it is not possible, or at least not straightforwardly possible, to obtain a good lower bound by stringing together a sequence of finite examples. Indeed, until very recently the best known asymptotic size was $(N\log N)^{1/3}$, that is, only a logarithmic improvement on the trivial bound of $cN^{1/3}$. This was obtained in a seminal paper of Ajtai, Komlós and Szemerédi [AKS] - seminal partly because of its interesting results, and partly because in it can be found the genesis of a major new technique in probabilistic combinatorics, now known as the Rödl nibble.

However, in 1998 this bound was substantially improved by Ruzsa [Ru4], who invented an astonishingly clever argument which gave a lower bound of $N^{\sqrt{2}-1+o(1)}$, the first time a power greater than 1/3 had been obtained. Although the correct answer is almost certainly that N^α is possible for every $\alpha < 1/2$, and although such a result is unlikely to be proved by Ruzsa's method, his paper is strongly recommended for its sheer beauty and ingenuity.

In the other direction, Erdős improved the upper bound of $N^{1/2}$ by a logarithmic factor: that is, if A is an infinite Sidon set then $|A \cap \{1, 2, \ldots, N\}| \leq (N/\log N)^{1/2}$ for infinitely many N. Thus, the trivial upper bound is not correct. Interestingly, a small modification to this result produces another famous open problem of Erdős.

Problem 11.14

Define a B_h-set to be a set containing no non-trivial solutions to the equation $x_1 + \cdots + x_h = y_1 + \cdots + y_h$ (so a Sidon set is a B_2-set). A simple counting argument like that for Sidon sets shows that the asymptotic size of a B_3-set cannot exceed $N^{1/3}$. However, unlike for Sidon sets, in this case it is not known whether there can be a set that achieves this trivial bound, to within a constant. There is an important technical difference between sums of two numbers and sums of three, which is that while there is a one-to-one correspondence between solutions of the equation $x + y = z + w$ and solutions of $x - y = z - w$, it is not possible to rewrite solutions to $u + v + w = x + y + z$ in terms of differences in a symmetrical way. In general, this makes problems about B_h-sets harder when h is odd.

This state of affairs can be compared with a well known problem from traditional additive number theory. While it is a classical fact that the asymptotic density of the set of sums of two squares is $c(\log N)^{-1/2}$, the problem of what happens with sums of three cubes is open. Moreover, it is conjectured that the answer is completely different, and that these numbers have positive density. (This fascinating problem does not count as 'combinatorial' in the sense in which I have been using the word, since it is clear that it will require advanced number-theoretic techniques for its solution.)

One can of course ask the corresponding finite problems for B_h-sets with $h > 2$, and for these the correct constants are no longer known. Until recently, almost all the best known upper bounds came from natural generalizations of the argument of Erdős and Turán for Sidon sets. For example, the largest B_4-subset of $\{1, 2, \ldots, N\}$ was shown by Lindström [Lin] 1969 to be of size at most $(8^{1/4} + o(1))N^{1/4}$. One exception was a complicated result of Graham [Gr] which improved the naturally occurring constant for B_3-sets from $4^{1/3}$ to $(4 - \frac{1}{228})^{1/3}$. However, Green, in a paper which will appear soon [Gre2], found a genuinely new way of looking at the problem which has improved all these bounds (including Graham's), not just by reducing the error estimates, but by actually decreasing the constant attached to the main term. In many cases these constants had stood still for over thirty years.

Bibliography.

[AKS] M. Ajtai, J. Komlós and E. Szemerédi, *A dense infinite Sidon sequence*, European J. Comb. **2** (1981), 1-11.

[BS] A. Balog and E. Szemerédi, *A Statistical Theorem of Set Addition*, Combinatorica **14** (1994), 263-268.

[Be] F. A. Behrend, *On sets of integers which contain no three in arithmetic progression*, Proc. Nat. Acad. Sci. **23** (1946), 331-332.

[BL] V. Bergelson and A. Leibman, *Polynomial extensions of van der Waerden's and Szemerédi's theorems*, J. Amer. Math. Soc. **9** (1996), 725-753.

[Bi] Y. Bilu, *Structure of sets with small sumset*, in Structure Theory of Set Addition, Astérisque **258** (1999), 77-108.

[Bo] N. N. Bogolyubov, *Sur quelques propriétés arithmétiques des presque-périodes*, Ann. Chaire Math. Phys. Kiev, **4** (1939), 185-194.

[Bou] J. Bourgain, *On triples in arithmetic progression*, Geom. Funct. Anal. **9** (1999), 968-984.

[C] M.-C. Chang, *A polynomial bound in Freiman's theorem*, submitted.

[CG] F. R. K. Chung and R. L. Graham, *Quasi-random subsets of* $\mathbb{Z}_N$, J. Combin. Theory Ser. A **61**, 64-86.

[E] G. Elekes, *On the number of sums and products*, Acta Arith. **81** (1997), 365-367.

[ET] P. Erdős and P. Turán, *On some sequences of integers*, J. London Math. Soc. **11** (1936), 261-264.

[ET2] P. Erdős and P. Turán, *On a problem of Sidon in additive number theory and on some related problems*, J. London Math. Soc. **16** (1941), 212-215.

[F1] G. R. Freiman, Foundations of a Structural Theory of Set Addition, (in Russian), Kazan Gos. Ped. Inst., Kazan (1966).

[F2] G. R. Freiman, Foundations of a Structural Theory of Set Addition, Translations of Mathematical Monographs **37**, Amer. Math. Soc., Providence, R. I., USA.

[F3] G. R. Freiman, *Structure theory of set addition*, Astérisque No. 258 (1999), 1-33.

[Fu] H. Furstenberg, *Ergodic behaviour of diagonal measures and a theorem of Szemerédi on arithmetic progressions*, J. Analyse Math. **31** (1977), 204-256.

[FKO] H. Furstenberg, Y. Katznelson and D. Ornstein, *The ergodic theoretical proof of Szemerédi's theorem*, Bull. Amer. Math. Soc. **7** (1982), 527-552.

[G] W. T. Gowers, *A new proof of Szemerédi's theorem*, Geometric and Functional Analysis, to appear.

[Gr] S. W. Graham, B_h *sequences*, in Analytic Number Theory Vol. I (Allerton Park, IL, 1995), 4331-449, Progress in Mathematics **138**, Birkhäuser, Boston MA 1996.

[Gre1] B. J. Green, *On arithmetic structures in dense sets of integers*, submitted.

[Gre2] B. J. Green, *The number of squares and* $B_h[g]$ *sets*, Acta Arith., to appear.

[H-B] D. R. Heath-Brown, *Integer sets containing no arithmetic progressions*, J. London Math. Soc. (2) **35** (1987), 385-394.

[L] B. Lindström, *A remark on B_4-sequences*, Journal of Comb. Th. **7** (1969), 276-277.

[N] M. B. Nathanson, Additive Number Theory: Inverse Problems and the Geometry of Sumsets, Graduate Texts in Mathematics 165, Springer-Verlag 1996.

[PSS] J. Pintz, W. L. Steiger and E. Szemerédi, *On sets of natural numbers whose difference set contains no squares*, J. London Math. Soc. (2) **37** (1988), 219-231.

[R] K. F. Roth, *On certain sets of integers*, J. London Math. Soc. **28** (1953), 245-252.

[Ru1] I. Z. Ruzsa, *Arithmetic progressions and the number of sums*, Periodica Math. Hungar. **25** (1992), 105-111.

[Ru2] I. Z. Ruzsa, *An application of graph theory to additive number theory*, Scientia, Ser. A **3** (1989), 97-109.

[Ru3] I. Z. Ruzsa, *Generalized arithmetic progressions and sumsets*, Acta Math. Hungar. **65** (1994), 379-388.

[Ru4] I. Z. Ruzsa, *An infinite Sidon sequence*, J. Number Theory **68** (1998), 63-71.

[S] A. Sárközy, *On difference sets of sequences of integers I*, Acta Math. ACad. Sci. Hungar. **15** (1984), 205-209.

[Sh] S. Shelah, *Primitive Recursive Bounds for van der Waerden Numbers*, J. Amer. Math. Soc. **1** (1988), 683-697.

[Si] J. Singer, *A theorem in finite projective geometry and some applications to number theory*, Trans. Amer. Math. Soc. **43** (1938), 377-385.

[So] J. Solymosi, *Note on a generalization of Roth's theorem*, preprint.

[Sz1] E. Szemerédi, *On sets of integers containing no four elements in arithmetic progression*, Acta Math. Acad. Sci. Hungar. **20** (1969), 89-104.

[Sz2] E. Szemerédi, *On sets of integers containing no k elements in arithmetic progression*, Acta Arith. **27** (1975), 299-345.

[Sz3] E. Szemerédi, *Integer sets containing no arithmetic progressions*, Acta Math. Hungar. **56** (1990), 155-158.

[ST] E. Szemerédi and W. T. Trotter, *Extremal problems in discrete geometry*, Combinatorica **3** (1983), 381-392.

[V] R. C. Vaughan, The Hardy-Littlewood Method (2nd ed.), Cambridge Tracts in Mathematics 125, CUP 1997.

[W] M. J. Walters, *Combinatorial proofs of the polynomial van der Waerden theorem and the polynomial Hales-Jewett theorem*, J. London Math. Soc. (2) **61** (2000), 1-12.

[We] H. Weyl, *Über die Gleichverteilung von Zahlen mod Eins*, Math. Annalen **77** (1913), 313-352.

[Wo] T. D. Wooley, *Large improvements in Waring's problem*, Ann. of Math. (2) **135** (1992), 131-164.

Current Developments in Mathematics, 2000
Pages 197 – 231

The Topology of Real Algebraic Varieties

János Kollár

Contents

1. Introduction

The aim of these notes is to give a survey of some of the recent results and directions in real algebraic geometry. We are mainly interested in the topological aspects of the theory.

The study of real algebraic varieties is classical, it even predates algebraic geometry over other fields. Despite this, in the last 50 years real algebraic geometry developed to a large extent independently of general algebraic geometry, notwithstanding the efforts of a few people active in both fields. There are even two quite different definitions of what a "real algebraic variety" should be. (See (21) for my conventions which come from algebraic geometry.)

It seems to me that this separation has been detrimental to both disciplines. Real algebraic geometry can gain by utilizing the methods of complex algebraic geometry and algebraic geometry can connect with many other fields through real questions. As an example, let me just mention the theory of hyperbolic PDE-s which leads to very interesting problems about real algebraic hypersurfaces and their deformations.

1. In studying the toplogy of real algebraic varieties, two main directions emerge:

(1.1) (Recognition problem) Which properties of the set of real points can be determined from algebraic geometry over $\mathbb{C}$. The simplest example is the dimension. If X is smooth then the real dimension of $X(\mathbb{R})$ (the set of real points) is the same as the complex dimension of $X(\mathbb{C})$ (the set of complex points), except that $X(\mathbb{R})$ may be empty.

(1.2) (Realization problem) Which topological data can be realized by real algebraic varieties. The answer here is especially interesting if the realization establishes a close connection between topology and algebra. Here the nicest example is the realization of semi–simple Lie groups by real algebraic matrix groups. Unfortunately, such close connection between topology and algebra is rather rare.

The first general result comparing topological properties of the real points with complex analytic properties is due to [**Harnack1876**], who proved the following (cf. [**Shafarevich72**, VII.4]).

THEOREM 2. *Let B be a smooth, projective, real algebraic curve. Then*

$$\#(\textit{connected components of } B(\mathbb{R})) \leq g(B) + 1.$$

Note the important feature of this result that we only have an inequality. Namely, we assert that if the complex curve is simple than the topology of the real points is also simple. In general we can not expect any implication the other way. Indeed, the set of real solutions of $(x^{2d} + y^{2d} = 1)$ is always a circle but the genus of the complex curve goes to infinity.

This leads to the following general question, which is the main topic of these notes.

MAIN PROBLEM 3. *Prove that the complexity of the set of real points $X(\mathbb{R})$ of a real algebraic variety can be bounded by invariants of the complex variety $X(\mathbb{C})$.*

1.1. Topological Methods. The simplest and prettiest theorems of this type are purely topological. Complex conjugation is an involution on $X(\mathbb{C})$ whose fixed point set is $X(\mathbb{R})$. It was noticed by [**Thom65**] and [**Sullivan71**] that general properties of involutions give the following:

THEOREM 4. *For any real algebraic variety X,*

$$\sum_i h^i(X(\mathbb{R}), \mathbb{Z}_2) \leq \sum_i h^i(X(\mathbb{C}), \mathbb{Z}_2), \quad \textit{and}$$
$$\sum_i h^i(X(\mathbb{R}), \mathbb{Z}_2) \equiv \sum_i h^i(X(\mathbb{C}), \mathbb{Z}_2) \mod 2.$$

By considering the degree as a basic invariant, [**Milnor64**] proved the following bound.

THEOREM 5. *Let $X \subset \mathbb{P}^n$ be a real algebraic variety of degree d. Then*

$$\sum_i h^i(X(\mathbb{R}), \mathbb{Z}_2) \leq d(2d-1)^{n-1}.$$

These results are quite strong in dimensions 1 and 2, since in these dimensions the Betti numbers pretty much determine the topology. The situation drastically changes starting with dimension 3, where homotopy theoretic information becomes indispensable.

A detailed survey of the homological study of real algebraic varieties is given in [**Degtyarev-Kharlamov00**].

It is also of interest to study other, more algebraic, invariants of X. One of the natural candidates is the Kodaira dimension of X, denoted by $\kappa(X)$ (22).

There is no way to bound the sum of the Betti numbers in terms of the Kodaira dimension, but one may hope to read off certain properties of $X(\mathbb{R})$ from the Kodaira dimension. Here we concentrate on the simplest case of varieties with Kodaira dimension $-\infty$.

This question is interesting, and largely unsolved, even for complex varieties. Using basic properties of Gromov–Witten invariants, [**Kollár98a**, 4.2.10] proves that the case of negative Kodaira dimension is an essentially topological property of the set of complex points:

THEOREM 6. *Let X be a smooth, complex, projective variety. Then one can decide from the underlying symplectic manifold $(X(\mathbb{C}),\omega)$ whether $\kappa(X) = -\infty$ or not.*

1.2. The Nash Conjecture. In the realization problem the first general result was proved in [**Nash52**] and later improved by [**Tognoli73**]. Their result says that every compact differentiable manifold is algebraic:

THEOREM 7. [**Nash52, Tognoli73**] *Let M^n be a compact differentiable manifold. Then there are real polynomials $f_i(x_1,\dots,x_N)$ such that their common zero set*

$$X(\mathbb{R}) := \{\mathbf{x}\in\mathbb{R}^N : f_i(\mathbf{x}) = 0\ \forall i\} \subset \mathbb{R}^N$$

is smooth and diffeomorphic to M^n.

Nash then went on to ask if this result can be improved by specifying various algebraic properties of the variety X. From (5) we see that X can not have bounded degree, in fact no bounded subset of algebraic varieties can represent all manifolds. Thus it is natural to ask if one can restrict the Kodaira dimension of X. It is easy to see that one can always make the canonical class of X ample:

EXAMPLE 8. Let $X \subset \mathbb{P}^n$ be a smooth real algebraic variety. Let $(x_0:\dots:x_n)$ be coordinates on $\mathbb{P}^n$ and $(s:t)$ coordinates on $\mathbb{P}^1$. Let $Y \subset X\times\mathbb{P}^1$ be the hypersurface defined by the equation

$$s^3(x_0^{2d}+\dots+x_n^{2d}) = t^3(x_0^{2d}+2x_0^{2d}+\dots+(n+1)x_n^{2d}).$$

Then $Y(\mathbb{R}) \to X(\mathbb{R})$ is a diffeomorphism and the canonical class of Y is very ample for $d \gg 1$.

There are, however, no general constructions which produce a variety of lower Kodaira dimension starting with a variety of higher Kodaira dimension, while keeping the dimension unchanged. (For any X, $\kappa(X\times\mathbb{P}^1) = -\infty$, but this does not yield anything interesting topologically.) Thus it is of interest to see if one can realize every differentiable manifold with varieties of small Kodaira dimension. Nash pointed out that the following would be the strongest possible result:

CONJECTURE 9. [**Nash52**, p.421] *Let M^n be a compact, connected, differentiable manifold. Then there is a smooth real algebraic variety X^n such that X is birational to $\mathbb{P}^n$ and $X(\mathbb{R})$ is diffeomorphic to M^n.*

This is obvious for curves. Unbeknownst to Nash, this question has been settled for surfaces much earlier. It is sign of the separation of various branches of mathematics that this seemed not to have been realized for quite a few years.

THEOREM 10. [**Comessatti14**] *Let S be a smooth real algebraic surface. Assume that S is birational to $\mathbb{P}^2$ and $S(\mathbb{R})$ is orientable.*

Then $S(\mathbb{R})$ is either a sphere or a torus.

Several attempts have been made to use the result of [**Comessatti14**] to get higher dimensional counter examples, for instance for threefolds of the form $S^1 \times$ (high genus surface) or for 4–folds which are the product of 2 high genus surfaces. This type of trivial dimension increase frequently works in complex algebraic geometry but it does not seem to work in this example.

[**Benedetti-Marin92**] showed that for every 3–manifold M^3 there is a *singular* real algebraic variety X birational to $\mathbb{P}^3$ such that $X(\mathbb{R})$ is homeomorphic to M^3. [**Akbulut-King91**] and [**Mikhalkin97**] show that the related "topological Nash conjecture" is true.

The 3–dimensional Nash conjecture has been settled recently in a series of papers [**Kollár98b, Kollár99a, Kollár99b, Kollár00a, Kollár00b**]. The first 4 papers provide a negative solution and the last one a positive solution. It turns out that the seemingly innocuous projectivity assumption drastically changes the nature of the problem. This is a completely new phenomenon of 3–fold geometry. The precise statements are the following. (The relevant concepts of 3–manifolds topology are recalled in (17).)

THEOREM 11 (The Nash conjecture fails in dimension 3).
Let X be a smooth, real, projective 3-fold birational to $\mathbb{P}^3$. Assume that $X(\mathbb{R})$ is orientable. Then every connected component of $X(\mathbb{R})$ is among the following:

1. *Seifert fibered,*
2. *connected sum of lens spaces,*
3. *torus bundle over S^1 or doubly covered by a torus bundle over S^1,*
4. *finitely many other possible exceptions, or*
5. *obtained from one of the above 3–manifolds by repeatedly taking connected sum with $\mathbb{RP}^3$ and $S^1 \times S^2$.*

THEOREM 12 (The Nash conjecture holds in dimension 3).
For every compact, connected, differentiable 3–manifold M there is a compact complex manifold X which can be obtained from $\mathbb{P}^3$ by a sequence of smooth, real blow ups and downs such that M is diffeomorphic to $X(\mathbb{R})$.

The two statements are not in contradiction since blowing down is not always possible among projective varieties. Thus we have to work in the slightly larger category of Moishezon manifolds or algebraic spaces, see section 4.

REMARK 13. The assumptions of (11) can be weakened considerably. Namely, (11) also holds if we assume the following much weaker conditions:

1. X has Kodaira dimension $-\infty$ (instead of assuming X to be birational to $\mathbb{P}^3$)
2. $X(\mathbb{R})$ does not contain any of the following types of embedded surfaces:
   (a) 2-sided $\mathbb{RP}^2$;
   (b) 1-sided torus;
   (c) 1-sided Klein bottle with nonorientable neighborhood,

(instead of assuming $X(\mathbb{R})$ to be orientable). In this case $X(\mathbb{R})$ may have several connected components and each satisfies the conclusions of (11). I do not know what happens if some components of $X(\mathbb{R})$ are orientable and some are not.

The proof of (11) establishes a close connection between certain algebraic properties of $X(\mathbb{C})$ and geometric structures of $X(\mathbb{R})$. In one case, which is known to contain only finitely many types of varieties, such a connection has not been proved, and this accounts for the finitely many unknown cases. I believe, however, that there are no exceptions:

CONJECTURE 14. *The cases (11.3–4) do not occur. (To be precise, if a torus bundle occurs then it is also Seifert fibered.)*

It is also of interest to note that the conditions (13.2.a–c) hold for every hyperbolic 3–manifold. Thus we obtain the following. (Again I conjecture that there are no exceptions.)

THEOREM 15. *There are only finiteley many hyperbolic 3-manifolds (orientable or not) among the $X(\mathbb{R})$ where X is a smooth, projective, real algebraic 3–fold with Kodaira dimension $-\infty$.*

A similar result was obtained in all dimensions by Viterbo, using stronger conditions on rational curves. (See [**Kharlamov00**] for a discussion of this and related topics.)

THEOREM 16. [**Viterbo98**] *Let X be a smooth, projective, real algebraic variety of dimension $n \geq 3$. Assume that $H_2(X(\mathbb{C}), \mathbb{Z}) \cong \mathbb{Z}$ and that $X(\mathbb{C})$ is covered by rational curves C_λ such that $[C_\lambda] \in H_2(X(\mathbb{C}), \mathbb{Z})$ is a generator.*

Then $X(\mathbb{R})$ does not carry any metric with negative sectional curvature.

DEFINITION 17. For relatively prime $0 < q < p$ consider the action of $\mathbb{Z}_p$ on $S^3 \sim (|x^2| + |y^2| = 1) \subset \mathbb{C}^2$ given by $(x, y) \mapsto (e^{2\pi i/p}x, e^{2\pi iq/p}y)$. The quotient is a 3–manifold called the *lens space* $L_{p,q}$.

A 3–manifold M is called *Seifert fibered* if there is a C^∞ morphism $f : M \to F$ to a topological surface such that every fiber is a circle. It turns out that M is an S^1-bundle except over finitely many points where the local structure is given by one of the normal forms

$$f_{c,d} : S^1 \times D^2 \to D^2 \quad \text{where} \quad f_{c,d}(u, z) = u^c z^d,$$

$S^1 \subset \mathbb{C}$ is the unit circle with coordinate u, $D^2 \subset \mathbb{C}$ the closed unit disc with coordinate z and c, d are relatively prime integers satisfying $0 < c < d$.

1.3. Moduli of Real Varieties. There has also been lot of work in describing all topological types represented by real algebraic varieties of a specific type. The case of plane curves is one of the Hilbert problems and quite a lot is known about them.

In the surface case, the first significant result was the classification of real cubic surfaces by [**Schläfli1863**]. [**Zeuthen1874, Klein1876**] studied plane curves of degree 4 and their bitangents, which is essentially equivalent to degree 2 Del Pezzo surfaces (though they related them to cubic surfaces only). The first systematic study was done in [**Comessatti13**] and [**Comessatti14**]. These papers presented a breakthrough whose extent has not been appreciated for over 50 years. The monograph [**Segre42**] treats cubic surfaces in detail.

The prototype of classification results that one would like is the following. In some form it has been known to [**Comessatti14**].

THEOREM 18. *Let F_1, F_2 be smooth, projective, real algebraic surfaces which are rational over $\mathbb{R}$. Then F_1 can be deformed to F_2 through a family of smooth, projective, real algebraic surfaces iff*

1. *$F_1(\mathbb{C})$ and $F_2(\mathbb{C})$ are homeomorphic, and*

2. $F_1(\mathbb{R})$ *and* $F_2(\mathbb{R})$ *are homeomorphic.*

The case of surfaces which are rational over $\mathbb{C}$ is quite a bit harder. First of all, there is an easy exception given by the two surfaces

$$F_1 = Q^{3,0} \times \mathbb{P}^1, \quad F_2 = Q^{3,0} \times Q^{3,0}$$

where $Q^{3,0}$ is the empty conic $(x^2+y^2+z^2=0) \subset \mathbb{P}^2$. Thus we have to work with the more precise topological information given by the action of complex conjugation on the set of complex points. It turns out that the analog of (18) holds in all cases.

THEOREM 19. [**Degtyarev-Kharlamov01**] *The deformation types of real algebraic surfaces which are rational over* $\mathbb{C}$ *are described by topological data, namely, by the action of complex conjugation on the set of complex points.*

Similar results have been obtained for other classes of surfaces:

THEOREM 20. *The deformation types of real algebraic surfaces are described by topological data in the following cases:*

1. *(Abelian surfaces) In this case the deformation type is determined by* $F(\mathbb{R})$ *alone. This is equivalent to the easy classification of real curves of genus 2.*
2. *(Hyperelliptic surfaces)* [**Catanese-Frediani00**] *In this case the deformation type is determined by the (algebraic) fundamental group.*
3. *(K3 surfaces)* [**Nikulin79, Kharlamov78**] *In this case the deformation type is determined by the action of complex conjugation on* $H^2(F(\mathbb{C}),\mathbb{Z})$.
4. *(Enriques surfaces)* [**DIK00**] *In this case the deformation type is determined by the actions of complex conjugation and of the fundamental group on* $H^2(F(\mathbb{C}),\mathbb{Z})$ *where* F *is the K3–cover of the Enriques surface.*

The proof of (20.4) is especially long and we end up with over 200 cases.

While these results are very nice, one should not expect that the topology determines the deformation type in many cases. This fails already for plane curves. Nonetheless, one can hope that similar results remain true for special classes of varieties in higher dimensions as well. Aside from the easy case of Abelian varieties, nothing is known.

NOTATION 21. A *real algebraic variety* is an algebraic variety defined by real equations. Sometimes I write $X_{\mathbb{R}}$ instead of X to emphasize that we are over $\mathbb{R}$. This is consistent with the usage in current algebraic geometry books (for instance [**Hartshorne77, Shafarevich72**]). This is, however, different from the definition of [**BCR87**] which essentially considers the germ of a variety along its real points.

If X is a variety over $\mathbb{R}$ then $X(\mathbb{R})$ denotes its set of real points with the Euclidean topology and $X(\mathbb{C})$ denotes its set of complex points as a complex manifold.

I try to systematically distinguish the latter from $X_{\mathbb{C}} := X \times_{\operatorname{Spec}\mathbb{R}} \operatorname{Spec}\mathbb{C}$, which is the complexification of X. One should also keep in mind the distinction between $X_{\mathbb{R}}$ and $X(\mathbb{R})$. The first is a scheme over $\mathbb{R}$, thus among its closed points we find all conjugate pairs of complex points of $X_{\mathbb{C}}$, but $X(\mathbb{R})$ is the collection of the real points only.

$\mathbb{P}^n$ stands for projective n-space (as a variety over $\mathbb{Q}$). Thus $\mathbb{P}^n_{\mathbb{R}}$ is projective n-space as a scheme over $\mathbb{R}$ and $\mathbb{P}^n(\mathbb{R})$ is the set of real points, customarily denoted by $\mathbb{RP}^n$.

When talking about a point, curve or vector bundle etc. on a real variety X, these are, by definiton, objects defined over $\mathbb{R}$. I will always explicitly mention if we are considering a point, curve or vector bundle etc. on $X_{\mathbb{C}}$.

If $Z \subset X_{\mathbb{C}}$ is a subscheme, then $\bar{Z} \subset X_{\mathbb{C}}$ denotes the subscheme defined by conjugate equations. Then $Z + \bar{Z} \subset X_{\mathbb{C}}$ can be defined by real equations and it can be viewed as a subscheme of $X_{\mathbb{R}}$. That is, there is a unique subscheme $Z_{\mathbb{R}} \subset X_{\mathbb{R}}$ such that $Z_{\mathbb{R}} \times_{\operatorname{Spec} \mathbb{R}} \operatorname{Spec} \mathbb{C} = Z + \bar{Z}$.

A property of X (irreducible, normal, smooth etc.) always refers to the scheme theoretic property of X. Thus if X is smooth then it is smooth not just at all points of $X(\mathbb{R})$, but also at all complex points. When talking about irreducibility (normality etc.) of $X_{\mathbb{C}}$, I say X is "geometrically irreducible" or "irreducible over $\mathbb{C}$". In some cases this does not matter (for instance X is normal or smooth iff $X_{\mathbb{C}}$ is) but in other cases the two versions are different (X may be irreducible but not geometrically irreducible).

The logical continuation of this terminology is to say that two irreducible $\mathbb{R}$-schemes X, Y are *birational* iff there is a map $f : X \dashrightarrow Y$ defined over $\mathbb{R}$ which is birational. Similarly, X is rational iff it is birational to $\mathbb{P}^n_{\mathbb{R}}$ over $\mathbb{R}$. Many authors, however, use the word "rational" to mean "geometrically rational". To avoid confusion, I will sometimes use the expressions *rational over* $\mathbb{R}$ and *rational over* $\mathbb{C}$.

Let $X \subset \mathbb{P}^n$ be a variety over $\mathbb{C}$ and $\bar{X}$ the variety defined by conjugate equations. The disjoint union of X and $\bar{X}$ is invariant under conjugation, and so there is a real variety $Y_{\mathbb{R}}$ such that $Y_{\mathbb{C}} \cong X \cup \bar{X}$. Such real varieties are not particularly interesting since the theory of $Y_{\mathbb{R}}$ over $\mathbb{R}$ is equivalent to the theory of X over $\mathbb{C}$. Thus it is reasonable to restrict our attention to real varieties which are geometrically irreducible.

Of course, during a proof we may run into a subvariety of $Y_{\mathbb{R}}$ which is geometrically reducible, and these have to be dealt with appropriately. Thus we can not ignore such varieties completely.

DEFINITION 22 (Kodaira dimension).

Let X be a smooth projective variety and let K_X denote its canonical bundle. That is, K_X is the highest exterior power of the cotangent bundle. The *Kodaira dimension* of X, denoted by $\kappa(X)$ measures the growth rate of the spaces of global sections $H^0(X, K_X^{\otimes m})$. To be precise, we set

$$\kappa(X) = -\infty \Leftrightarrow H^0(X, K_X^{\otimes m}) = 0 \quad \forall m > 0.$$

If $H^0(X, K_X^{\otimes m}) \neq 0$ for some $m > 0$ then, as it turns out, there is a unique integer $0 \leq \kappa(X) \leq \dim X$ such that

$$0 < \limsup \frac{\dim H^0(X, K_X^{\otimes m})}{m^{\kappa(X)}} < \infty.$$

(In fact, one expects that for all nonzero values of $H^0(X, K_X^{\otimes m})$ the above quotient converges to a rational number. This number should be a very basic measure of the variety but even its existence is unknown in dimensions ≥ 4.)

It is also conjectured that $\kappa(X) = -\infty$ iff X is uniruled (that is, there is a generically finite map $\mathbb{P}^1 \times Y \dashrightarrow X$ where $\dim Y = \dim X - 1$). This is known only in dimension ≤ 3.

2. Minimal Models of Real Algebraic Surfaces

This section is a warm up to the proof of (11). We study the topology of real algebraic surfaces, using methods that are applicable in dimension 3. The main result is the following strengthening of (10). (This is not to hard obtain from topological considerations but the present proof shows the way to the 3–dimensional results in sections 3 and 4.)

THEOREM 23. *Let F be a smooth, projective surface over $\mathbb{R}$ such that $F_{\mathbb{C}}$ is rational. Then one of the following holds:*

0. $F(\mathbb{R}) = \emptyset$.
1. $F(\mathbb{R}) \sim S^1 \times S^1$.
2. $F(\mathbb{R}) \sim \#r_1\mathbb{RP}^2 \uplus \cdots \uplus \#r_m\mathbb{RP}^2$ *for some* $r_1, \dots, r_m \geq 0$.

All these cases do occur. (Note that $S^2 = \#0\mathbb{RP}^2$ by convention.)

We start with the first steps of the general minimal model program.

DEFINITION 24. Let X be a variety over a field k. A *1–cycle* on X is a formal linear combination $C = \sum c_iC_i$, where the $C_i \subset X$ are irreducible, reduced and proper curves. A 1–cycle is called *effective* if $c_i \geq 0$ for every i.

Two 1–cycles C, C' are *numerically equivalent* if $(C \cdot D) = (C' \cdot D)$ for every Cartier divisor D on X. 1–cycles with real coefficients modulo numerical equivalence form a vectorspace, denoted by $N_1(X)$. $N_1(X)$ is finite dimensional by the Theorem of the base of Néron–Severi (cf. [**Hartshorne77**, p.447]). Its dimension, denoted by $\rho(X)$, is called the *Picard number* of X.

Effective 1–cycles generate a cone $NE(X) \subset N_1(X)$. Its closure in the Euclidean topology $\overline{NE}(X) \subset N_1(X)$ is called the *cone of curves* of X.

If K_X is Cartier (or at least some multiple of K_X is Cartier) then set

$$\overline{NE}(X)_{K\geq 0} := \{z \in \overline{NE}(X) | (z \cdot K_X) \geq 0\}.$$

Let $V \subset \mathbb{R}^n$ be a closed convex cone. For $v \in V$, a ray $\mathbb{R}^{\geq 0}v \subset V$ is called *extremal* if $u, u' \in V$, $u + u' \in \mathbb{R}^{\geq 0}v$ implies that $u, u' \in \mathbb{R}^{\geq 0}v$. Intuitively: $\mathbb{R}^{\geq 0}v$ is an edge of V.

An extremal ray $\mathbb{R}^{\geq 0}z \subset \overline{NE}(X)$ is called K_X*-negative* if $(z \cdot K_X) < 0$. This does not depend on the choice of z in the ray.

We use the following description, due to [**Mori82**], of the cone of curves of smooth varieties over $\mathbb{C}$. See also [**Kollár96, Kollár-Mori98**] for proofs.

THEOREM 25 (Cone Theorem). *Let X be a smooth projective variety over an algebraically closed field. Then there are rational curves $C_i \subset X$ such that*

$$\overline{NE}(X) = \overline{NE}(X)_{K\geq 0} + \sum \mathbb{R}^{\geq 0}[C_i],$$

and the $\mathbb{R}^{\geq 0}[C_i]$ are K_X-negative extremal rays of $\overline{NE}(X)$.

Let now X be a smooth projective variety over $\mathbb{R}$. If C is a 1–cycle on $X_{\mathbb{C}}$ then $C + \bar{C}$ is a 1–cycle on X, and every 1–cycle on X arises this way, at least with rational or real coefficients. Thus (25) immediately gives:

THEOREM 26 (Cone Theorem over $\mathbb{R}$). *Let X be a smooth projective variety over $\mathbb{R}$. Then there are rational curves $C_i \subset X_{\mathbb{C}}$ such that*

$$\overline{NE}(X) = \overline{NE}(X)_{K\geq 0} + \sum \mathbb{R}^{\geq 0}[C_i + \bar{C}_i],$$

and the $\mathbb{R}^{\geq 0}[C_i + \bar{C}_i]$ are K_X-negative extremal rays of $\overline{NE}(X)$.

Proof. There is one point that we need to be careful about. Namely, it happens frequently that C_i gives an extremal ray but $C_i + \bar{C}_i$ does not. So we have to throw away some of the C_i appearing in (25). □

THEOREM 27 (Contraction theorem). *Let X be a smooth projective variety over $\mathbb{R}$ and $R = \mathbb{R}^{\geq 0}[C_i + \bar{C}_i]$ a K_X-negative extremal ray of $\overline{NE}(X)$. Then there is a unique morphism, called the contraction of R, $\operatorname{cont}_R : X \to Y$ such that*

1. *Y is normal and cont_R has connected fibers.*
2. *An irreducible curve $D \subset X$ is mapped to a point in Y iff $[D] \in R$.*

The above theorems are sufficient to complete the minimal model program for surfaces. The key point is the classification of all contractions. This is stated after a definition.

DEFINITION 28. Let S be a smooth projective surface over a field k. S is called a *Del Pezzo surface* if S is geometrically irreducible and $-K_S$ is ample. It is called *minimal* (over k) if $\rho(S) = 1$.

S, together with a morphism to a smooth curve $f : S \to B$ is called a *conic bundle* if every fiber is isomorphic to a plane conic. A conic bundle is called *minimal* if $\rho(S) = 2$.

The geometric description and meaning of the extremal rays occurring in (26) is given in the next result:

THEOREM 29. *Let F be a smooth projective geometrically irreducible surface over $\mathbb{R}$ and $R = \mathbb{R}^{\geq 0}[C_i + \bar{C}_i] \subset \overline{NE}(F)$ a K_F-negative extremal ray. Then R can be contracted $f : F \to F'$, and we obtain one of the following cases:*

(B) *(Birational) F' is a smooth projective surface over $\mathbb{R}$ and $\rho(F') = \rho(F) - 1$. F is the blow up of F' at a closed point P. We have two cases:*
   (a) *$P \in F'(\mathbb{R})$, or*
   (b) *P is a pair of conjugate points.*

(C) *(Conic bundle) $B := F'$ is a smooth curve, $\rho(F) = 2$ and $F \to B$ is a conic bundle. The fibers $f^{-1}(P) : P \in B(\mathbb{C})$ are smooth, except for an even number of ponts $P_1, \dots, P_{2m} \in B(\mathbb{R})$. $(K_F^2) = 8(1 - g(B)) - 2m$.*

(D) *(Del Pezzo surface) F' is a point, $\rho(F) = 1$, $-K_F$ is ample.*

Proof. The adjunction formula says that $2g(C) - 2 = (C^2) + (C \cdot K_F)$. In our case $(C \cdot K_F) < 0$, so $-2 < (C^2)$. We start with the low values of (C^2).

If $(C^2) = -1$ and $C = \bar{C}$ then the contraction of C in $F_{\mathbb{C}}$ is defined over $\mathbb{R}$, thus F is the blow up of a surface at a real point. (This is Castelnuovo's contraction theorem, cf. [**Hartshorne77**, V.5.7].)

If $(C^2) = -1$, C and $\bar{C}$ are disjoint, then we can contract them simultaneously over $\mathbb{R}$ to obtain $f : F \to F'$ which is an isomorphism near $F(\mathbb{R})$.

If $(C^2) = -1$ and $(C \cdot \bar{C}) = 1$ or $(C^2) = (C \cdot \bar{C}) = 0$, then $C + \bar{C}$ has selfintersection 0. It is relatively easy to see that $C + \bar{C}$ is a fiber of a conic bundle.

Over $\mathbb{C}$ we can contract one of the components of every singular fiber to obtain a minimal ruled surface. The selfintersection number of the canonical class of a minimal ruled surface is $8(1 - g(B))$, and each singular fiber drops this number by 1. We see in (33) that the number of singular fibers is even.

We are left with the cases when the selfintersection of $(C + \bar{C})$ is positive. By an a easy lemma (cf. [**Kollár-Mori98**, 1.21]), this implies that $\overline{NE}(F)$ is 1-dimensional, hence $-K_F$ is ample. □

As a consequence we obtain the minimal model program (MMP for short) for real algebraic surfaces:

THEOREM 30 (MMP for real surfaces). *Let F be a smooth projective geometrically irreducible surface over $\mathbb{R}$. Then there is a sequence of morphisms*

$$F = F_0 \stackrel{f_0}{\to} F_1 \to \cdots F_{m-1} \stackrel{f_{m-1}}{\to} F_m = F^*$$

such that each $f_i : F_i \to F_{i+1}$ is a birational contraction as in (29.B) and F^* satisfies precisely one of the following properties:

(M) *(Minimal model) K_{F^*} is nef. (That is, it has nonnegative intersection number with every curve in F^*.)*

(C) *(Conic bundle) F^* is a conic bundle over a curve $f : F^* \to B$. In particular, $\rho(F^*) = 2$*

(D) *(Del Pezzo surface) $\rho(F^*) = 1$ and $-K_{F^*}$ is ample.*

Proof. We do the steps of (29.B) as long as we can. $\rho(F_{i+1}) = \rho(F_i) - 1$, so eventually we reach $F^* = F_m$ where we can not perform a contraction as in (29.B). If K_{F^*} is nef then we have a minimal model.

If K_{F^*} is not nef, then by (29) we can perform a contraction as in (29.C–D). This gives our last two cases. □

REMARK 31. In the above proofs we had to establish several times that certain line bundles on $F_{\mathbb{C}}$ are defined over $\mathbb{R}$. This is frequently a quite subtle point.

Let X be a scheme over $\mathbb{R}$ and L a line bundle on X. Then $L_{\mathbb{C}}$ is a line bundle on $X_{\mathbb{C}}$ and $L_{\mathbb{C}} \cong \bar{L}_{\mathbb{C}}$. Thus if M is a line bundle on $X_{\mathbb{C}}$ and $M \not\cong \bar{M}$, then M is not the complexification of a real line bundle.

On the other hand, if $M \cong \bar{M}$ then M is the complexification of a real line bundle, provided $X(\mathbb{R}) \neq \emptyset$. The latter condition is necessary. (For instance, consider the curve $C \cong (x^2 + y^2 + z^2) \subset \mathbb{P}^2$ over $\mathbb{R}$ and let M be a degree 1 line bundle on $C_{\mathbb{C}}$. Then $M \cong \bar{M}$ but M is not the complexification of a real line bundle.)

Let X be a scheme over $\mathbb{R}$ and M a line bundle on $X_{\mathbb{C}}$ such that $M \cong \bar{M}$. Then $M^{\otimes 2}$ is always the complexification of a real line bundle.

In is quite straightforward to follow the MMP for real surfaces from the topological point of view. The main results are already in [**Comessatti14**] (though many people found some of his proofs very hard to follow).

NOTATION 32. $M \uplus N$ denotes the disjoint union of M and N. $\uplus rN$ denotes the disjoint union of r copies of N. $M\#N$ denotes the connected sum of two manifolds M and N (which are assumed to have the same dimension). $\#rN$ denotes the connected sum of r copies of N. (By definition, $\#0M = S^{\dim M}$.) $M \sim N$ denotes that M and N are homeomorphic.

One can give a complete topological description of the various contractions in (29):

THEOREM 33. *Let F be a smooth projective geometrically irreducible surface over $\mathbb{R}$ and $R \subset \overline{NE}(F)$ a K_F-negative extremal ray. The following is the topological description of the corresponding contraction:*

(B) *(Birational) F is the blow up of F' at a closed point P. We have two cases:*

(a) *If* $P \in F'(\mathbb{R})$ *then* $F(\mathbb{R}) \sim F'(\mathbb{R})\#\mathbb{RP}^2$.
(b) *If* P *is a pair of conjugate points then* $F(\mathbb{R}) \sim F'(\mathbb{R})$.

(C) *(Conic bundle)* $f : F \to B$ *is a conic bundle with singular fibers* $f^{-1}(P_1), \dots, f^{-1}(P_{2m})$. *Then*

$$F(\mathbb{R}) \sim \uplus m S^2 \uplus N_1 \uplus \cdots \uplus N_b,$$

where b *is less than or equal to the number of connected components of* $B(\mathbb{R})$ *which do not contain any of the points* P_i *and each* N_i *is either a torus or a Klein bottle.*

(D) *(Del Pezzo surface) There are 4 cases:*
(a) *If* $(K_F^2) = 9$ *then* $F(\mathbb{R}) \sim \mathbb{RP}^2$.
(b) *If* $(K_F^2) = 8$ *then* $F(\mathbb{R}) \sim S^2$.
(c) *If* $(K_F^2) = 2$ *then* $F(\mathbb{R}) \sim \uplus 4S^2$.
(d) *If* $(K_F^2) = 1$ *then* $F(\mathbb{R}) \sim \mathbb{RP}^2 \uplus 4S^2$.

Proof. Blowing up replaces a point with all tangent directions through that point. So we remove a disc and put in an interval bundle over S^1 whose boundary is connected. This is a Möbius strip and so $F(\mathbb{R}) \sim F'(\mathbb{R})\#\mathbb{RP}^2$.

In the conic bundle case, let $M \sim S^1$ be a connected component of $B(\mathbb{R})$. If none of the P_i lie on M then $F(\mathbb{R}) \to B(\mathbb{R})$ is an S^1-bundle over M, this gives either a torus, a Klein bottle or the empty set. If k of the points $P_1, \dots, P_k \in M \sim S^1$ correspond to singular fibers then, after reindexing, they divide M into k intervals $[P_i, P_{i+1}]$ (subscript $\mod k$). $F(\mathbb{R})$ is alternatingly empty or a copy of S^2 over the intervals. Thus k is even.

In the Del Pezzo case, let us prove first that the value of (K_F^2) is one of those listed.

These cases appear in the proof of (30) in 3 different ways:

1. $(C^2) = -1$ and $r := (C \cdot \bar{C}) \geq 2$,
2. $(C^2) = 0$ and $r := (C \cdot \bar{C}) \geq 1$, or
3. $(C^2) > 0$.

In the first case

$$((C + \bar{C})^2) = -2 + 2(C \cdot \bar{C}) = 2r - 2 > 0.$$

By an a easy lemma (cf. [**Kollár-Mori98**, 1.21]), this implies that $\overline{NE}(F)$ is 1-dimensional, hence $-K_F \equiv a(C + \bar{C})$ for some $a > 0$. $(-K_F \cdot (C + \bar{C})) = 2$, thus

$$(C + \bar{C}) \equiv (1 - r)K_F \quad \text{and} \quad 2r - 2 = (1 - r)^2 (K_F^2).$$

This gives the possibilities $r = 2, (K_F^2) = 2$ or $r = 3, (K_F^2) = 1$.

If $(C^2) = 0$ and $r := (C \cdot \bar{C}) \geq 1$ then a computation as above gives that $8 = r(K_F^2)$, which allows too many cases. It is better to consider this geometrically.

By (25), C is a fiber of a $\mathbb{P}^1$-bundle $g : F_{\mathbb{C}} \to D$ over $\mathbb{C}$ and $\bar{C}$ is a (possibly multiple) section of g. Thus D is rational. By the classification of minimal ruled surfaces, either $F_{\mathbb{C}} \cong \mathbb{P}^1 \times \mathbb{P}^1$ and we are done by (34), or g has a unique section E with negative selfintersection. E is then defined over $\mathbb{R}$, thus $\rho(F) = 2$, a contradiction.

We are left with the case when $(C^2) > 0$. This is actually mostly a complex question and the theory gives us that $F_{\mathbb{C}} \cong \mathbb{CP}^2$. Let L be a line in $F_{\mathbb{C}} \cong \mathbb{CP}^2$. Then $\bar{L}$ is another line and L and $\bar{L}$ intersect in a unique point, which is therefore real. We can get another real point, and so also a real line. Thus $\mathcal{O}_F(1)$ is defined over $\mathbb{R}$ and $F \cong \mathbb{RP}^2$.

We are also done with (33.D.a–b). The cases $(K_F^2) = 2, 1$ are considerably harder. They follow from (92) and (95). □

In the next lemma, set

$$Q^{p,q} := (x_1^2 + \cdots + x_p^2 = x_{p+1}^2 + \cdots + x_{p+q}^2) \subset \mathbb{P}^{p+q-1}.$$

LEMMA 34. *Let F be a smooth projective surface over $\mathbb{R}$ such that $F_{\mathbb{C}} \cong \mathbb{P}^1 \times \mathbb{P}^1$. Then one of the following holds:*

1. $F \cong Q^{2,2} \cong Q^{2,1} \times Q^{2,1}$, $\rho(F) = 2$ *and* $F(\mathbb{R}) \sim S^1 \times S^1$,
2. $F \cong Q^{3,1}$, $\rho(F) = 1$ *and* $F(\mathbb{R}) \sim S^2$,
3. $F \cong Q^{4,0} \cong Q^{3,0} \times Q^{3,0}$, $\rho(F) = 2$ *and* $F(\mathbb{R}) = \emptyset$,
4. $F \cong Q^{3,0} \times \mathbb{P}^1$, $\rho(F) = 2$ *and* $F(\mathbb{R}) = \emptyset$.

Proof. Let $C \subset F_{\mathbb{C}}$ be one of the rulings. Then $\bar{C}$ is another ruling, thus either $(C \cdot \bar{C}) = 0$ or $(C \cdot \bar{C}) = 1$.

If $(C \cdot \bar{C}) = 0$ then the linear system $|C + \bar{C}|$ is defined over $\mathbb{R}$ and maps F onto a conic. Similarly for the other rulings, thus F is the product of two conics. All 3 possibilities are listed.

If $(C \cdot \bar{C}) = 1$ then $\mathcal{O}_F(C + \bar{C})$ is a line bundle on F which is of type $\mathcal{O}_{F_{\mathbb{C}}}(1,1)$ over $\mathbb{C}$. Thus its global sections embed F as a quadric. $Q^{3,1}$ is the only quadric not yet accounted for. □

Using (33) it is easy to determine which 2-manifolds occur as $F(\mathbb{R})$ for geometrically rational surfaces F.

35 (Proof of (23)). Apply the MMP over $\mathbb{R}$ to get $F = F_1 \to F_2 \to \cdots$. We prove the theorem by induction on the number of blow ups in the sequence. If $F_i \to F_{i+1}$ is the inverse of the blowing up of a real point, then $F_i(\mathbb{R}) \sim F_{i+1}(\mathbb{R}) \# \mathbb{RP}^2$. If $F_i \to F_{i+1}$ is the inverse of the blowing up of a pair of conjugate points, then $F_i(\mathbb{R}) \sim F_{i+1}(\mathbb{R})$. The induction works since $(S^1 \times S^1) \# \mathbb{RP}^2 \sim \# 3\mathbb{RP}^2$.

Thus we are reduced to one of the following two cases:

1. F has a conic bundle structure $F \to B$, or
2. F is Del Pezzo and $\rho(F) = 1$.

In the first case, $B_{\mathbb{C}} \cong \mathbb{CP}^1$ since $F_{\mathbb{C}}$ is rational. Thus either $B(\mathbb{R}) = \emptyset$ and so $F(\mathbb{R}) = \emptyset$, or $B \cong \mathbb{RP}^1$. Thus $F(\mathbb{R})$ is the torus, the Klein bottle or the empty set if there are no singular fibers and $F(\mathbb{R}) \sim \uplus m S^2$ if there are $2m > 0$ singular fibers by (33.C).

In the second case we use (33.D). □

REMARK 36. The hardest part of the proof of (33) is the still unfinished business of identifying the topology of Del Pezzo surfaces with $(K^2) = 1, 2$. It is very easy to establish that these cases give only finitely many topological types. Thus if we want to know the above theorems up to finitely many possible exceptions, then we are already done.

A similar situation accounts for the finitely many exceptions in (11). In 3–dimensions I do not see how to settle the missing cases.

3. The Minimal Model Program for Real 3–folds

In the surface case, the contraction of an extremal ray always results in a smooth variety. This is unfortunately not the case in higher dimensions. One of the main conceptual developments in higher dimensional birational geometry is the realization that we have to live with singularities. The smallest class that we need is called *terminal singularities.* A precise definition is given later (57), here let us accept that this class exists and go on stating the minimal model program.

Very roughly, the contractions of extremal rays can be classified into 3 types.

DEFINITION 37. Let $g : X \to Y$ be an extremal contraction.

We say that g is of *fiber type* if $\dim Y < \dim X$.

We say that g is a *divisorial* contraction if the exceptional set $\mathrm{Ex}(g)$ is the support of Cartier divisor. In this case $\mathrm{Ex}(g)$ is irreducible over K.

We say that g is a *small* contraction if $\dim \mathrm{Ex}(g) \leq \dim X - 2$.

REMARK 38. A fiber type contraction is an end of a minimal model program. In effect we reduced the study of X to the study of Y and the fibers of g. Notethat $\dim Y = 0$ is allowed, in which case we may not have achieved much. However, in this case $-K_X$ is ample, thus we need to develop a theory of these rather special class of varieties.

A divisorial contraction produces a new variety Y of the same dimension. It turns out that the singularities are usually not worse than for X and we can continue the program with Y. The Picard number drops by 1, so this step can not be repeated forever.

A small contraction produces a very badly singular variety Y, and something radically new needs to be done. This operation, called a flip, is defined next.

DEFINITION 39. Let $f : X \to Y$ be a small K_X-negative extremal contraction. A variety X^+ together with a proper birational morphism $f^+ : X^+ \to Y$ is called a *flip* of f if

1. K_{X^+} is $\mathbb{Q}$-Cartier,
2. K_{X^+} is f^+-ample, and
3. the exceptional set $\mathrm{Ex}(f^+)$ has codimension at least two in X^+.

By a slight abuse of terminology, the rational map $\phi : X \dashrightarrow X^+$ is also called a flip. A flip gives the following diagram:

$$\begin{array}{ccccc} & X & \overset{\phi}{\dashrightarrow} & X^+ & \\ -K_X \text{ is } f\text{-ample} & & \searrow \quad \swarrow & & K_{X^+} \text{ is } f^+\text{-ample} \\ & & Y & & \end{array}$$

It is not hard to see that a flip is unique and the main question is its existence.

We are ready to state the 3-dimensional MMP over an arbitrary field:

THEOREM 40 (MMP over K). *Let X be a smooth projective 3-fold defined over a field K (of characteristic zero). Then there is a sequence*

$$X = X_0 \overset{f_0}{\dashrightarrow} X_1 \dashrightarrow \cdots \dashrightarrow X_i \overset{f_i}{\dashrightarrow} X_{i+1} \dashrightarrow \cdots \overset{f_{n-1}}{\dashrightarrow} X_n =: X^*$$

with the following properties

1. *Each X_i is a projective 3-fold over K with terminal singularities (57).*

2. *Each f_i is either a K_X-negative divisorial extremal contraction or the flip of a K_X-negative small extremal contraction.*
3. *One of the following holds for X^*:*
   (a) *K_{X^*} is nef (that is $(C \cdot K_{X^*}) \geq 0$ for any curve $C \subset X^*$), or*
   (b) *there is a fiber type extremal contraction $X^* \to Z$.*

The most important result about the finer workings of the MMP we need is the classification of terminal 3-fold singularities over nonclosed fields, established in [**Kollár98b**], building on the earlier complex classification. First let us fix our notation.

NOTATION 41. For a field $K = \mathbb{R}$ or $K = \mathbb{C}$ let $K\{x_1, \dots, x_n\}$ denote the ring of those formal power series which converge in some neighborhood of the origin.

For a power series F, F_d denotes the degree d homogeneous part. The *multiplicity*, denoted by $\operatorname{mult}_0 F$, is the smallest d such that $F_d \neq 0$. If we write a power series as $F_{\geq d}$ then it is assumed that its multiplicity is at least d.

For $F \in \mathbb{R}\{x_1, \dots, x_n\}$ let $(F = 0)$ denote the germ of its zero set in $\mathbb{C}^n$ with its natural real structure. I always think of it as a complex analytic germ with a real structure and not just as a real analytic germ in $\mathbb{R}^n$.

$(F = 0)/\frac{1}{n}(a, b, c, d)$ means the quotient of the hypersurface singularity $(F = 0)$ by the group action $(x, y, z, t) \mapsto (\epsilon^a x, \epsilon^b y, \epsilon^c z, \epsilon^d t)$ where ϵ is a primitive nth root of unity. In all of our cases the quotient has a natural real structure.

THEOREM 42. [**Kollár98b**] *Let X be a real algebraic or analytic 3-fold and $0 \in X(\mathbb{R})$ a real point. Then X has a terminal singularity (57) at 0 iff a neighborhood of $0 \in X$ is real analytically equivalent to one of the following:*

name	*equation*
cA_0	$(t = 0)$
cA_1	$(x^2 + y^2 \pm z^2 \pm t^m = 0)$
$cA^+_{>1}$	$(x^2 + y^2 + g_{\geq 3}(z, t) = 0)$
$cA^-_{>1}$	$(x^2 - y^2 + g_{\geq 3}(z, t) = 0)$
cD_4	$(x^2 + f_{\geq 3}(y, z, t) = 0)$, *where* $f_3 \neq l_1^2 l_2$ *for linear forms* l_i
$cD_{>4}$	$(x^2 + y^2 z + f_{\geq 4}(y, z, t) = 0)$,
cE_6	$(x^2 + y^3 + y g_{\geq 3}(z, t) + h_{\geq 4}(z, t) = 0)$, *where* $h_4 \neq 0$
cE_7	$(x^2 + y^3 + y g_{\geq 3}(z, t) + h_{\geq 5}(z, t) = 0)$, *where* $g_3 \neq 0$
cE_8	$(x^2 + y^3 + y g_{\geq 4}(z, t) + h_{\geq 5}(z, t) = 0)$, *where* $h_5 \neq 0$
cA_0/n	$(t = 0)/\frac{1}{n}(r, -r, 1, 0)$ *where* $n \geq 2$ *and* $(n, r) = 1$
$cA_1/2$	$(x^2 + y^2 \pm z^n \pm t^m = 0)/\frac{1}{2}(1, 1, 1, 0)$ *where* $\min\{n, m\} = 2$
$cA^+_{>1}/2$	$(x^2 + y^2 + f_{\geq 3}(z, t) = 0)/\frac{1}{2}(1, 1, 1, 0)$
$cA^-_{>1}/2$	$(x^2 - y^2 + f_{\geq 3}(z, t) = 0)/\frac{1}{2}(1, 1, 1, 0)$
cA/n	$(xy + f(z, t) = 0)/\frac{1}{n}(r, -r, 1, 0)$ *where* $n \geq 3$ *and* $(n, r) = 1$
$cAx/2$	$(x^2 \pm y^2 + f_{\geq 4}(z, t) = 0)/\frac{1}{2}(0, 1, 1, 1)$
$cAx/4$	$(x^2 \pm y^2 + f_{\geq 2}(z, t) = 0)/\frac{1}{4}(1, 3, 1, 2)$
$cD/2$	$(x^2 + f_{\geq 3}(y, z, t) = 0)/\frac{1}{2}(1, 1, 0, 1)$
$cD/3$	$(x^2 + f_{\geq 3}(y, z, t) = 0)/\frac{1}{3}(0, 1, 1, 2)$ *where* $f_3(0, 0, 1) \neq 0$
$cE/2$	$(x^2 + y^3 + f_{\geq 4}(y, z, t) = 0)/\frac{1}{2}(1, 0, 1, 1)$

In order to understand the birational steps of the real MMP, let us first look at some examples of contractions in 3–manifold theory.

43 (The topology of contractions of 3–manifolds).

Let M be a compact 3–manifold and $S \subset M$ a closed 2–dimensional subcomplex. Let $\pi : M \to N = N(M, S)$ denote the map that collapses S to a point and is a homeomorphism outside S. Thus N may have a single nonmanifold point at $P := f(S)$. We would like to know how to reconstruct M from knowing N and S.

(43.1) For every 3–manifold M there is a 2–dimensional subcomplex $S \subset M$ such that $N(M, S)$ is a bouquet of spheres. We can even choose S to be a union of spheres intersecting transversally.

(43.2) For every 3–manifold M and every $g \gg 1$ there is a surface $S \subset M$ of genus g such that $N(M, S)$ is a union of two copies of T_g where T_g is the connected sum of g solid tori with the boundary collapsed. Thus the precise knowledge of N and of S tells us nothing about M.

(43.3) Assume that S has only finitely many nonmanifold points and $N(M, S)$ is the bouquet of 3-manifolds. Then a rather straightforward topological argument (cf. [**Kollár99a**, 5.7]) shows that M can be obtained from N by repeated application of the following operations:

1. separating the components of N,
2. taking connected sums of connected components,
3. taking connected sum with $S^1 \times S^2$,
4. taking connected sum with $S^1 \tilde{\times} S^2$,
5. taking connected sum with $\mathbb{RP}^3$.

If we think of $M \to N$ as the real part of a morphism of 3–folds with exceptional set E then $S = E(\mathbb{R})$ has only finitely many nonmanifold points if E is a normal surface. Unfortunately, there does not seem to be any method to factor any birational morphism into steps which contract only one normal surface at a time. Thus we end up deriving very precise descriptions of the steps of the MMP and reading off normality of the exceptional surface at the end. In some cases normality fails, but then we have good enough control of the situation to see that everyhting is nice topologically.

From the technical point of view the key point is to control the singularities that occur during a minimal model program.

44 (Which singularities are complicated?).

The singularities can cause trouble in two ways.

First, a singularity can be complicated algebraically. It is by now well understood that the most intricate features of the 3–dimensional MMP occur only in the presence of quotient singularities. Thus if we can exclude these, the MMP is much more manageable. Hence we should exclude the last 10 types of singularitites on the list (42).

Second, a singularity can be complicated topologically. An isolated singularity on a real algebraic threefold is locallly a cone over a surface, called the *link* of the singularity. As long as this surface is the union of spheres, we are dealing essentially with a manifold and the usual methods of manifold theory work well. The situation becomes quite complicated if the link is a high genus surface. In this case the real variety is very far from a manifold and it becomes nearly impossible to follow the minimal model program topologically.

Singularities which are very similar algebraically can have very different topological behaviour. For instance, the link of

$$x^2 + y^2 + \prod_{i=1}^{n}(z - a_i t) = 0$$

is a union of n spheres and the link of

$$x^2 - y^2 + \prod_{i=1}^{n}(z - a_i t) = 0$$

is a surface of genus n. These two singularities are complex analytically equivalent.

Assume that we start with a smooth real algebraic threefold X and we run the minimal model program. Let X_i be one of the intermediate steps. One can think of X as a resolution of singularities of X_i. This is not completely correct since the flips in the program are not everywhere defined, thus $X \dashrightarrow X_i$ is not always a morphism. This however effects only finitely many curves and for our current purposes it does not matter much. Thus, at least philosophycally we are asking the following:

QUESTION 45. Which of the 3–dimensional, real, terminal singularities admit a resolution with orientable real part.

REMARK 46. One can ask similar questions about different topological properties. For instance, the proof of (13) requires the study of those real, terminal singularities that admit a resolution whose real part satisfies the conditions (13.2.a–c).

One can also ask for stronger conditions. For instance it is of interest to study those resolutions where the real part has vanishing homology groups.

In order to understand (45) let us study the analogous surface question first.

47 (Resolutions of surface singularities).

The topological study of resolutions of surface singularities is made quite easy by the fact that every surface singularity has a unique minimal resolution. That is, every other resolution is obtained from the minimal one by blowing up points (which is connected sum by $\mathbb{RP}^2$ at the topological level). Thus a surface singularity has an orientable resolution iff the minimal resolution is orientable. Thus we have a concrete model to check.

48 (Resolutions of threefold singularities).

The topological study of resolutions of threefold singularities is more complicated since we do not have a unique minimal resolution. To be precise, starting with a threefold X, there is a nearly unique partial resolution $X^t \to X$ where X^t has terminal singularities. The resolutions of X^t are, however, far from being unique. This does not help us at all, since we start with a threefold with terminal singularities and the first step $X^t \to X$ is the identity. Thus we need a way to compare two resolutions. One way is the following.

DEFINITION 49. Let X be a variety, $f : Y \to X$ a proper birational morphism and $E \subset Y$ an exceptional divisor. Let $g : Z \to X$ be another proper birational morphism. The composite $h := g^{-1} \circ f : Y \dashrightarrow Z$ is birational and the image $h(E)$ is either a subset of codimension at least 2, or it is a subset of codimension 1 and $E \dashrightarrow h(E)$ is a birational map. In the latter case we say that E *appears on* Z and that E and $h(E)$ are *equivalent exceptional divisors*.

Assume now that everything is defined over $\mathbb{R}$ and that $E(\mathbb{R})$ represents a nontrivial homology class in $Y(\mathbb{R})$. Then there is a loop $L = \mathrm{im}[S^1 \to Y(\mathbb{R})]$ which produces a nonzero intersection number with $E(\mathbb{R})$. The loop can be moved away from the codimension 2 set where h is not defined, thus $h(E(\mathbb{R}))$ and $h(L)$ have nonzero intersection number again. (We have to be a little more careful if h collapses some divisors to curves or points.) Thus we may be able to carry information about the homology of one resolution to the homology of another resolution.

This is of course useless if every exceptional divisor on Y is collapsed by h. Thus we are led to the following:

QUESTION 50. Which divisors appear on every resolution?

Let us illustrate this first with the case of smooth points. Here the simplest resolution is the identity, thus we should ask about all nontrivial birational maps $f : Y \to X$.

EXAMPLE 51 (Blow ups of smooth varieties). If $\dim X = 2$ and f^{-1} is not defined at a point $P \in X$ then f factors through the blow up $B_P X$. Thus we have a unique nontrivial minimal blow up.

If $\dim X = 3$ then we have many more choices. For instance, we can blow up a point P or we can blow up any smooth curve through P. These already give infinitely many different choices but they correspond to just 2 different types.

Thus we can still hope to get a complete list of proper birational morphisms $g_\lambda : Z_\lambda \to X$ such that every proper birational morphism $f : Y \to X$ factors through at least one of them as

$$f : Y \xrightarrow{h} Z_\lambda \xrightarrow{g_\lambda} X.$$

As a first attempt, we should restrict ourself to the smooth case. As we saw, the list of $Z_\lambda \to X$ should include the blow ups of points and smooth curves.

There are, however, further possibilities. For instance, choose local coordinates (x, y, z) at P and blow up a curve $P \in C \subset X$ with local equations

$$(z = x^2 \pm y^2 + \text{(higher degree terms)} = 0).$$

We get $X_1 \to X$ where X_1 has a unique singular point. Next we blow up the singular point. We get $X_2 \to X_1$ with exceptional divisor $F \subset X_2$. It turns out that X_2 is smooth. Let $E \subset B_P X$ be the exceptional divisor of the ordinary blow up. By explicit computation we see that E and F are equivalent. $E \cong \mathbb{P}^2$ and F is a quadric, so they are not isomorphic. It is easy to see that X_2 does not dominate any of the blow ups considered first.

There are many more such examples, and we do not even have a conjecture about what a complete list $Z_\lambda \to X$ should look like. It is quite probable that the list is enormous, nearly as big as the list of all birational morphisms $Z \to X$. Thus the list is unlikely to be useful.

There are two related ideas that seem to work better.

52 (Last steps of the MMP).

Given a variety X, we may ask for a list of all divisorial contractions $Z_\lambda \to X$. This would give some information on all resolutions of X since the MMP starting with any resolution would go through one of the $Z_\lambda \to X$. This is a quite interesting question. Unfortunately, the answer is not known for most terminal singularities.

A list like this should be useful in many other questions and finding it is one of the interesting open problems of 3–dimensional birational geometry.

The case when X is smooth was solved very recently by [**Kawakita00**].

A second, less ambitious option is to ask for a list of exceptional divisors that appear on every resolution.

53 ("Minimal" exceptional divisors).

The right formulation of the question is the following. Given a variety X, find all exceptional divisors E_i which appear on every resolution $f : Y \to X$ whose exceptional set is a divisor.

(Many of the singularities on the list (42) admit a resolution whose exceptional set is a curve. In the MMP, this problem is usually dealt with by requiring all singularities to be $\mathbb{Q}$-factorial. This is a technical condition which takes care of the problem completely. It has no effect on the general framework of the proof, so we ignore it for the rest of these notes.)

The following is a prototypical result in this direction.

THEOREM 54. [**Sun95**] *Let $f : Y \to X$ be a birational morphism between smooth threefolds.*

1. *If f^{-1} is not defined along a curve $C \subset X$ then the exceptional divisor of $B_C X \to X$ appears on Y.*
2. *If f^{-1} is not defined at finitely many points $P_i \in X$ then the exceptional divisor of $B_{P_i} X \to X$ appears on Y.*

For toric varieties a complete description of these divisors is given in [**Bouvier98**].

Our aim is to develop a version of this result for the real terminal singularities occurring in (42). To do this, we have to have a way of measuring how soon exceptional divisors appear on blow ups.

55 (The hierarchy of exceptional divisors).

Let X be any normal variety and assume that its canonical class K_X is a Cartier divsior, or at least some multiple of K_X is Cartier. This holds, for instance, for all hypersurface singularities or for all quotients of hypersurface singularities. These cover all the cases in (42).

Let $f : Y \to X$ be any proper birational map. Then we can write f^*K_X as

$$f^*K_X = K_Y + \sum_i a(E_i, X) E_i$$

where the summation runs through all exceptional divisors. The (rational) number $a(E_i, X)$ is called the *discrepancy* of E_i with respect to X. In practice we want to look at the discrepancies of exceptional divisors on all proper birational maps $Y \to X$. It is easy to see that equivalent divisors have the same discrepancy.

EXAMPLE 56. Let $f : Y \to X$ be a birational morphism between smooth varieties. Then $a(E, X)$ is the order of vanishing of the Jacobian of f along E. Thus $a(E, X) \geq 1$ for every exceptional divisor. It is also not hard to see that $a(E, X) = 1$ iff $f(E) \subset X$ has codimension 2 and E is equivalent to the exceptional divisor of the blow up of $f(E)$.

DEFINITION 57. We say that X has terminal singularities if the discrepancy of every exceptional divisor is positive.

In the terminal case, there is a useful sufficient condition which guarantees that an exceptional divisor appears on every resolution.

PROPOSITION 58. *Let X be a variety with terminal singularities and E an exceptional divisor*

1. *If $a(E,X) < 1$ then E appears on every resolution of X (whose exceptional set is a divisor).*
2. *If $a(E,X) = 1$ then*
   (a) *either E appears on every resolution of X (whose exceptional set is a divisor), or*
   (b) *E is the exceptional divisor of the blow up of a codimension 2 subvariety $V \subset X$.*

Proof. Let $g : Z \to X$ be any resolution of singularities. $h : g^{-1} \circ f : Y \dashrightarrow Z$ is a rational map. We are done if E appears on Z. Otherwise, E is an exceptional divisor of h and we have

$$\begin{aligned} f^*K_X &= h^*K_Z + \sum_i a(E_i, Y)E_i \\ &= h^*(g^*K_X) + h^*(\sum_j a(F_j, X)F_j) + \sum_i a(E_i, Y)E_i, \end{aligned}$$

where E_i are the exceptional divisors of h and F_j the exceptional divisors of g. E appears among the E_i and its discrepancy is at least 1 by (56). E also may appear in the effective divsior $h^*(\sum_j a(F_j, X)F_j)$, necessarily with positive coefficient. This already shows (1).

If $a(E,X) = 1$ then $h(E)$ is not a subset of any exceptional divisor of g and E corresponds to the blow up of a codimension 2 subvariety $V \subset Y$ by (56). Thus E also corresponds to the blow up of the codimension 2 subvariety $g(V) \subset X$. (Here we use that the excptional set has pure codimension 1.) □

Now our task is relatively straightforward:

59 (Main computation).

For any of the singularities in (42), [**Kollár99a**] finds all exceptional divisors with discrepancy ≤ 1 and computes if the presence of these divisors contradicts the orientability of the real part or not.

The algebraic part is closely related to the earlier works of [**Markushevich96**] and [**Hayakawa99**].

For most cases this is rather easy to do and one can exclude many singularities this way. A number of singularities are, however, not excluded by divisors of discrepancy ≤ 1. Then one needs secondary computations of divisors with discrepancy ≤ 2 and in one case one even needs to understand divisors with discrepancy ≤ 3.

In these cases one has to take the real structure carefully into consideration. Also, the divisors with discrepancy > 1 do not need to appear on every resolution, so additional considerations are needed.

At the end we get the following:

THEOREM 60. *Let X be a smooth, projective, real algebraic 3-fold and assume that $X(\mathbb{R})$ satisfies the 3 conditions (13.2.a–c).*

Let X_i be any of the intermediate steps of the MMP over $\mathbb{R}$ starting with X and $0 \in X_i(\mathbb{R})$ a real point. Then a neighborhood of $0 \in X_i$ is real analytically equivalent to one of the following standard forms:

1. *(cA_0) Smooth point.*
2. *($cA^+_{>0}$) $(x^2+y^2+g_{\geq 2}(z,t)=0)$, where g is not everywhere negative in a punctured neighborhood of* 0.
3. *(cE_6) $(x^2+y^3+(z^2+t^2)^2+yg_{\geq 4}(z,t)+g_{\geq 6}(z,t)=0)$.*

The above points of type cE_6 form a codimension 6 family in the space of all cE_6 singularities. They all do occur. Points of type $cA^+_{>0}$ occur for many choices of g.

REMARK 61. It should be emphasized that our computations limit only the real singular points on the X_i. There are no restrictions on the complex singularities, except that they appear in conjugate pairs. Thus the full complexity of the 3–dimesional MMP appears but it does not influence the real part much.

It is also very convenient to have a quick way of determining the orientability of the real part using algebraic data.

62 (Degrees of line bundles over $\mathbb{R}$ and $\mathbb{C}$).

Let B be a smooth projective curve over $\mathbb{C}$ and L a line bundle on B. Let s be a nonzero meromorphic section of L. The number of zeros minus the number of poles of s on B (counted with multiplicity) is called the *degree of* L. Let Y be a smooth projective variety over $\mathbb{C}$ and L a line bundle on Y. For any curve $B\subset Y$ the degree of $L|_B$ is defined. It is also called the *intersection number* of B and L and denoted by $(B\cdot L)$.

Let $A\sim S^1$ be a compact 1–dimensional manifold and L a real line bundle on M. Let s be a nonzero section of L. The number of zeros of s on A (counted with multiplicity) makes sense only mod 2. If M is a compact manifold and L a real line bundle on M then for any 1-cycle $A\subset M$ we obtain the $\mathbb{Z}_2$-valued *intersection number* of A and L. It is denoted by $(A\cap L)$. (To be precise, I should write $(A\cap w_1(L))$ where $w_1(L)$ stands for the first Stiefel–Whitney class of L. This is a class in $H^1(X(\mathbb{R}),\mathbb{Z}_2)$ analogous to the first Chern class of a complex line bundle, cf. [**Milnor-Stasheff74**, Sec.4].)

Let now X be a smooth projective variety over $\mathbb{R}$, $C\subset X$ a curve and L a line bundle on X. We obtain two numbers:

$$(C(\mathbb{R})\cap L(\mathbb{R}))\quad\text{and}\quad(C_{\mathbb{C}}\cdot L_{\mathbb{C}}).$$

What is the relationship between them?

To answer this, take a real meromorphic section s of L which has only finitely many zeros and poles on C. When we count the real zeros and poles of s on $C(\mathbb{R})$, we miss the complex zeros and poles of s on $C_{\mathbb{C}}$. Since s is real, the complex zeros and poles come in conjugate pairs. Thus we conclude that

$$(C(\mathbb{R})\cap L(\mathbb{R}))\equiv(C_{\mathbb{C}}\cdot L_{\mathbb{C}})\mod 2,$$

which is best possible since the left hand side is defined only mod 2 anyhow.

63 (Orientability of $X(\mathbb{R})$ and the canonical class).

Let M be a differentiable manifold, $0\in M$ a point and $x_1,\dots,x_n$ local coordinates. A *local orientation* of M at 0 is a choice of an n-form $f(x)dx_1\wedge\cdots\wedge dx_n$ with $f(0)\neq 0$ up to multiplication by a positive function. An *orientation* of M is a nowhere zero global n-form on M, up to multiplication by a positive function. n-forms are sections of the real line bundle $\det T^*_M$. If $S^1\sim A\subset M$ is a loop then one can choose a consistent orientation of M along $A \Leftrightarrow \det T_M^*$ has a nowhere zero section along $A \Leftrightarrow (\det T_M^* \cap A) = 0$.

If X is a smooth variety over $\mathbb{R}$ then n-forms appear as sections of the canonical line bundle. This proves that

$$\det T_{X(\mathbb{R})}^* \cong K_X(\mathbb{R}).$$

In many cases this gives a way to decide if $X(\mathbb{R})$ is orientable or not.

COROLLARY 64. *Let X be a smooth projective variety over $\mathbb{R}$. Assume that there is a curve $C \subset X$ such that $(C \cdot K_X)$ is odd. Then $X(\mathbb{R})$ is not orientable.*

Proof. We have proved above that

$$(C(\mathbb{R}) \cap \det T_{X(\mathbb{R})}^*) \equiv (C \cdot K_X) \equiv 1 \mod 2.$$

$C(\mathbb{R})$ may have several components, but along one of them $\det T_{X(\mathbb{R})}^*$ has odd degree, so we can not choose a consistent orientation along that component. □

If $(C \cdot K_X)$ is even, then it can happen that $X(\mathbb{R})$ is not orientable along an even number of components of $C(\mathbb{R})$. In some cases we are still able to conclude orientability of $X(\mathbb{R})$ using stronger assumptions:

LEMMA 65. *Let X be a smooth projective variety over $\mathbb{R}$. Assume that $K_X \cong L^{\otimes 2}$ for a real line bundle L. Then $X(\mathbb{R})$ is orientable.*

EXAMPLE 66. $\mathbb{RP}^n$ is orientable iff n is odd. Let $X \subset \mathbb{P}^n$ be a smooth hypersurface of degree d. Then $X(\mathbb{R})$ is orientable if $n - d$ is odd. $X(\mathbb{R})$ is not orientable if n and d are both odd. If n and d are both even, then $X(\mathbb{R})$ may or may not be orientable.

Once the singularities of the intermediate stages of the MMP are limited, a rather similar computation comes up with a list of possible steps of the MMP.

THEOREM 67. *Let X be a smooth, projective, real algebraic 3-fold such that $X(\mathbb{R})$ satisfies the conditions (13.2.a–c).*

Let $f_i : X_i \dashrightarrow X_{i+1}$ be any of the intermediate steps of the MMP over $\mathbb{R}$ starting with X. Then the induced map $f_i : X_i(\mathbb{R}) \to X_{i+1}(\mathbb{R})$ is everywhere defined and the following is a complete list of possibilities for f_i:

1. *($\mathbb{R}$-trivial) f_i is an isomorphism in a (Zariski) neighborhood of the set of real points.*
2. *($\mathbb{R}$-small) $f_i : X_i(\mathbb{R}) \to X_{i+1}(\mathbb{R})$ collapses a 1-complex to points.*
3. *(smooth point blow up) f_i is the inverse of the blow up of a smooth point $P \in X_{i+1}(\mathbb{R})$.*
4. *(singular point blow up) f_i is the inverse of a (weighted) blow up of a singular point $P \in X_{i+1}(\mathbb{R})$. There are two cases:*
   (a) *($cA_{>0}^+$, $\operatorname{mult}_0 g$ even) Up to real analytic equivalence near P, $X_{i+1} \cong (x^2 + y^2 + g_{\geq 2m}(z,t) = 0)$ where $g_{2m}(z,t) \neq 0$ and X_i is the weighted blow up $B_{(m,m,1,1)}X_{i+1}$.*
   (b) *($cA_{>0}^+$, $\operatorname{mult}_0 g$ odd) Up to real analytic equivalence near P, $X_{i+1} \cong (x^2 + y^2 + g_{\geq 2m+1}(z,t) = 0)$ where $z^{2m+1} \in g$ and $z^i t^j \notin g$ for $2i + j < 4m + 2$. X_i is the weighted blow up $B_{(2m+1,2m+1,2,1)}X_{i+1}$.*

68 (End of the proof).

The previous consideration assure us that we understand the change of the topology of the real part through the birational steps of the MMP. If $\kappa(X) = -\infty$, then the last step is a fiber type contraction. Hence we still need to understand the topology of a variety Y in the following 3 cases.

1. There is an extremal contraction $f : Y \to S$ to a surface. Here the general fiber is $\mathbb{P}^1$.
2. There is an extremal contraction $f : Y \to C$ to a curve. Here the general fiber is a Del Pezzo surface.
3. There is an extremal contraction $f : Y \to P$ to a point. Here $-K_Y$ is ample.

In the first case the general fiber of $Y(\mathbb{R}) \to S(\mathbb{R})$ is S^1 and it is reasonable to hope that $Y(\mathbb{R})$ is Seifert fibered. From the topological point of view this is quite delicate since the singular fibers can have a very substantial effect on the topology. Still, after careful computations in [**Kollár99b**] we obtain that $Y(\mathbb{R})$ is either Seifert fibered or a connected sum of lens spaces.

In the second case the general fiber is either a torus or a union of spheres. The tricky part is again to control the singular fibers. This was done in [**Kollár00a**].

Finally we are left with the case when Y is a Fano variety, that is, $-K_Y$ is ample. Smooth Fano threefolds are completelly classified, but we can have conjugate pairs of complex quotient singularities. Fortunately, even these singular Fano threefolds form a bounded family by a theorem of [**Kawamata92**]. This is the main source of the finitely many exceptions.

A complete list of singular Fano threefolds is not known and there are no reasonable bounds about the number of families.

Even if the complex variety is well known, the topology of the real part may be rather hard to determine. Here is the simplest unsolved question:

CONJECTURE 69. *Let $X \subset \mathbb{P}^4$ be a real hypersurface of degree 4. Then $X(\mathbb{R})$ s not hyperbolic.*

4. The Nash Conjecture for Nonprojective Threefolds

In this section we consider the Nash conjecture for nonprojectivite 3–folds. Allowing quasi projective varieties instead of projective ones does not help at all. Indeed, if Y is quasi projective and $Y(\mathbb{R})$ is compact then $\bar{Y}(\mathbb{R}) = Y(\mathbb{R})$ for every smooth compactification $\bar{Y} \supset Y$. To get something interesting, we should look at compact complex manifolds which can be obtained from $\mathbb{P}^3$ by a sequence of smooth blow ups and blow downs. This class of manifolds has been studied by many authors.

DEFINITION 70. [**Artin68, Moishezon67**] A compact complex manifold Y is called a *Moishezon manifold* or a (smooth) *algebraic space* if it is bimeromorphic to a projective variety.

A *real Moishezon manifold* or a *real algebraic space* is an algebraic space Y with an antiholomorphic involution $\tau : Y \to Y$.

It is not hard to see that if (Y, τ) is a real algebraic space then there is a projective real algebraic variety (Y', τ') and a conjugation invariant bimeromorphic map $\phi : Y \dashrightarrow Y'$.

By a result of Chow and Kodaira, every smooth Moishezon surface is projective (cf. [**BPV84**, IV.5]). The first smooth nonprojective examples in dimension 3 were found by Hironaka (see [**Hartshorne77**, App.B.3]).

Moishezon manifolds are quite close to projective varieties. In fact, as a general philosophy, if a property of projective varieties does not obviously involve projectivity then it should also be true for Moishezon manifolds. Thus it was quite a surpize to me that the Nash conjecture is true for Moishezon manifolds in a strong form.

Before explaining the method of the proof, we give a quick review of surgeries of 3–manifolds. As a general reference, see [**Rolfsen76**, Chap.2].

71 (Surgery on 3–manifolds).

Let D^2 be the unit disc $(|u| \leq 1) \subset \mathbb{C}_u$ and S^1 the unit circle $(|v| = 1) \subset \mathbb{C}_v$. A three–manifold T diffeomorphic to $D^2 \times S^1$ is called a *solid torus*. The boundary of a solid torus

$$\partial T \sim (|u| = 1) \times (|v| = 1) \sim S^1 \times S^1$$

is a *torus*. Any simple closed curve on the torus $S^1 \times S^1$ is isotopic to one of the form

$$C_{a,b} := \operatorname{im}[(|z| = 1) \to S^1 \times S^1] \quad \text{given by} \quad z \mapsto (z^a, z^b) \quad \text{for } (a,b) = 1.$$

The *meridian* of a solid torus is any curve isotopic to $C_{\pm 1,0}$. Note that these curves are generators of $\ker[\pi_1(\partial T) \to \pi_1(T)]$, so their isotopy class is well defined.

The correspondence

$$\begin{pmatrix} a & b \\ c & d \end{pmatrix} \mapsto [(u,v) \mapsto (u^a v^b, u^c v^d)]$$

gives an isomorphism

$$GL(2,\mathbb{Z}) \cong \frac{\text{(diffeomorphisms of the torus)}}{\text{(modulo isotopy)}}.$$

Up to isotopy, the diffeomorphisms of a solid torus are given by

$$(u,z) \mapsto (uz^m, z^{\pm 1}) \quad \text{or} \quad (u,z) \mapsto (\bar{u}z^m, z^{\pm 1}), \quad \text{where} \quad m \in \mathbb{Z}.$$

They correspond to the subgroup

$$B(2,\mathbb{Z}) := \begin{pmatrix} \pm 1 & * \\ 0 & \pm 1 \end{pmatrix} \subset GL(2,\mathbb{Z}).$$

Let M be a 3–manifold and $L \subset M$ a *knot*. (That is, an embedded copy of S^1.) Assume that M is orientable along L. L has a tubular neighborhood which is a solid torus T with bundary $\partial T \sim S^1 \times S^1$. Set $N := M \setminus \operatorname{Int} T$. We can think of M as being glued together from these two pieces $M = T \cup N$. Let $\phi : \partial T \to \partial T$ be any orientation preserving diffeomorphism. We can glue together T and N using ϕ to obtain another 3–manifold $M_\phi := T \cup_\phi N$. The operation that creates M_ϕ from M is called a *surgery* along L. (If ϕ is orientation reversing, we can compose ϕ with an orientation reversing diffeomorphism of T to see that we do not get anything new.) If we fix a diffeomorphism $h : T \sim D^2 \times S^1$ then a diffeomorphism $\phi : \partial T \to \partial T$ corresponds to an element of $SL(2,\mathbb{Z})$.

Two diffeomorphisms of the torus that differ by a diffeomorphism of the solid torus give diffeomorphic surgeries. Computing representatives of the cosets $GL(2,\mathbb{Z})/B(2,\mathbb{Z})$ we obtain the following:

PROPOSITION 72. *The image of the meridian of T under ϕ determines the surgery up to diffeomorphism.*

We are especially interested in surgeries that can be realized algebraically. From the algebraic point of view the simplest surgery–type operations are flops. It turns out that we need only the simplest flop.

73 (Flops of $(-1,-1)$-curves).

Let X be a complex 3-manifold and $\mathbb{CP}^1 \cong C \subset X$ a curve. Assume that the normal bundle of C in X is isomorphic to $\mathcal{O}_C(-1) + \mathcal{O}_C(-1)$. Let us blow up C in X. We obtain $p_1 : X_1 \to X$ with exceptional divisor $Q \cong \mathbb{CP}^1 \times \mathbb{CP}^1$. Moreover, the normal bundle N_{Q,X_1} has intersection number -1 with both rulings of Q.

Thus $Q \subset X_1$ is symmetrical with respect to the two factors of Q and by a result of [**Nakano70**] it can be contracted (if we allow X' to be an algebraic space) in the other direction to obtain $p_2 : X_1 \to X'$. The bimeromorphic map $\phi := p_2 \circ p_1^{-1} : X \dashrightarrow X'$ is called the *flop of C*. Frequently X' is also called a flop of X. Note that X and X' have a completely symmetrical role in a flop.

Assume next that X has a real structure and C is also real and isomorphic to $\mathbb{P}^1$ over $\mathbb{R}$. Thus $C(\mathbb{R}) \sim S^1$. A tubular neighborhood of $S^1 \sim C(\mathbb{R}) \subset X(\mathbb{R})$ is a trivial disc bundle over S^1, but the algebraic direct sum decomposition of the normal bundle gives that

$$N_{C(\mathbb{R})|X(\mathbb{R})} \cong \mathcal{O}_{C(\mathbb{R})}(-1) + \mathcal{O}_{C(\mathbb{R})}(-1)$$

which is the sum of two Möbius bands.

We can identify $N_{C(\mathbb{R})|X(\mathbb{R})}$ with a tubular neighborhood $T \supset C(\mathbb{R})$, then any of the direct summands $\mathcal{O}_{C(\mathbb{R})}(-1)$ corresponds to an embedded Möbius band $B \subset T$ with boundary $\partial B \subset \partial T$.

$X_1(\mathbb{R})$ is obtained from $X(\mathbb{R})$ by blowing up $C(\mathbb{R})$ which is contained in B. Thus the birational transform $(p_1)_*^{-1}B$ is isomorphic to B and it intersects $Q(\mathbb{R})$ in a fiber of p_2. Hence the birational transform of B in X' is a disc which intersects $C'(\mathbb{R})$ transversally. Thus $X(\mathbb{R}) \mapsto X'(\mathbb{R})$ is a surgery along $C(\mathbb{R})$ where ∂B becomes the new meridian.

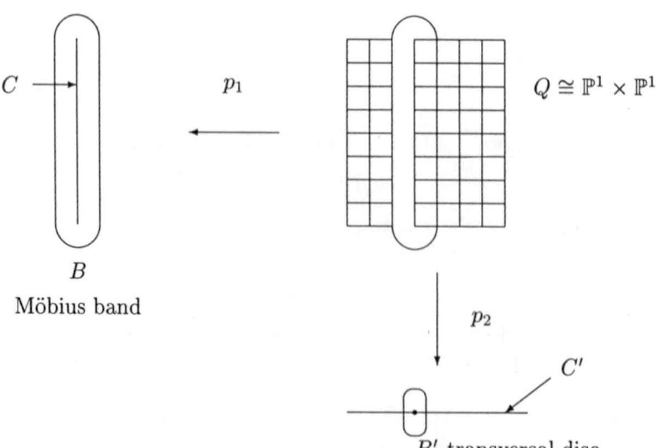

DEFINITION 74. Let M be a 3–manifold and $B \subset M$ an embedded Möbius band with center L. Assume that M is orientable along L. By (72) B defines a unique surgery $M \mapsto M'$. We call M' the *topological flop* of M along B.

The diffeomorphism class of M' is determined by the pair (M, B) up to isotopy.

The relationship between algebraic and topological flops is summarized in the next result which is a direct consequence of our previous discussions.

PROPOSITION 75. *Let X be a 3-dimensional, smooth, real, algebraic space and $\mathbb{P}^1 \cong C \subset X$ a real curve. Assume that $N_{C|X} \cong \mathcal{O}_C(-1) + \mathcal{O}_C(-1)$. Let $B \subset X(\mathbb{R})$ be a Möbius band corresponding to one of the $\mathcal{O}_C(-1)$ summands and $X \dashrightarrow X'$ the flop of B. Then $X'(\mathbb{R})$ is diffeomorphic to the topological flop of $X(\mathbb{R})$ along B.*

Now we can explain the main steps of the proof of (12).

76 (Idea of the proof of (12)).

1. First we find a normal form for 3–manifolds up to topological flops. This is a purely topological result which has been accomplished in [**Benedetti-Marin92**].
2. We realize each of the above normal forms by a 3–manifold $X(\mathbb{R})$ for some real projective algebraic variety X, birational to $\mathbb{P}^3$. In all but one case this was done in [**Benedetti-Marin92**]. In the last case this probably can not be done and we need a nonprojective example.
3. We realize every topological flop as a sequence of blow ups followed by an algebraic flop. Of course we need to choose the blow ups in such a way that they do not change the topology of the real points.

The first step of (76) in the orientable case relies on the following very nice result.

THEOREM 77. [**Benedetti-Marin92**] *Let M_1, M_2 be two compact orientable 3-manifolds. Then M_2 can be obtained from M_1 by a sequence of topological flops iff $h_1(M_1, \mathbb{Z}_2) = h_1(M_2, \mathbb{Z}_2)$.*

Instead of explaining the proof, let me illustrate the method in a key step: how to untangle knots by flops.

Let $L \subset \mathbb{R}^3$ be a knot. It is known that each knot has a *Seifert surface*, that is, an embedded surface S whose boundary is exactly L.

First we claim that we may assume S to be nonorientable. Indeed, if we slide the handle on the left hand side of the picture to the edge of the Seifert surface (with 2 new holes) on the right, we get a new nonorientable Seifert surface of the knot L.

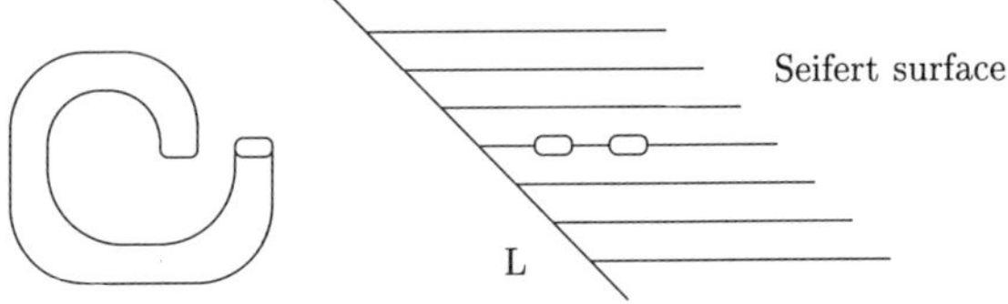

Every nonorientable surface with disk boundary is the connected sum of projective planes $\#k\mathbb{RP}^2$ with a disc removed. Let $C_1, \dots, C_k$ be lines in the projective planes. The neighborhood of a line in $\mathbb{RP}^2$ is a Möbius band, so each $C_i \subset S$ corresponds to a topological flop of $\mathbb{R}^3$. Performing the flops contracts each C_i to a point, thus the image of S after all these flops is a disc. Hence L becomes unknotted.

COROLLARY 78. *Let X_k denote a rational variety obtained from the quadric $(x_0^2 + x_1^2 + x_2^2 + x_3^2 = x_4^2) \subset \mathbb{P}^4$ by blowing up k real points. Then every compact orientable 3-manifold M can be obtained from $X_k(\mathbb{R})$ (with $k = h_1(M, \mathbb{Z}_2)$) by a sequence of topological flops.*

Proof. $X_k(\mathbb{R})$ is diffeomorphic to
$S^3\#k\mathbb{RP}^3$, hence it is orientable and $h_1(X_k(\mathbb{R}), \mathbb{Z}_2) = k$. We are done by (77). □

In the general case, we use the following consequence of the classification, which takes care of the first two steps in (76). The construction of the exceptional case is by a series of explicitly described blow ups and downs, see [**Kollár00b**, Sec.6] for details.

THEOREM 79. [**Benedetti-Marin92**, Thm.B and Ex.C.5] *Any compact 3-manifold M can be obtained by a sequence of topological flops from a 3-manifold N such that either*

1. *N can be obtained form S^3 by repeatedly blowing up points and simple closed curves, or*
2. *N is a torus bundle over S^1 with monodromy* $\begin{pmatrix} 0 & 1 \\ 1 & 1 \end{pmatrix} \in GL(2, \mathbb{Z})$.

A Möbius band $B \subset M$ with center L can be thought of as a subline bundle B' of the normal bundle $N_{L|M}$. Thus, the algebraic version of a Möbius band should be a smooth, real, rational curve $\mathbb{P}^1 \cong C \subset X$ and a real subline bundle (of odd degree) of the normal bundle $N_{C|X}$.

Taking into account the ultimate aim of producing algebraic flops, we would like to get a real subline bundle of degree -1 of the normal bundle $\mathcal{O}_C(-1) \hookrightarrow N_{C|X}$. The next theorem says that this is indeed possible.

THEOREM 80. *Let X be a compact, smooth, real, algebraic space which is birational to $\mathbb{P}^3$. Let $B \subset X(\mathbb{R})$ be an embedded Möbius band. Then there is*

1. *a smooth, real, rational curve $\mathbb{P}^1 \cong C \subset X$ such that $C \cdot K_X \leq 0$, and*
2. *a real subline bundle $\mathcal{O}_C(-1) \cong B' \subset N_{C|X}$*

such that the corresponding Möbius band is isotopic to $B \subset X(\mathbb{R})$.

Idea of the proof. Any continuous function on S^1 can be approximated by trigonometric polynomials. Thus any map $S^1 \to \mathbb{R}^3$ can be approximated by trigonometric polynomials, and we can even assume this to be a C^1-approximation. The trigonometric polynomials are exactly the polynomials $p(x, y)$ restricted to the circle. This way we see that any $S^1 \subset \mathbb{R}^3$ can be approximated by the real points of a rational algebraic curve obtained as the image of the plane conic $x^2 + y^2 = 1$. By carefully following the projective version, [**Bochnak-Kucharz99**] proves that with X as above there are plenty of morphisms $f : \mathbb{P}^1 \to X$ such that $f(\mathbb{RP}^1)$ is a very good approximation of L. A precise technical consequence of "plenty" is that f^*T_X can be assumed to be as ample as we want. This implies that there will be many subline bundles $\mathcal{O}_{\mathbb{P}^1}(-1) \hookrightarrow N_{C|X}$, and we can realize $B \subset X(\mathbb{R})$ by the real points of a subline bundle $\mathcal{O}_{\mathbb{P}^1}(-1) \cong B' \hookrightarrow N_{C|X}$.

A precise formulation of this is a bit tricky since there is no canonical choice for the trivialization of a tubular neighborhood in general. Thus some artificial choice has to be made and everything measured accordingly.

81 (Construction of algebraic flops).

Assume now that we have our real subline bundle $\mathcal{O}_{\mathbb{P}^1}(-1) \cong B' \hookrightarrow N_{C|X}$. Let Q' be the quotient $Q' := N_{C|X}/B'$. $\deg N_{C|X} = -K_X \cdot C - 2 \geq -2$ by (80.1), thus $\deg Q' \geq -1$. In order to flop, we need to achieve $\deg Q' = -1$. This will be done by a blow up.

Let $D \subset X$ be a smooth, real, algebraic curve such that

1. $D \cap C$ is precisely r points $P_1, \dots, P_r$, and
2. the tanget vector of D at every point P_i lies in $B' \subset N_{C|X}$.

Let us blow up D to obtain $p : X_1 := B_D X \to X$ and set $C_1 := p_*^{-1} C$. There is an exact sequence

$$0 \to B' \to N_{C_1|X_1} \to Q'(-\sum P_i) \to 0$$

If $\deg Q'(-\sum P_i) = \deg Q' - r = -1$, the sequence splits and

$$N_{C_1|X_1} \cong \mathcal{O}_{C_1}(-1) + \mathcal{O}_{C_1}(-1).$$

Thus $C_1 \subset X_1$ can be flopped to get X_2. Topologically we want to flop a curve in $X(\mathbb{R})$, and this is achieved as long as $X_1(\mathbb{R})$ is diffeomorphic to $X(\mathbb{R})$. This can happen only if $D(\mathbb{R}) = \emptyset$. There are plenty of real curves without real points, so this should be easy to achieve. The only additional problem is that in this case $D \cap C$ consists of conjugate point pairs, hence r is even. Thus our construction works only if $\deg Q' = -K_X \cdot C - 1$ is odd.

$X(\mathbb{R})$ is orientable along $C(\mathbb{R})$, thus $w_1(X(\mathbb{R})) \cdot C(\mathbb{R}) = 0$ where w_1 denotes the first Stiefel–Whitney class (cf. [**Milnor-Stasheff74**]).

$$c_1(X(\mathbb{C})) \cdot C \equiv w_1(X(\mathbb{R})) \cdot C(\mathbb{R}) \mod 2$$

by (62), hence $-K_X \cdot C = c_1(X(\mathbb{C})) \cdot C$ is even. Thus the only remaining question is the choice of D. This is actually trickier than it seems.

82 (General position curves in algebraic spaces).

If X is a quasi projective variety, it is easy to construct a curve D as above by taking the intersection of suitable hypersurfaces. On an algebraic space there may not exist any base point free linear system, so intersections of hypersurfaces are unlikely to produce the needed curve D. In fact, there are singular algebraic spaces with a unique singular point $P \in X$ such that every curve in X passes through P. No such *smooth* examples are known and it is conjectured that this can not happen for smooth algebraic spaces. (See [**Kollár91**, 5.2.6] for the case when $\dim X \leq 3$.)

In our case we deal with Moishezon spaces birational to $\mathbb{P}^3$, and so they contain plenty of rational curves. The deformation spaces of rational curves are very well understood and standard techniques (see, for instance [**Kollár-Mori98**]) allow us to construct D.

5. The Topology of Real Del Pezzo Surfaces

In this section we describe the topological types of all real Del Pezzo surfaces. This is the last step needed to finish the proofs of (33) and of (23).

We start with the description of some explicit isomorphisms and birational maps.

DEFINITION 83. Let F be a smooth real algebraic surface. A surface obtained from F by blowing up a real points and b pairs of conjugate complex points (possibly infinitely near) is denoted by $(F, a, 2b)$.

LEMMA 84. *We have the following elementary birational equivalences between the minimal models in (29).*

1. $(\mathbb{P}^2, 2, 0) \cong (Q^{2,2}, 1, 0)$.
2. $(\mathbb{P}^2, 0, 2) \cong (Q^{3,1}, 1, 0)$.
3. $(Q^{4,0}, 0, 2)$ *is isomorphic to the blow up of* $Q^{3,0} \times \mathbb{P}^1$ *at a pair of conjugate points on the same section* $Q^{3,0} \times P$, $P \in \mathbb{RP}^1$.
4. *Any minimal conic bundle over a rational curve with* 2 *singular fibers is isomorphic to* $(Q^{3,1}, 0, 2)$.

Proof. In the first two cases we blow up the 2 points in $\mathbb{P}^2$ and then contract the line through them to get a quadric.

$Q^{4,0} \cong Q^{3,0} \times Q^{3,0}$, let π_1 be the first projection. The pencil of planes through the 2 points gives a map $p : Q^{4,0} \dashrightarrow \mathbb{P}^1$.

$$(\pi_1, p) : Q^{4,0} \dashrightarrow Q^{3,0} \times \mathbb{P}^1 \quad \text{is birational}$$

and becomes a morphism after blowing up the 2 points.

Finally assume that $F \to B$ is a minimal conic bundle over a rational curve with 2 singular fibers. By (29.C), $B(\mathbb{R}) \neq \emptyset$, thus $B \cong \mathbb{P}^1$. $F_{\mathbb{C}}$ is the blow up of a minimal ruled surface F'' at 2 points. We can even assume that F'' has a section E with negative selfintersection $(E^2) = -k$ and the two points are not on E. If $k \geq 2$ then all other sections of F'' have selfintersection at least 2, so $E \subset F$ is the unique section with negative selfintersection. Thus E is defined over $\mathbb{R}$ and $F \to B$ is not minimal.

Thus $k = 1$ and there is a unique section $E' \subset F''$ such that $(E'^2) = 1$ and E' passes through the two blown up points. Let $\bar{E} \subset F$ be the birational transform. Then E and $\bar{E}$ have to be conjugate. Contracting them gives the quadric $Q^{3,1}$. □

PROPOSITION 85. *Let F be a smooth real Del Pezzo surface with $F(\mathbb{R}) = \emptyset$ whose minimal model has degree 8 or 9. Set $d := (K_F^2)$. Then $d \in \{8, 6, 4, 2\}$.*

If $d = 8$ then F is isomorphic to either $Q^{3,0} \times Q^{3,0}$ or to $Q^{3,0} \times \mathbb{P}^1$.

If $d < 8$ then F is isomorphic to $Q^{3,0} \times \mathbb{P}^1$ blown up in $\frac{1}{2}(8 - d)$ pairs of conjugate points.

Proof. Apply the MMP over $\mathbb{R}$ to obtain $F \to \cdots \to F^*$. If F is Del Pezzo then so is F^*, and $(K_{F^*}^2) \geq (K_F^2)$. Hence in our case F^* is either $Q^{3,0} \times Q^{3,0}$ or $Q^{3,0} \times \mathbb{P}^1$. By (84.3) any blow up of $Q^{3,0} \times Q^{3,0}$ at a pair of conjugate points is also a blow up of $Q^{3,0} \times \mathbb{P}^1$. □

PROPOSITION 86. *Let F be a smooth real Del Pezzo with $F(\mathbb{R}) \neq \emptyset$ whose minimal model has degree 8 or 9. Set $d := (K_F^2)$. Then $9 \geq d \geq 1$ and we have one of the following cases:*

If $d = 9$ then F is isomorphic to $\mathbb{P}^2$.

If $d = 8$ then F is isomorphic to either $Q^{3,1}$ or to $Q^{2,2}$ or to $\mathbb{P}^2$ blown up at a real point.

If $d < 8$ then F is isomorphic to one of the following:

1. $\mathbb{P}^2$ *blown up at* $a \geq 0$ *real points and* $b \geq 0$ *pairs of conjugate points for some* $a + 2b = 9 - d$. *Thus* $F(\mathbb{R}) \sim \#(a+1)\mathbb{RP}^2$.
2. $Q^{3,1}$ *blown up at* $b = \frac{1}{2}(8-d)$ *pairs of conjugate points (so d is even). Thus* $F(\mathbb{R}) \sim S^2$.
3. $Q^{2,2}$ *blown up at* $b = \frac{1}{2}(8-d)$ *pairs of conjugate points (so d is even). Thus* $F(\mathbb{R}) \sim S^1 \times S^1$.

Proof. The minimal model of such a surface is either $\mathbb{P}^2$, $Q^{3,1}$ or $Q^{2,2}$. By (84.1–2) any blow up of $Q^{3,1}$ or to $Q^{2,2}$ at a real point is also a blow up of $\mathbb{P}^2$. □

The two propositions above account for all Del Pezzo surfaces of degrees $d \geq 5$. The results are summarized in the next statement:

COROLLARY 87. *The following table lists all topological types of the real points of Del Pezzo surfaces of degrees* $9 \geq d \geq 5$.

degree	*topological types*
9	$\mathbb{RP}^2$
8	S^2 *or* $S^1 \times S^1$ *or* $\mathbb{RP}^2 \# \mathbb{RP}^2$ *or* $\emptyset$
7	$\mathbb{RP}^2$ *or* $\#3\mathbb{RP}^2$
6	S^2 *or* $S^1 \times S^1$ *or* $\mathbb{RP}^2 \# \mathbb{RP}^2$ *or* $\#4\mathbb{RP}^2$ *or* $\emptyset$
5	$\mathbb{RP}^2$ *or* $\#3\mathbb{RP}^2$ *or* $\#5\mathbb{RP}^2$

The following result shows that odd degree Del Pezzo surfaces over $\mathbb{R}$ are relatively easy to understand:

LEMMA 88. *Every degree* $2d - 1$ *Del Pezzo surface* F *over* $\mathbb{R}$ *with* $\rho(F) \geq 2$ *is the blow up of a degree* $2d$ *Del Pezzo surface at a real point.*

Proof. Since (K_F^2) is odd, F is not a minimal conic bundle. Thus F is either the blow up of a degree $2d$ Del Pezzo surface at a real point or the blow up of a degree $2d + 1$ Del Pezzo F' surface at a conjugate pair of complex points $P + \bar{P}$.

If $F' \cong \mathbb{P}^2$ then let $L \subset \mathbb{P}^2$ be the line through the two points. Its birational transform on F is a line.

Otherwise, F' is again not minimal by (29), hence F' contains a line L over $\mathbb{R}$ by induction. $P, \bar{P} \notin L$ since otherwise the birational transform of L on F would have a nonnegative intersection number with K_F. Thus L gives a real line on F.

Contracting a real line on F we get a degree $2d$ Del Pezzo surface. □

This shows that the study of degree 3 Del Pezzo surfaces is reduced to the study of degree 4 cases. The classification of these two classes is summarized next. These results were obtained by [**Schläfli1863**], who actually worked directly with cubic surfaces.

COROLLARY 89. *The following table lists all topological types of the real points of Del Pezzo surfaces of degrees 4 and 3.*

degree	*topological types*
4	S^2 *or* $S^1 \times S^1$ *or* $\mathbb{RP}^2 \# \mathbb{RP}^2$ *or* $\#4\mathbb{RP}^2$ *or* $\emptyset$ *or* $S^2 \uplus S^2$
3	$\mathbb{RP}^2$ *or* $\#3\mathbb{RP}^2$ *or* $\#5\mathbb{RP}^2$ *or* $\#7\mathbb{RP}^2$ *or* $S^2 \uplus \mathbb{RP}^2$

Proof. As we noted above, it is sufficient to describe all degree 4 Del Pezzo surfaces.

If a degree 4 Del Pezzo surface F is obtained from a higher degree surface by blowing up then we are reduced to (87). Otherwise F is a conic bundle over $\mathbb{P}^1$ with 4 singular fibers, thus $F(\mathbb{R}) \sim S^2 \uplus S^2$. □

Next we move to the harder cases of Del Pezzo surfaces of degrees 2 and 1.

NOTATION 90. Let $D \subset \mathbb{P}^2$ be a degree 4 smooth real curve. $D(\mathbb{R})$ divides $\mathbb{RP}^2$ into connected open sets and precisely one of these is nonorientable (denoted by U_D). We choose an equation $f(x,y,z) \in \mathbb{R}[x,y,z]$ of D such that f is negative on U_D.

We can associate two different degree 2 Del Pezzo surfaces to D. One is $F_D^+ := (u^2 = f(x,y,z) \subset \mathbb{P}^3(1,1,1,2)$ and the other $F_D^- := (u^2 = -f(x,y,z) \subset \mathbb{P}^3(1,1,1,2)$.

The correspondence $F_D^+ \leftrightarrow F_D^-$ is a natural involution on the space of degree 2 real Del Pezzo surfaces.

D has 28 bitangents over $\mathbb{C}$ and over each bitangent of D we get a pair of lines on $F_D^{\pm}$. This gives a total of 56 lines.

The topological classification of degree 4 plane curves over $\mathbb{R}$ is very old, it is already contained in [**Plücker1839**]. (See [**Viro90**] for a recent survey of the study of low degree real plane curves.) This implies the topological classification of degree 2 real Del Pezzo surfaces. The following proposition summarizes these results.

PROPOSITION 91. *There are 6 topological types of degree 4 smooth real plane curves. Correspondingly there are 12 topological types of degree 2 real Del Pezzo surfaces. The following table gives the complete list. The types in the same row correspond to each other under* $D \leftrightarrow F_D^+ \leftrightarrow F_D^-$.

$D(\mathbb{R})$	$F_D^+(\mathbb{R})$	$F_D^-(\mathbb{R})$
○○○○	$\uplus 4S^2$	$\#8\mathbb{RP}^2$
○○○	$\uplus 3S^2$	$\#6\mathbb{RP}^2$
○○	$S^2 \uplus S^2$	$\#4\mathbb{RP}^2$
○	S^2	$\#2\mathbb{RP}^2$
$\emptyset$	$\emptyset$	$\mathbb{RP}^2 \uplus \mathbb{RP}^2$
◎	$S^1 \times S^1$	$S^2 \uplus \#2\mathbb{RP}^2$

[**Zeuthen1874**] studied the bitangents of degree 4 plane curves. He proved the equivalence of (92.1) and (92.5). He understood the relationship between degree 4 plane curves and cubic surfaces. (Projecting a cubic surface from one of its points, the branch curve is a plane quartic. Equivalently, blowing up the cubic at a point we get a degree 2 Del Pezzo surface.) This is, however, not the natural thing to do from the modern viewpont. Most of (92) is proved in [**Comessatti13**].

THEOREM 92. *Let* $D \subset \mathbb{P}^2$ *be a degree 4 smooth real curve. The following are equivalent:*

1. *All 28 bitangents of* D *are real.*
2. *All 56 lines of* F_D^- *are real.*
3. F_D^- *is isomorphic to* $\mathbb{P}^2$ *blown up in 7 real points.*
4. $F_D^-(\mathbb{R}) \sim \#8\mathbb{RP}^2$.
5. $D(\mathbb{R}) \sim \uplus 4S^1$.
6. $F_D^+(\mathbb{R}) \sim \uplus 4S^2$.

7. *F_D^+ has Picard number 1 over $\mathbb{R}$.*

Proof. (1) $\Rightarrow$ (2): A neighborhood of a line in $\mathbb{RP}^2$ is not orientable, thus any bitangent is contained in U_D (except for the points of tangency). f is negative on any bitangent and so $u^2 = -f$ has real solutions, giving 56 real lines on F_D^-.

(2) $\Rightarrow$ (3): Over $\mathbb{C}$, F_D^- is the blow up of $\mathbb{P}^2$ at 7 points, hence it has 7 disjoint lines. If all lines are real, we have 7 disjont real lines. Contracting these we get a Del Pezzo surface of degree 9 over $\mathbb{R}$. By (29.D) it is $\mathbb{P}^2_{\mathbb{R}}$.

(3) $\Rightarrow$ (4): Topologically, each blowing up is connected sum with $\mathbb{RP}^2$.

(4) $\Rightarrow$ (5): This follows from (91).

(5) $\Rightarrow$ (6): This also follows from (91).

(6) $\Rightarrow$ (7): Assume to the contrary that F_D^+ has Picard number ≥ 2 over $\mathbb{R}$. By (29) we have one of 2 cases:

1. F_D^+ is a minimal conic bundle with 6 singular fibers. In this case $F_D^+(\mathbb{R}) \sim \uplus 3S^2$, a contradiction.
2. F_D^+ is the blow up of a Del Pezzo surface of degree 3 or 4 over $\mathbb{R}$. By (89) $F_D^+(\mathbb{R})$ has at most 2 connected components, a contradiction.

(7) $\Rightarrow$ (1): Assume that D has a complex bitangent L. Its conjugate $\bar{L}$ is again a bitangent. Let $C \subset F_D^+$ be a complex line over L. Its conjugate $\bar{C}$ lies over $\bar{L}$. Then $(C \cdot \bar{C}) \leq 1$ (the only possible intersection point lies over $L \cap \bar{L}$). Thus F_D^+ has either a disjoint pair of conjugate lines or a conic bundle structure, a contradiction. □

93 (Degree 1 Del Pezzo surfaces).

Let F be a degree 1 Del Pezzo surface over any field k. $|-K_F|$ is a pencil with exactly one base point. So this is a k-point and $F(k) \neq \emptyset$. $|-2K_F|$ is base point free and exhibits F as a double cover of a quadric cone $Q \subset \mathbb{P}^3$, ramified along a curve $D \subset Q$ which is a complete intersection of Q with a cubic surface with equation $(f = 0)$. D does not pass through the vertex of the cone.

$F_{\bar{k}}$ contains 240 lines (that is -1-curves); cf. [**Manin72**, IV.4.3]. We obtain these as follows. Take a plane $H \subset \mathbb{P}^3$ which is tangent to D at 3 points. The preimage of $H \cap Q$ in F has 2 irreducible components, each is a line. Thus we conclude that there are 120 planes which are tangent to D at 3 points.

Assume now that $k = \mathbb{R}$. Since $Q(\mathbb{R}) \neq \emptyset$, we can write $Q = (x^2 + y^2 = 1)$ in suitable affine coodinates (x, y, z) on $\mathbb{A}^3$. That is, $Q(\mathbb{R})$ is a cylinder with a singular point at infinity.

As in the degree 2 case, for each (nonhomogeneous) cubic $f(x, y, z)$ we obtain two degree 1 Del Pezzo surfaces, given by affine equations

$$F_f^{\pm} := (x^2 + y^2 - 1 = u^2 \mp f(x, y, z) = 0) \subset \mathbb{A}^4.$$

There are 2 types of simple closed loops on a cylinder: null homotopic ones (I call them ovals) and those homotopic to a plane section (I call them big circles).

Since $D(\mathbb{R})$ is the intersection of a cylinder with a cubic, it has 3 or 1 intersection points with any ruling line of the cylinder. Thus $D(\mathbb{R})$ contains either 3 big circles (and no ovals) or 1 big circle. D has genus 4, hence by Harnack's theorem (2), $D(\mathbb{R})$ has at most 5 connected components. An oval can not be inside another oval since this would give 4 points on a ruling. Furthermore, we can not have an oval on one side the big circle and at least two ovals on the other side. Indeed, choosing points P_1, P_2, P_3 inside the 3 ovals, there is a plane H through them. Then H intersects each oval in at least 2 points, and also the big circle. So $(H \cdot D) \geq 8$, but D has degree 6, a contradiction.

If all the ovals are on the same side of the big circle, we can normalize f so that it is positive on the other side. The other cases are symmetrical and it makes little sense to normalize f.

We can summarize these results in the following table:

PROPOSITION 94. *There are 7 topological types of degree 6 smooth real complete intersection curves on the cylinder $(x^2 + y^2 = 1)$, not passing through the point at infinity. Correspondingly there are 11 topological types of degree 1 real Del Pezzo surfaces. The following table gives the complete list. The types in the same row correspond to each other under $D = (f = 0) \cap Q \leftrightarrow F_f^+ \leftrightarrow F_f^-$.*

$D(\mathbb{R})$	$F_f^+(\mathbb{R})$	$F_f^-(\mathbb{R})$
1 big circle + 4 ovals	$\mathbb{RP}^2 \uplus 4S^2$	$\#9\mathbb{RP}^2$
1 big circle + 3 ovals	$\mathbb{RP}^2 \uplus 3S^2$	$\#7\mathbb{RP}^2$
1 big circle + 2 ovals	$\mathbb{RP}^2 \uplus 2S^2$	$\#5\mathbb{RP}^2$
1 big circle + 1 oval	$\mathbb{RP}^2 \uplus S^2$	$\#3\mathbb{RP}^2$
1 big circle + 0 oval	$\mathbb{RP}^2$	$\mathbb{RP}^2$
1 big circle + 1+1 ovals	$\#3\mathbb{RP}^2 \uplus S^2$	$\#3\mathbb{RP}^2 \uplus S^2$
3 big circles	$\mathbb{RP}^2 \uplus \#2\mathbb{RP}^2$	$\mathbb{RP}^2 \uplus \#2\mathbb{RP}^2$

The following theorem, due to [**Comessatti13**], is the degree 1 version of (92). I thank F. Russo for checking the arguments of Comessatti.

THEOREM 95. *Let $D = (f = 0) \subset Q$ be a degree 6 smooth real complete intersection curve on the cylinder $Q = (x^2 + y^2 = 1)$. The following are equivalent:*

1. *All 120 triple tangents of D are real and f is negative on all of them.*
2. *All 240 lines of F_f^- are real.*
3. *F_f^- is isomorphic to $\mathbb{P}^2_{\mathbb{R}}$ blown up in 8 real points.*
4. *$F_f^-(\mathbb{R}) \sim \#9\mathbb{RP}^2$.*
5. *$D(\mathbb{R}) \sim \uplus 5S^1$.*
6. *$F_f^+(\mathbb{R}) \sim \mathbb{RP}^2 \uplus 4S^2$.*
7. *F_f^+ has Picard number 1 over $\mathbb{R}$.*

Proof. (1) $\Rightarrow$ (2): If f is negative on a triple tangent then $u^2 = -f$ has real solutions, giving a pair of real lines on F_f^-.

(2) $\Rightarrow$ (3): Over $\mathbb{C}$, F_f^- is the blow up of $\mathbb{P}^2$ at 8 points. Thus it has 8 disjoint lines. If all lines are real, we have 8 disjont real lines. Contracting these we get a Del Pezzo surface of degree 9 over $\mathbb{R}$. By (29.D) it is $\mathbb{P}^2_{\mathbb{R}}$.

(3) $\Rightarrow$ (4): Topologically, each blowing up is connected sum with $\mathbb{RP}^2$.

(4) $\Rightarrow$ (5): This follows from (94).

(5) $\Rightarrow$ (6): This also follows from (94).

(6) $\Rightarrow$ (7): Assume to the contrary that F_f^+ has Picard number ≥ 2 over $\mathbb{R}$. F_f^+ can not be a minimal conic bundle since (K^2) is odd. Thus F_f^+ is the blow up of Del Pezzo surface of degree 2 or 3 over $\mathbb{R}$. By (91, 89) $F_f^+(\mathbb{R})$ has at most 4 connected components, a contradiction.

(7) $\Rightarrow$ (1): If D has a complex triple tangent, we can argue as in (92.(7) $\Rightarrow$ (1)) that F_f^+ contains a conjugate pair of lines $C, \bar{C}$ such that $(C \cdot \bar{C}) \leq 2$. $(C + \bar{C}) \equiv rK$ for some $r \in \mathbb{Z}$, thus $2(C \cdot \bar{C}) - 2 = r^2$. This is impossible.

If there is a real triple tangent such that f is positive on it then as in (1) $\Rightarrow$ (2) we get real lines on F_f^+. $\square$

REMARK 96. It is not obvious that Del Pezzo surfaces with the same topological type form a connected family. Let us look for instance at cubics which are obtained from $\mathbb{P}^2$ by blowing up 6 real points. These correspond to 6 points in $\mathbb{P}^2$, no 3 on a line, not all on a conic. The possible configurations do not form a connected set.

On the other hand, a cubic surface like this is obtained as a blow up of $\mathbb{P}^2$ in many different ways, so the set of all such cubics still may be a connected set.

It is in fact true that all real Del Pezzo surfacs with the same degree and the same topological type form a connected family, with the exception of degree 8 surfaces without real points. For a modern proof see [**DIK00**, 17.3].

ACKNOWLEDGMENTS . I thank L. Bonavero, S. Cynk, S. Endrass, S. Kharlamov and W. Kucharz for numerous comments and improvements. A. Marin directed me to several 19th century references. F. Russo checked the arguments of Comessatti about degree 1 and 2 Del Pezzo surfaces and pointed out some misunderstandings on my part. Partial financial support was provided by the NSF under grant number DMS-0096268.

References

Akbulut-King91. S. Akbulut and H. King, Rational structures on 3–manifolds, Pacific J. Math. 150 (1991) 201-204

Artin68. M. Artin, The implicit function theorem in algebraic geometry, in: Algebraic geometry, Bombay, Oxford Univ. Press 1968, 13–34

BPV84. W. Barth, C. Peters and A. Van de Ven, Compact Complex Surfaces, Springer, 1984

Benedetti-Marin92. R. Benedetti and A. Marin, Déchirures de variétés de dimension trois, Comm. Math. Helv. 67 (1992) 514–545

BCR87. J. Bochnak, M. Coste and M-F. Roy, Géométrie algébrique réelle, Springer 1987; revised English translation: Real algebraic geometry, Springer, 1999

Bochnak-Kucharz99. J. Bochnak and W. Kucharz, The Weierstrass approximation theorem for maps between real algebraic varieties, Math. Ann. 314 (1999) 601–612.

Bouvier98. C. Bouvier, Diviseurs essentiels, composantes essentielles des variétés toriques singulières. Duke Math. J. 91 (1998) 609–620

Catanese-Frediani00. F. Catanese and P. Frediani, Real hyperelliptic surfaces and the orbifold fundamental group, preprint (2000)

Comessatti13. A. Comessatti, Fondamenti per la geometria sopra superfizie razionali dal punto di vista reale, Math. Ann. 73 (1913) 1-72

Comessatti14. A. Comessatti, Sulla connessione delle superfizie razionali reali, Annali di Math. 23(3) (1914) 215-283

DIK00. A. Degtyarev, I. Itenberg and V. Kharlamov, Real Enriques Surfaces, Lecture Notes in Mathematics no. 1746. Springer-Verlag, Berlin, 2000

Degtyarev-Kharlamov00. A. Degtyarev and V. Kharlamov, Topological properties of real algebraic varieties: Rokhlin's way, preprint

Degtyarev-Kharlamov01. A. Degtyarev and V. Kharlamov, Real rational surfaces are quasi–simple, preprint

Harnack1876. Über die Vieltheiligkeit der ebenen algebraischen Kurven, Math. Ann. 10 (1876) 189-198

Hartshorne77. R. Hartshorne, Algebraic Geometry, Springer, 1977

Hayakawa99. T. Hayakawa, Blowing ups of 3-dimensional terminal singularities, Publ. Res. Inst. Math. Sci. 35 (1999) 515–570

Hempel76. J. Hempel, 3-manifolds, Princeton Univ. Press, 1976

Iskovskikh67. V. A. Iskovskikh, Rational surfaces with a pencil of rational curves, Math. USSR Sb. 3 (1967) 563-587

Iskovskikh80. V. A. Iskovskikh, Minimal models of rational surfaces over arbitrary fields, Math. USSR Izv. 14 (1980) 17-39

Kawakita00. M. Kawakita, Divisorial contractions in dimension three which contract divisors to smooth points, (preprint) 2000

Kawamata92. Y. Kawamata, Boundedness of $\mathbb{Q}$-Fano threefolds, Proc. Int. Conf. Algebra, Contemp. Math. vol. 131 (1992) 439-445

Kharlamov78. V. M. Kharlamov, Real algebraic surfaces (Russian) Proceedings of the International Congress of Mathematicians (Helsinki, 1978) pp. 421–428, Acad. Sci. Fennica, Helsinki, 1980.

Kharlamov00. V. M. Kharlamov, Variétés de Fano réelles, Sém. Bourbaki, No. 872, March 2000

Klein1876. F. Klein, Über eine neuer Art von Riemannschen Flächen, Math. Ann. 10 (1876) 398–416, Reprinted in : F. Klein, Gesammelte Mathematische Abhandlungen, Springer, 1922, vol. II.

Kollár91. J. Kollár, Flips, Flops, Minimal Models, etc., Surv. in Diff. Geom. 1 (1991) 113-199

Kollár96. J. Kollár, Rational Curves on Algebraic Varieties, Springer Verlag, Ergebnisse der Math. vol. 32, 1996

Kollár98a. J. Kollár, Low degree polynomial equations, in: European Congress of Math. Birkhäuser, 1998, 255-288

Kollár98b. J. Kollár, Real Algebraic Threefolds I. Terminal Singularities, Collectanea Math. (FERRAN SERRANO, 1957-1997) 49 (1998) 335-360

Kollár98c. J. Kollár, The Nash conjecture for algebraic threefolds, ERA of AMS 4 (1998) 63–73

Kollár99a. J. Kollár, Real Algebraic Threefolds II. Minimal Model Program, Jour. AMS 12 (1999) 33–83

Kollár99b. J. Kollár, Real Algebraic Threefolds III. Conic Bundles, J. Math. Sci. (New York) 94 (1999) 996–1020

Kollár00a. J. Kollár, Real Algebraic Threefolds IV. Del Pezzo Fibrations, in: Complex analysis and algebraic geometry, de Gruyter, Berlin, 2000, 317–346,

Kollár00b. J. Kollár, The nonprojective Nash conjecture (preprint)

Kollár-Mori98. J. Kollár and S. Mori, Birational geometry of algebraic varieties, English edition: Cambridge Univ. Press, 1998; Japanese edition: Iwanami Shoten, 1998

Kollár-Smith97. J. Kollár and K. Smith, Rational and Non-rational Algebraic Varieties (e-prints: alg-geom/9707013)

Manin66. Yu. I. Manin, Rational surfaces over perfect fields (in Russian), Publ. Math. IHES 30 (1966) 55-114

Manin72. Yu. I. Manin, Cubic forms (in Russian), Nauka, 1972. English translation: North-Holland, 1974, second expanded edition, 1986

Markushevich96. D. G. Markushevich, Minimal discrepancy for a terminal cDV singularity is 1, J. Math. Sci. Univ. Tokyo 3 (1996) 445-456

Mikhalkin97. G. Mikhalkin, Blow up equivalence of smooth closed manifolds, Topology, 36 (1997) 287–299

Milnor64. J. Milnor, On the Betti numbers of real varieties. Proc. Amer. Math. Soc. 15 (1964) 275–280

Milnor-Stasheff74. J. Milnor and J. Stasheff, Characteristic classes, Princeton Univ. Press, 1974

Moishezon67. B. Moishezon, On n-dimensional compact varieties with n algebraically independent meromorphic functions, Amer. Math. Soc. Transl. 63 (1967) 51-177

Mori82. S. Mori, Threefolds whose Canonical Bundles are not Numerically Effective, Ann. of Math. 116 (1982) 133-176

Nash52. J. Nash, Real algebraic manifolds, Ann. Math. 56 (1952) 405-421

Nakano70. S. Nakano, On the inverse of monoidal transformations, Publ. Res. Inst. Math. Sci. Kyoto, 6 (1970) 483–503

Narasimhan68. R. Narasimhan, Analysis on real and complex manifolds, North–Holland, 1968

Nikulin79. V. V. Nikulin, Integer symmetric bilinear forms and some of their geometric applications. (Russian) Izv. Akad. Nauk SSSR Ser. Mat. 43 (1979) 111–177

Plücker1839. J. Plücker, Theorie der algebraischen Kurven, Bonn, 1839

Reid85. M. Reid, Young person's guide to canonical singularities, in Algebraic Geometry, Proc. Symp. Pure Math. vol.46, pp. 345-414

Rolfsen76. D. Rolfsen, Knots and links, Publish or Perish, 1976

Schläfli1863. L. Schläfli, On the distribution of surfaces of the third order into species, Phil. Trans. Roy. Soc. London, 153(1863)193-241. Reprinted in : L. Schläfli, Gesammelte Mathematische Abhandlungen, Birkhäuser, 1953, vol. II.

Scott83. P. Scott, The geometries of 3–manifolds, Bull. London Math. Soc., 15 (1983) 401-487

Segre42. B. Segre, The non-singular cubic surfaces, Clarendon Press, 1942

Segre51. B. Segre, The rational solutions of homogeneous cubic equations in four variables, Notae Univ. Rosario 2 (1951) 1-68

Shafarevich72. R. I. Shafarevich, Basic Algebraic Geometry (in Russian), Nauka, 1972. English translation: Springer, 1977, second expanded edition, 1994

Silhol84. R. Silhol, Real algebraic surfaces with rational or elliptic fibering, Math. Zeitschr. 186(1984) 465-499

Silhol89. R. Silhol, Real algebraic surfaces, Springer Lecture Notes vol. 1392, 1989

Sullivan71. D. Sullivan, Combinatorial invariants of analytic spaces, Springer Lecture Notes vol. 192, (1971) 165-168

Sun95. X. Sun, A regularity theorem on birational morphisms, J. Algebra 178 (1995) 919–927

Thom65. R. Thom, Sur l'homologie des variétés algébriques réelles, in: Differential and combinatorial topology, Princeton Univ. Press, 1965, 255-265

Tognoli73. A. Tognoli, Su una congettura di Nash, Ann. Sci. Norm. Sup. Pisa 27 (1973) 167-185

Viro90. O. Ya. Viro, Real algebraic plane curves, Leningrad Math. J. 1 (1990) 1059-1134

Viterbo98. C. Viterbo, Symplectic real algebraic geometry, to appear

Zeuthen1874. H.G. Zeuthen, Sur les différentes formes des courbes du quatrième ordre, Math. Ann. 7(1874) 410-432

Princeton University, Princeton NJ 08544-1000
kollar@math.princeton.edu

Current Developments in Mathematics, 2000
Pages 233 – 253

Scaling limits of random processes and the outer boundary of planar Brownian motion

Oded Schramm

1. Scaling limits and the Stochastic Löwner Evolution

1.1. Scaling Limits. Often Scaling limits exhibit more symmetry and are more canonical than the discrete models.

Example. Consider the simple random walk on the grid $\mathbb{Z}^2$. If you rescale the grid (and rescale time appropriately), then the simple random walk converges to Brownian motion.

In simple random walk (SRW), at each step you are at some vertex and walk to one of the neighbors at random with equal probability.

The scaling is as follows. Let $X_s(t) := X(s^2t)/s$, where $X(t)$ is the position of the SRW at time t. Then $X_s(t)$ converges to Brownian motion, in the appropriate topology, as $s \to \infty$.

Brownian motion is invariant under rotations, although the simple random walk is not.

Actually, in two dimensions, BM is conformally invariant (up to time-change).

1.2. Conformal invariance. A conformal map $f : D \to D'$ is a homeomorphism from D onto D' which is complex-differentiable.

Equivalently, f preserves angles.

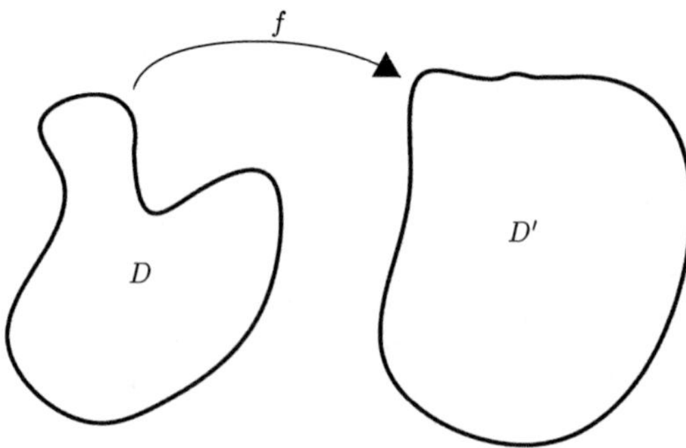

A random process is conformally invariant, if the image of the process under a conformal map $f : D \to D'$ has the same distribution as the process in D'.

Example. Run Brownian motion B from $p \in D$ until you hit the boundary of D. Take the trace (image).

One may consider $f(B)$ as a path from $f(p)$ to $\partial D'$. Conformal invariance means that up to reparameterization, $f(B)$ is the same as BM in D' from $f(p)$ to $\partial D'$.

The above is not a precise definition of what conformal invariance means. Rather, it should serve as a guiding principle. It is perhaps more instructive at this point not to present an exact definition.

1.3. Percolation. Here is one of several models for critical percolation.

Each hexagon is white (open) with probability 1/2, independently. The connected components of the white regions are studied.

More generally, given $p \in [0,1]$, we may consider Bernoulli(p) percolation, where each hexagon is white with probability p.

There are various other percolation models. A favorite is Bernoulli(p) bond percolation where each edge of the grid in $\mathbb{Z}^d$ is kept with probability p and removed with probability $1-p$.

Harris [**Har60**] showed that with probability 1 Bernolli(1/2) bond percolation on $\mathbb{Z}^2$ has no infinite components, while Kesten [**Kes80**] showed that when $p > 1/2$ there are infinite components with probability 1. The same statements are also known for the percolation model discussed above. The critical parameter $p_c = p_c(d)$ is defined as the supremum of $p \in [0,1]$ such that with probability 1 there are no infinite components for Bernoulli(p) bound percolation on $\mathbb{Z}^d$. For $d > 2$, the value of $p_c(d)$ is not known precisely.

There are many interesting works on percolation in 2 and higher dimensions. One problem is to show that there are no infinite components for critical (i.e., $p = p_c$) Bernoulli bond percolation in $\mathbb{Z}^d$. This is known only for $d = 2$ (Kesten [**Kes80**]) and $d \geqslant 19$ (Hara and Slade [**HS90**]).

1.4. Cardy's formula (Carleson's version). Take an equilateral triangle with unit length edges. Let $x \in [0,1]$. From the upper corner, mark an arc of length x along the right edge. Put a grid of very small hexagons, sample a critical percolation configuration, and consider the event that inside the triangle there is a crossing from the lower edge to the marked arc inside white hexagons. It is a conjecture that for every $x \in [0,1]$ the probability of this event tends to x as the mesh of the grid goes to zero.

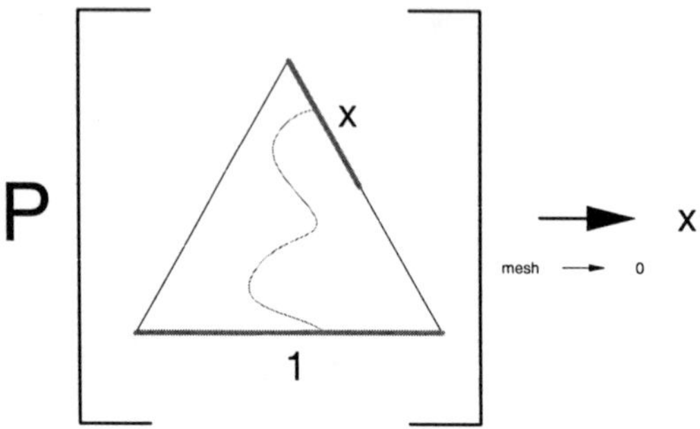

Mathematically speaking, this is a conjecture (Cardy is a physicist). Cardy's reasoning [**Car92**] uses Conformal field theory and the Virasoro algebra. Cardy assumes the conjectured conformal invariance of percolation. (Aizenmann; Langlands, Pouliot, Saint-Aubin [**LPSA94**]).

Cardy's formula was for the crossing of a rectangle. Carleson observed that assuming conformal invariance it is equivalent to the above. Cardy's original formula involves hypergeometric functions.

Added in revision: For critical percolation on the triangular lattice Cardy's formula was recently proven by Stanislav Smirnov [**Smi**], who also showed that the scaling limit exists and is conformally invariant.

Langlands et al. attempted to understand Cardy's reasoning from a mathematical point of view, unsuccessfully, they say. However they presented the very appealing conformal invariance conjecture which they attribute to Aizenmann. This drew several mathematicians to this subject, myself included.

1.5. Uniform spanning trees (UST). Consider a random-uniform spanning tree of an $n \times n$ square in the grid $\mathbb{Z}^2$.

A spanning tree of a graph is a subgraph that contains all the vertices and where there is precisely one simple path joining any pair of distinct vertices.

A UST on a finite graph is a random spanning tree selected according to the uniform measure on the collection of all spanning trees of the graph. (That is, each spanning tree has the same probability of being selected.)

There is also an emerging theory of UST on infinite graphs, with contributions from Aizenman, Benjamini, Burchard, Häggström, Kenyon, Lyons, Newman, Pemantle, Peres, Schramm, Wilson and others.

One nice thing about the UST is that it connects well to classical mathematics, such as potential theory, and random walks.

1.6. Loop-erased random walk. If you fix two vertices a, b in a finite graph G, then the UST path joining them is loop-erased random walk (LERW), from a to b.

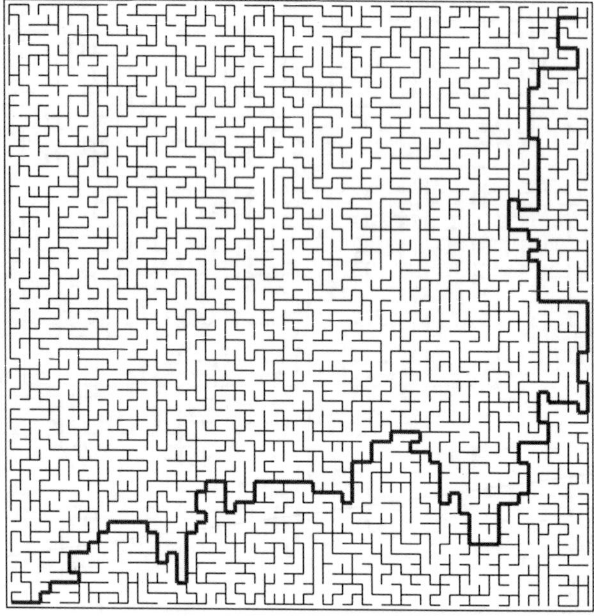

1.7. The definition of LERW. The LERW is obtained by performing simple random walk (SRW), and removing loops as they are created.

In other words, in the loop-erasure of a path γ, at each step you go from a vertex v along the last edge of γ incident with v. The notion of LERW was introduced by Greg Lawler [**Law80**]. The UST relation was first discovered by Aldous [**Ald90**] and Broder [**Bro89**].

Following is an algorithm due to Wilson [**Wil96**] for generating the UST of a finite graph G using the LERW. Pick an arbitrary ordering $\{v_1, \ldots, v_n\}$ of the vertices of G. Inductively, we build a subtree T_j, as follows. Set $T_1 = \{v_1\}$. Given T_j, $j < n$, let L_{j+1} be the loop-erasure of the simple random walk started from v_{j+1} which stops when it hits T_j. (If $v_{j+1} \in T_j$, then set $L_{j+1} = \{v_{j+1}\}$.) Set $T_{j+1} := T_j \cup L_{j+1}$. Wilson proved that T_n has the same distribution as the UST on G.

1.8. The Peano curve associated with the UST. The complement of the UST in the plane is another UST (on a dual grid). Between the UST and its dual winds the Peano path, which is shown in black in the figure.

Note that the UST Peano curve is space-filling, it visits every vertex in the appropriate grid.

1.9. The Big Conjecture. Conjecture. Percolation, UST, LERW and the Peano curve are conformally invariant in the scaling limit.

Special to 2 dimensions.

Added in revision: For percolation on the triangular grid, this has been recently established by Stanislav Smirnov [**Smi**].

In order to state this more precisely, one should define each of these processes in a domain. For example, for the LERW, let $D \subset \mathbb{R}^2$ be a domain and p a basepoint in D. Take a fine square grid in the plane. Let p' be a vertex in the grid closest to p. Perform SRW on the grid starting from p' until the walk hits ∂D. Let γ be the loop-erasure of this random walk. The conjectured conformal invariance applies to the (weak) limit of γ as the mesh of the grid tends to zero, as a set of points, say, or as an unparameterized path.

Richard Kenyon [**Ken00b, Ken00a, Ken00c, Ken**] has shown that some properties of LERW and UST are conformally invariant in the scaling limit. His work is based on the relation with domino tilings.

There is much Monte-Carlo type and theoretical support for the conformal invariance conjecture.

1.10. The Scaling limit of LERW. Consider the loop-erasure of a simple random walk starting at $a \in D$, which stops when ∂D is first hit.

We may consider that as a random closed subset of $\overline{D}$.

The scaling limit is the limit of the law of this closed subset.

Theorem ([Sch00]). Any subsequential scaling limit of LERW is almost surely a simple path.

As we shall see below, there are natural situations where the scaling limit of a random simple path is not a simple path.

1.11. On the LERW scaling limit. Assuming the conformal-invariance of the LERW scaling limit, one can construct the limit directly. We now describe it.

Consider the LERW scaling limit γ in the unit disk $\mathbb{U}$ starting from 0. Consider a terminal segment γ_t of the path γ from some point $\gamma(t)$ to $\partial\mathbb{U}$. We may uniformize the complement of the segment in the unit disk, using Riemann's mapping theorem. That is, consider the conformal map $g_t : \mathbb{U} \setminus \gamma_t \to \mathbb{U}$ normalized by $g_t(0) = 0$ and $g_t'(0) > 0$. This map is unique. The image of the tip $\gamma(t)$ under the conformal map g_t is some point $\zeta(t)$ on the unit circle.

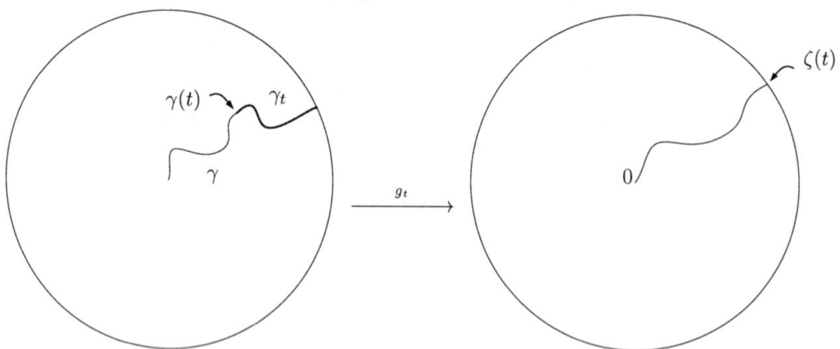

1.12. Theorem ([Sch00]). Assuming the conformal-invariance of LERW, $\zeta(t)$ is (1-dimensional) Brownian motion on the unit circle, run at twice the standard speed. (With the appropriate capacity parameterization of γ.)

The capacity parameterization of γ is the unique parameterization $\gamma : [0, \infty] \to \mathbb{U}$ such that $g_t'(0) = \exp t$.

The theorem allows a reconstruction of γ, since by Löwner's theorem, γ can be reconstructed from ζ:

Corollary ([Sch00]). Assuming that LERW has a conformally invariant scaling limit, the scaling limit of LERW from 0 to $\partial\mathbb{U}$ is the path

$$\gamma(t) = f_t(\zeta(t)),$$

where

$$\zeta(t) = B(-2t),$$

$B(t)$ is BM on $\partial\mathbb{U}$, and f_t is defined by Löwner's equation with parameter ζ:

$$\frac{\partial}{\partial t} f_t(z) = z f_t'(z) \frac{\zeta(t) + z}{\zeta(t) - z}, \qquad f_0(z) = z.$$

Remark: $f_t = g_t^{-1}$.

1.13. Sketch of the proof. The reasoning proceeds roughly as follows. LERW has the following combinatorial property (even before we pass to the scaling limit). If we condition on γ_t, the terminal part of the path from $\gamma(t)$ onward, the other arc, $\gamma \setminus \gamma_t$, is just LERW in the domain $D \setminus \gamma_t$ conditioned to hit the boundary at $\gamma(t)$. Using the assumed conformal invariance, we may then map $\mathbb{U} \setminus \gamma_t$ to the unit disk, and we are back at the start, except that the endpoint of the path is fixed. This identity of distribution gives the Markov property for ζ, and also shows that ζ is stationary. It is then not hard to conclude that ζ is Brownian motion run at some constant speed.

To see that the speed is 2, we calculate the asymptotics as $\epsilon \to 0$ of the variance of the winding number of the path around zero from the circle of radius ϵ to $\partial\mathbb{U}$, and compare with the analogous computation that Richard Kenyon [**Ken00b**] performed for LERW.

1.14. Other parameters? When $\zeta(t) = B(\kappa|t|)$, the solution of Löwner's differential equation with driving parameter ζ will be called Stochastic Löwner Evolution with parameter κ, or SLE_κ. SLE_2 (conjecturally) gives the scaling limit of LERW. What happens with other constants κ?

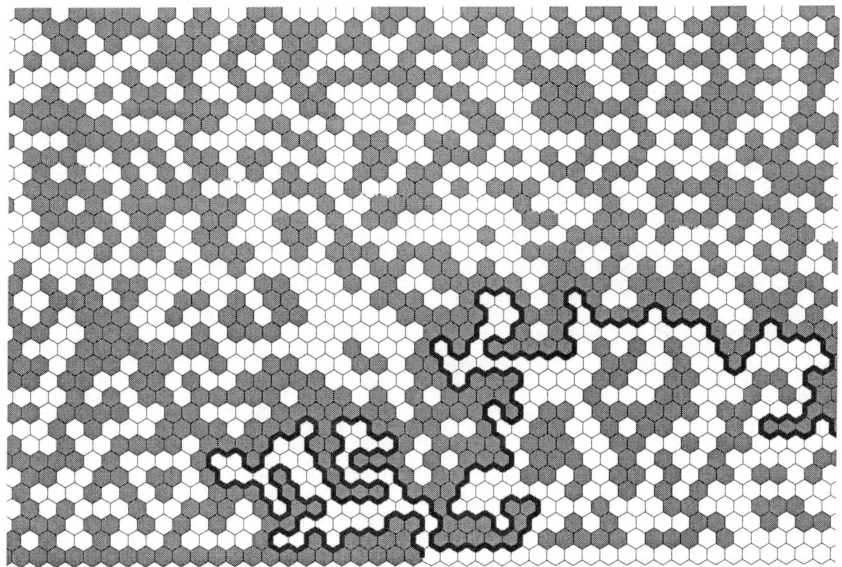

1.15. Critical percolation boundary path. In the figure, each of the hexagons is colored black with probability 1/2, independently, except that the hexagons intersecting the positive real ray are all white, and the hexagons intersecting the negative real ray are all black. There is a boundary path β, passing through 0 and separating the black and the white connected components adjacent to 0. The curve β is a random path in the upper half-plane $\mathbb{H}$ connecting the boundary points 0 and ∞.

1.16. Critical percolation boundary path scaling limit. Theorem (S). Assuming the conformal invariance conjecture for critical percolation, the scaling limit of the path β has the same law as $f_t(\zeta(t))$, where $\zeta(t) = B(6t)$, B is Brownian motion in $\mathbb{R}$ starting at $B(0) = 0$, and f_t is the solution of a Löwner-like equation with parameter ζ:

$$\frac{\partial}{\partial t} f_t(z) = \frac{2f_t'(z)}{\zeta(t) - z},$$

with $f_0(z) = z$.

Remark. It can be shown that the scaling limit of the percolation boundary curve is **not** a simple path.

1.17. Two versions of SLE. Note that the LERW path is a path from an interior point in the domain to a boundary point, while the percolation boundary curve is a path joining two endpoints of the domain (0 and ∞, in this case).

This requires somewhat different treatment. The normalization of the conformal map changes, and hence the differential equation changes too.

The version where two points on the boundary are joined is called chordal SLE while the version where a point in the interior is joined to the boundary is called radial SLE.

It turns out that radial and chordal SLE are essentially the same.

1.18. UST Peano. At $\kappa = 2$, we "got" LERW. At $\kappa = 6$, we "got" percolation. Assuming that LERW has a conformally invariant scaling limit, it follows that chordal SLE_8 is the scaling limit of the UST Peano curve (with partially wired boundary).

In the figure, the square boundary is wired on the lower and left edges. The dual graph is then wired on the right and top edges. The Peano curve then joins the lower right and the upper left corners.

1.19. Phases of SLE. Theorem (Rohde-Schramm [RS]). For all $\kappa \geqslant 0$, $\kappa \neq 8$, SLE produces a continuous path. It is a simple path iff $\kappa \leqslant 4$. It is space filling iff $\kappa > 8$.

The SLE path is defined by $f_t(\zeta(t))$. Continuity is nontrivial, since it is not a priori clear that f_t extends continuously to the boundary.

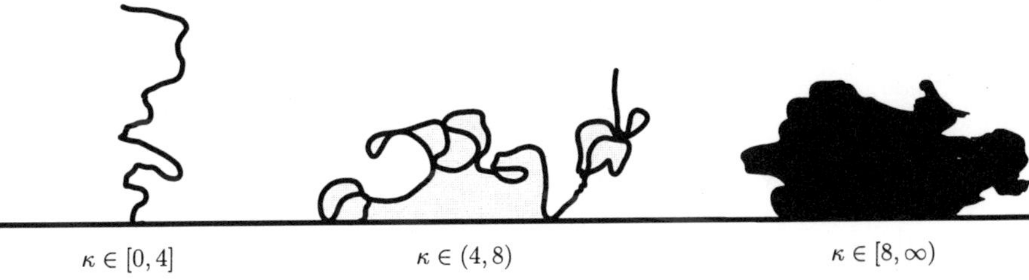

In the phase $\kappa \in (4,8)$, the SLE path makes loops "swallowing" parts of the domain. However, it never crosses itself.

The Hausdorff dimension of the SLE path is conjectured to be $1 + \kappa/8$ when $\kappa \leqslant 8$.

1.20. Computing with SLE. One can compute using the SLE. Typically, problems about SLE convert to PDE questions.

Cardy's formula has been proven for SLE_6 (Lawler-Schramm-Werner [**LSW1**]). This gives a proof of Cardy's formula modulo the conformal invariance conjecture.

Also, various generalizations and variations on Cardy's formula have been proven for SLE.

1.21. Example calculation sketch: Cardy's formula. Recalling the definition of the percolation boundary path, it is clear what Cardy's formula for SLE should be. It is a formula describing the probability that $I := (-\infty, -1)$ is hit by the SLE before $J := (s, \infty)$, as a function of s, $s > 0$. (Conformal invariance is used here.)

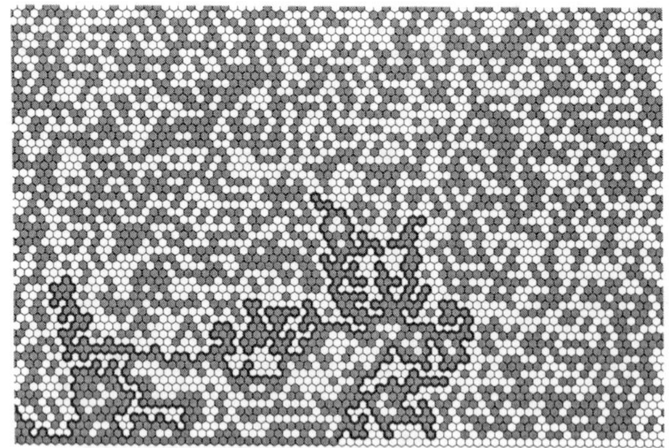

The differential equation for the (normalized) map going from the half-plane slitted by the SLE to the half-plane is

$$\partial_t g_t(z) = \frac{2}{g_t(z) - \zeta(t)} .$$

We run this random flow from $g_0(z) = z$ and look for the probability that $g_t(s)$ hits $\zeta(t)$ before $g_t(-1)$ does.

This is so because when t increases to the time when s is "swallowed" by the SLE, $g_t(s) - \zeta(t) \to 0$. This is a little exercise in conformal maps.

Note that one only needs to keep track of $g_t(s)$, $g_t(-1)$ and $\zeta(t)$, the behaviour of g at other points can be ignored.

Let $F(a, b)$ be the probability that in this flow $g_t(a)$ hits $\zeta(t)$ before $g_t(b)$, where $\zeta(0) = 0$. It is clear that $F\big(g_t(a) - \zeta(t), g_t(b) - \zeta(t)\big)$ is a (local) martingale. This gives a PDE for F. Moreover, scale invariance shows that $F(\lambda a, \lambda b) = F(a, b)$ for $\lambda > 0$. This reduces the PDE to an ODE. Cardy's formula in the upper half plane is the solution of the ODE.

1.22. SLE$_6$ **and the outer boundary of planar BM.** There is a version of SLE in the whole plane started from 0 and going to ∞, which we have not discussed. Call it full-plane SLE.

Theorem (Lawler-Schramm-Werner). Let F be the outer boundary of planar BM started from 0 and stopped on hitting the unit circle. Let F' be the outer boundary of full plane SLE$_6$ stopped on hitting the unit circle. Then F and F' have the same distribution.

The outer boundary of a closed bounded set $X \subset \mathbb{R}^2$ is its intersection with the closure of the unbounded connected component of the complement of X.

Note: there is no conjecture assumed, because conformal invariance is known for BM.

1.23. Summary. SLE_κ is conjectured to be the scaling limit of LERW at $\kappa = 2$, of the percolation boundary path at $\kappa = 6$, and of the UST Peano path at $\kappa = 8$. There are also conjectures for $\kappa = 4$ and for all $\kappa \in [0, 8]$.

The outer boundary of SLE$_6$ is essentially identical with the outer boundary of planar BM.

The study of properties of SLE_κ is in progress.

2. Applications to planar Brownian motion

2.1. What is Brownian motion? Define simple random walk on $\mathbb{Z}$ starting at 0 by setting $X(0) = 0$ and for each $n \in \mathbb{N}$, given $X(1), \ldots, X(n)$, let $X(n+1)$ be $X(n) - 1$ or $X(n) + 1$, each with probability $1/2$. For $t \in [n, n+1)$ set $X(t) = X(n)$.

BM on $\mathbb{R}$ is the limit as $\delta \to 0$ of the process $t \mapsto \delta X(t/\delta^2)$.

It has amazing properties. For one, if B_1 and B_2 are independent BM's starting at 0, then (B_1, B_2) is rotationally invariant. It is called 2-dimensional BM.

2.2. Conformal invariance of BM. In fact, 2-dimensional BM is also conformally invariant if we ignore the time parameterization. That is, suppose that $f : D \to D'$ is a conformal homeomorphism between domains in $\mathbb{R}^2$ and $f(0) = 0$. Let B be BM starting from zero stopped when it hits ∂D, and let B' be BM starting from zero stopped when it hits $\partial D'$. Then $f(B)$ and B' have the same distribution, as unparameterized paths.

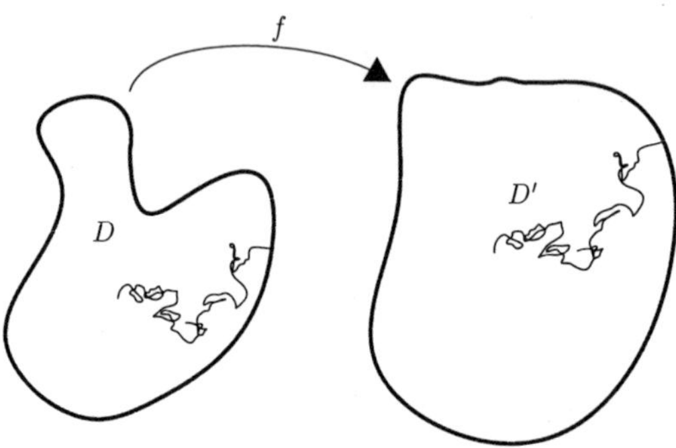

2.3. Brownian intersection exponents. Consider BM in the plane. The simplest BM exponent is $\xi(1,1)$:

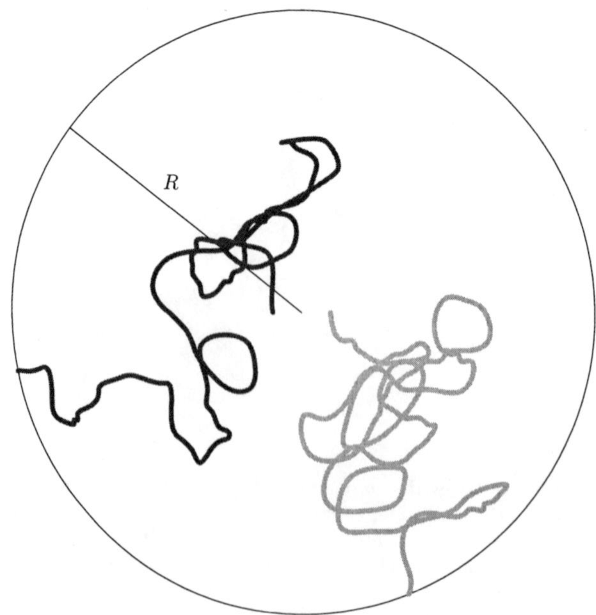

Run two brownian motions in the plane, "green" B starts from 1, and "red" B' starts from -1. Consider the probability that the paths do not intersect until they hit the circle of radius R about 0. It is not hard to see that this probability decays like a power of R as $R \to \infty$. Thus, the intersection exponent $\xi(1,1)$ is defined by

$$\mathbf{P}[\mathbf{B} \cap \mathbf{B}' = \emptyset] = \mathbf{R}^{-\xi(\mathbf{1},\mathbf{1})+\mathbf{o}(\mathbf{1})}, \qquad \mathbf{R} \to \infty.$$

2.4. More general exponents.

$$P[\text{different colors stay disjoint}] = R^{-\xi(2,1,3)+o(1)}$$

Similarly, other intersection exponents can be defined. For example, suppose you had two independent red BM's, one blue BM and 3 green BM, then the probability that paths of different color do not intersect until they hit the circle of radius R decays like $R^{-\xi(2,1,3)+o(1)}$.

2.5. Significance of the exponents. The exponents encode much information about BM and SRW. For example, the probability that two SRW paths of n steps each starting from zero will not intersect again decays like $n^{-\xi(1,1)/2}$ (Burdzy-Lawler [**BL90**]).

The Hausdorff dimension of the set of cut points of $B[0,1]$ is almost surely $2 - \xi(1,1)$ (Lawler [**Law96b**]).

A cut point of a connected set X is a point $p \in X$ such that $X \setminus \{p\}$ is disconnected.

2.6. Determination of the exponents. The values of the exponents $\xi(1,1,\ldots,1)$ have been conjectured by Duplantier-Kwon [**DK88**]. The following theorem proves a generalization of this conjecture.

Theorem (Lawler-Schramm-Werner [LSW3]).

$$\xi(n_1, n_2, \ldots, n_k) = \frac{\left(\sqrt{24n_1+1} + \cdots + \sqrt{24n_k+1} - k\right)^2 - 4}{48}.$$

Corollary [LSW2]. The Hausdorff dimension of the set of cut points of $B[0,1]$ is $3/4$ almost surely.

2.7. Heuristic. Here's an imprecise explanation for the relation between the dimension of the set of cut points and the exponent $\xi(1,1)$.

In order for p to be a cut point, it first has to lie on the path. But the probability that in a disk of radius ϵ there will be a point on the path decays slowly (logarithmically) as $\epsilon \to 0$.

Given that p is on the path, in order for it to be a cut point we need the "past" (the part before hitting p) and the "future" (the part after hitting p) to be disjoint. By reversability of BM, the "probability" for that is essentially the same as the probability that two independent BM's starting from p will not intersect. The decay for this event is governed by $\xi(1,1)$.

The 2 in the formula $2 - \xi(1,1)$ comes from the 2-dimensionality of $\mathbb{R}^2$. The $-\xi(1,1)$ is measures the improbability that two independent BM's will not intersect.

The proof follows along these lines.

2.8. Disconnection exponents. Let $f_k(R)$ be the probability that k independent BM paths in $\mathbb{R}^2$ starting from 1 will hit the circle of radius R about 0 before disconnecting 0 from ∞. The disconnection exponent $\eta(k)$ is defined by the relation

$$f_k(R) = R^{-\eta(k)+o(1)}, \qquad R \to \infty.$$

The event defining $\eta(3)$.

Theorem (Lawler-Schramm-Werner [LSW4]).

$$\eta(k) = \frac{\left(\sqrt{24k+1}-1\right)^2-4}{48}.$$

2.9. Brownian frontier. Corollary (Lawler-Schramm-Werner [LSW4]). The Hausdorff dimension of the outer boundary of $B[0,1]$ is almost surely $4/3$. (As conjectured by Mandelbrot.)

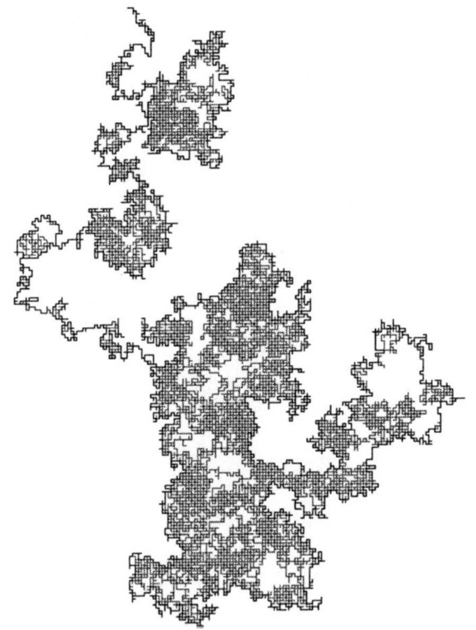

Proof: This follows from $\eta(2) = 2/3$ and a result of Lawler [**Law96a**] showing that the hausdorff dimension of the outer boundary is equal to $2 - \eta(2)$.

Exercise: Figure out the heuristic for the result of Lawler that $2 - \eta(2)$ should be the dimension of the outer boundary of planar BM.

2.10. Universality. Suppose that to every simply connected domain $D \subset \mathbb{C}$, we have a random set $X = X_D$ in the closure $\overline{D}$. X may also depend on one or two points in the interior or boundary of D. (Call those anchor points.) E.g., BM, LERW, or percolation boundary curve.

Lawler and Werner [**LW99, LW00**] showed that if the process X satisfies

- conformal invariance,
- locality, and
- other stuff,

then the intersection exponents of BM can be determined from those of X. This they called universality.

Conformal invariance means that when $f : D \to D'$ is conformal, the distribution of $f(X_D)$ is the same as that of $X_{D'}$, with the anchor points mapped by f.

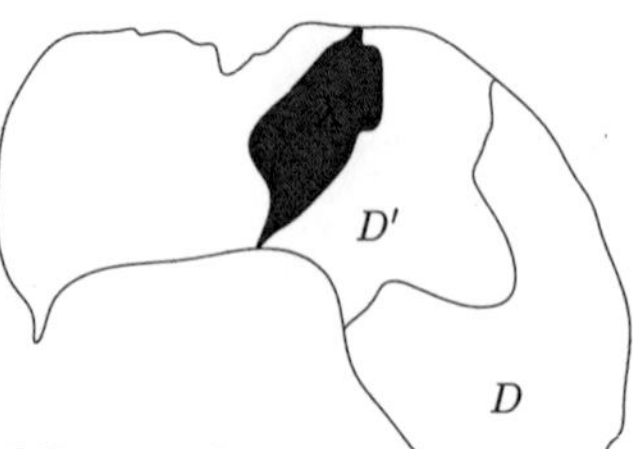

Locality means the following. Suppose that $D' \subset D$. Then the restriction of $X_{D'}$ to the event $X_{D'} \cap \partial D' \cap D = \emptyset$ has the same distribution as the restriction of X_D to the event $X_D \cap \partial D' \cap D = \emptyset$.

2.11. The universality argument (LW). Suppose that X has a single anchor point p in the interior of the domain D, and X is a random set containing p and having a single point on the boundary ∂D.

Then conformal invariance implies that (when D is simply connected) $X \cap \partial D$ has the distribution of harmonic measure on ∂D. That is, it is the pull-back of uniform measure on the unit circle under the Riemann map from D to $\mathbb{U}$ taking p to 0.

Let B and B' be two independent BM's starting from p and q, respectively, in $R\mathbb{U}$. Let X start from q.

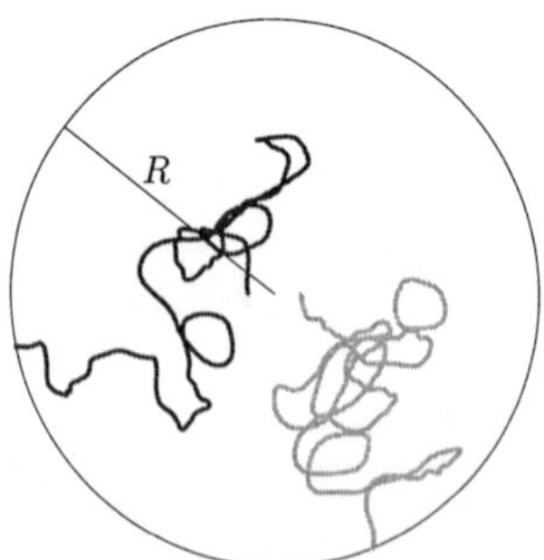

For a set $Y \subset R\overline{\mathbb{U}}$, let $h_q(Y)$ denote the harmonic measure from q of Y in $R\mathbb{U} \setminus Y$.

The following shows that one may replace B' with X when calculating the non-intersection probability.

$$\begin{aligned} P[B \cap B' = \emptyset] &= E[P[B \cap B' = \emptyset | B]] \\ &= E[1 - h_q(B)] \\ &= E[P[B \cap X = \emptyset | B]] \\ &= P[B \cap X = \emptyset]. \end{aligned}$$

In the second and last equalities, locality and conformal invariance for B' and X are used.

2.12. How SLE fits in. At roughly the same time, I defined *SLE* [**Sch00**] and conjectured that SLE_6 is the scaling limit of the percolation boundary curve.

SLE_κ can be defined in an arbitrary simply connected domain via conformal maps. It is then conformally invariant.

Assuming that SLE_6 is the scaling limit of the percolation boundary path, locality is obvious. Without the conjecture, locality was established [**LSW1**] with difficulty. The "other stuff" got handled differently.

Universality allows to replace the calculation of the exponents for BM with the exponents for SLE_6.

2.13. Exponents calculation for SLE_6. As the above universality calculation shows, it suffices to be able to determine the distribution of $h_0(\mathrm{SLE}_6)$. That is, in the SLE evolution, we need to keep track of the length of the image $g_t(\partial\mathbb{U})$. It turns out that it suffices to consider instead the decay of $|g_t'(1)|$. This can be converted to a PDE problem, since the infinitesimal evolution is known.

Why are exponents easier for SLE_6?

• Since SLE_6 does not cross itself, we only need to keep track of the exterior of the domain. By conformal invariance, this reduces to a (small) finite number of parameters.

• Reduction of the problem to 1-dimension (BM on $\mathbb{R}$).

2.14. When universality fails. The above is an outline for the calculation of the $\xi(n_1,\dots,n_k)$ exponents. However, the universality argument does not seem to work for the disconnection exponents.

2.15. Generalized exponents. The exponent $\xi(k,\lambda)$ can be naturally defined for $\lambda \geqslant 0$ which is not necessarily an integer, as follows.

Let $B_1,\dots,B_k$ be k independent BM's starting from $1 \in \mathbb{C}$ and stopped when they hit $R\partial\mathbb{U}$, and let B be an independent BM from 0, which is also stopped when it hits $R\partial\mathbb{U}$.

Given $B_1,\dots,B_k$, let $Z = Z(B_1,\dots,B_k)$ be the probability that B does not intersect $B_1 \cup \cdots \cup B_k$:

$$Z = \mathbf{P}\Big[B \cap (B_1 \cup \cdots \cup B_k) = \emptyset \Big| B_1,\dots,B_k\Big].$$

In other words, Z is the harmonic measure from 0 of $R\partial\mathbb{U}$ in $R\mathbb{U} \setminus (B_1 \cup \cdots \cup B_k)$.

Then $\xi(k,\lambda)$ is defined by the relation

$$\mathbf{E}[Z^\lambda] = R^{-\xi(k,\lambda)+o(1)}, \qquad R \to \infty.$$

Note that this coincides with the previous definition when $\lambda = 1, 2, \dots$.

2.16. Analyticity. Lawler [**Law00**] showed that $\eta(k) = \lim_{\lambda\searrow 0} \xi(k,\lambda)$.

The determination of the disconnection exponents $\eta(k)$ then proceeds by first determining $\xi(k,\lambda)$ when λ is large. In this range, universality can be used. Then we [**LSW4**] show that $\xi(k,\lambda)$ is an analytic function of λ. Hence, one gets the formula for $\eta(k)$ by analytic continuation.

There's hope for a more direct proof.

References

[AB99] M. Aizenman and A. Burchard. Hölder regularity and dimension bounds for random curves. *Duke Math. J.*, 99(3):419–453, 1999.

[ABNW99] Michael Aizenman, Almut Burchard, Charles M. Newman, and David B. Wilson. Scaling limits for minimal and random spanning trees in two dimensions. *Random Structures Algorithms*, 15(3-4):319–367, 1999. Statistical physics methods in discrete probability, combinatorics, and theoretical computer science (Princeton, NJ, 1997).

[Ald90] David J. Aldous. The random walk construction of uniform spanning trees and uniform labelled trees. *SIAM J. Discrete Math.*, 3(4):450–465, 1990.

[BL90] Krzysztof Burdzy and Gregory F. Lawler. Nonintersection exponents for Brownian paths. II. Estimates and applications to a random fractal. *Ann. Probab.*, 18(3):981–1009, 1990.

[BLPS01] I. Benjamini, R. Lyons, Y. Peres, and O. Schramm. Uniform spanning forests. *Ann. Probab.*, 29:1–65, 2001.

[Bro89] Andrei Broder. Generating random spanning trees. In *Foundations of Computer Science*, pages 442–447, 1989.

[Car92] John L. Cardy. Critical percolation in finite geometries. *J. Phys. A*, 25(4):L201–L206, 1992.

[DK88] Bertrand Duplantier and K.-H. Kwon. Conformal invariance and intersection of random walks. *Phys. Rev. Let.*, pages 2514–2517, 1988.

[Gri89] Geoffrey Grimmett. *Percolation.* Springer-Verlag, New York, 1989.

[Häg95] Olle Häggström. Random-cluster measures and uniform spanning trees. *Stochastic Process. Appl.*, 59(2):267–275, 1995.

[Häg98] Olle Häggström. Uniform and minimal essential spanning forests on trees. *Random Structures Algorithms*, 12(1):27–50, 1998.

[Har60] T. E. Harris. A lower bound for the critical probability in a certain percolation process. *Proc. Cambridge Philos. Soc.*, 56:13–20, 1960.

[HS90] Takashi Hara and Gordon Slade. Mean-field critical behaviour for percolation in high dimensions. *Comm. Math. Phys.*, 128(2):333–391, 1990.

[Ken00a] Richard Kenyon. Long-range properties of spanning trees. *J. Math. Phys.*, 41(3):1338–1363, 2000. Probabilistic techniques in equilibrium and nonequilibrium statistical physics.

[Ken00b] Richard Kenyon. Conformal invariance of domino tiling. *Ann. Probab.*, 28(2):759–795, 2000.

[Ken00c] Richard Kenyon. The asymptotic determinant of the discrete laplacian. *Acta Math.*, 185(2):239–286, 2000.

[Ken] Richard Kenyon. Dominos and the Gaussian free field, arXiv:math-ph/0002027.

[Kes80] Harry Kesten. The critical probability of bond percolation on the square lattice equals $\frac{1}{2}$. *Comm. Math. Phys.*, 74(1):41–59, 1980.

[Law80] Gregory F. Lawler. A self-avoiding random walk. *Duke Math. J.*, 47(3):655–693, 1980.

[Law96a] G. Lawler. The dimension of the frontier of planar Brownian motion. *Electron. Comm. Probab.*, 1:no. 5, 29–47, 1996.

[Law96b] Gregory F. Lawler. Hausdorff dimension of cut points for Brownian motion. *Electron. J. Probab.*, 1:no. 2, approx. 20 pp., 1996.

[Law00] Gregory F. Lawler. Strict concavity of the half plane intersection exponent for planar Brownian motion. *Electron. J. Probab.*, 5:no. 8, 33 pp., 2000.

[LPSA94] Robert Langlands, Philippe Pouliot, and Yvan Saint-Aubin. Conformal invariance in two-dimensional percolation. *Bull. Amer. Math. Soc. (N.S.)*, 30(1):1–61, 1994.

[LSW1] Gregory F. Lawler, Oded Schramm, and Wendelin Werner. Values of Brownian intersection exponents I: Half-plane exponents. *Acta Math.*, to appear. arXiv:math.PR/9911084.

[LSW2] Gregory F. Lawler, Oded Schramm, and Wendelin Werner. Values of Brownian intersection exponents II: Plane exponents. *Acta Math.*, to appear. arXiv:math.PR/0003156.

[LSW3] Gregory F. Lawler, Oded Schramm, and Wendelin Werner. Values of Brownian intersection exponents III: Two-sided exponents. *Acta Math.*, to appear. arXiv:math.PR/0005294.

[LSW4] Gregory F. Lawler, Oded Schramm, and Wendelin Werner. Analyticity of intersection exponents for planar Brownian motion, *Acta Math.*, to appear. arXiv:math.PR/0005295.

[LSW5] Gregory F. Lawler, Oded Schramm, and Wendelin Werner. The dimension of the planar Brownian frontier is 4/3. *Math. Res. Lett.*, to appear.

[LW99] Gregory F. Lawler and Wendelin Werner. Intersection exponents for planar Brownian motion. *Ann. Probab.*, 27(4):1601–1642, 1999.

[LW00] Gregory F. Lawler and Wendelin Werner. Universality for conformally invariant intersection exponents. *J. Eur. Math. Soc. (JEMS)*, 2(4):291–328, 2000.

[Pem91] Robin Pemantle. Choosing a spanning tree for the integer lattice uniformly. *Ann. Probab.*, 19(4):1559–1574, 1991.

[RS] Steffen Rohde and Oded Schramm. Basic properties of SLE. arXiv:math.PR/0106036.

[Sch00] Oded Schramm. Scaling limits of loop-erased random walks and uniform spanning trees. *Israel J. Math.*, 118:221–288, 2000.

[Smi] Stanislav Smirnov. Critical percolation in the plane. I. Conformal invariance and Cardy's formula. II. Continuum scaling limit. Preprint.

[Wer00] Wendelin Werner. Critical exponents, conformal invariance and planar Brownian motion, 2000, arXiv:math.PR/0007042.

[Wil96] David Bruce Wilson. Generating random spanning trees more quickly than the cover time. In *Proceedings of the Twenty-eighth Annual ACM Symposium on the Theory of Computing (Philadelphia, PA, 1996)*, pages 296–303, New York, 1996. ACM.

Current developments in mathematics